LINEAR CIRCUITS

M. E. VAN VALKENBURG

Department of Electrical Engineering
University of Illinois at Urbana-Champaign

B. K. KINARIWALA

Department of Electrical Engineering
University of Hawaii at Manoa

PRENTICE-HALL, INC., *Englewood Cliffs, New Jersey* 07632

Library of Congress Cataloging in Publication Data

VAN VALKENBURG, M. E. (MAC ELWYN) *date—*
    Linear circuits.

    Includes index.
    1. Electric circuits, Linear.
I. Kinariwala, B. K. (Bharat K.) II. Title.
TK454.V35   621.319′2   81-17809
ISBN 0-13-536722-0   AACR2

Editorial/production supervision
and interior design by *Virginia Huebner*
Manufacturing buyer: *Joyce Levatino*
Cover design by *Lee Cohen*

Printed in the United States of America

10 9 8 7 6 5 4 3 2

ISBN 0-13-536722-0

Prentice-Hall International, Inc., *London*
Prentice-Hall of Australia Pty. Limited, *Sydney*
Prentice-Hall of Canada, Ltd., *Toronto*
Prentice-Hall of India Private Limited, *New Delhi*
Prentice-Hall of Japan, Inc., *Tokyo*
Prentice-Hall of Southeast Asia Pte. Ltd., *Singapore*
Whitehall Books Limited, *Wellington, New Zealand*

CONTENTS

PREFACE   xi

I   **INTRODUCTION**   I

     1-1   Charge and Current   *1*
     1-2   SI Units and Prefixes   *4*
     1-3   Energy and Voltage   *6*
     1-4   Batteries and Measurements   *8*
     1-5   Resistors and Ohm's Law   *12*
     1-6   Reference Directions   *15*
     1-7   Circuit Connections   *17*
     1-8   Short and Open Circuits   *18*
     1-9   Series and Parallel Connections   *20*
     1-10   Summary   *22*

2   **THE KIRCHHOFF LAWS**   26

     2-1   Kirchhoff's Voltage Law   *26*
     2-2   Some KVL Applications   *32*
     2-3   The Voltage-Divider Circuit   *36*
     2-4   Kirchhoff's Current Law   *39*
     2-5   The Current-Divider Circuit   *42*
     2-6   Energy and Power in Resistor Circuits   *45*
     2-7   Generalization   *49*

3   CIRCUIT SIMPLIFICATION TECHNIQUES   58

3-1   Series and Parallel Circuits   *58*
3-2   Source Transformation   *63*
3-3   Source Shifting   *68*
3-4   Thévenin and Norton Equivalent Circuits   *73*
3-5   Equivalent versus Input Resistance   *84*

4   CIRCUIT ANALYSIS METHODS   92

4-1   Node Analysis   *92*
4-2   Loop Analysis   *98*
4-3   Circuits with Mixed Sources   *103*
4-4   Ladder Circuits   *106*
4-5   Superposition   *110*
4-6   Power Transfer   *113*
4-7   Effect of Feedback on Resistance   *116*

5   OP AMP–RESISTOR CIRCUITS   124

5-1   Two-Port Circuits   *124*
5-2   The Op Amp   *125*
5-3   Op Amp–Resistor Circuits   *128*
5-4   Voltage Follower (or Isolator)   *131*
5-5   Analog Addition and Subtraction   *133*

6   RESISTOR–CAPACITOR CIRCUITS   146

6-1   The Capacitor   *147*
6-2   Energy and Power   *153*
6-3   Linearity and Superposition   *154*
6-4   Series and Parallel Combinations   *155*
6-5   *RC* Circuits   *157*
6-6   Driven *RC* Circuits   *162*
6-7   Simple Capacitor–Op Amp Circuits   *169*

7   RESISTOR–INDUCTOR CIRCUITS   178

7-1   Inductors   *178*
7-2   *LR* Circuits   *184*
7-3   Coupled Coils and Transformers   *189*

8 RISE TIME, RINGING, AND SUSTAINED OSCILLATIONS   199

8-1 Step Response of First-Order Differential Equations   *199*
8-2 Damped Oscillations (Ringing)   *203*
8-3 Step Response for a Second-Order Circuit   *206*
8-4 Overshoot and Settling Time   *208*
8-5 Sustained Oscillations   *209*
8-6 Transient versus Steady-State Analysis   *211*

9 SINUSOIDS AND PHASORS   214

9-1 The Sinusoid   *214*
9-2 The Addition of Sinusoids of the Same Frequency   *218*
9-3 Sinusoidal Steady State   *220*
9-4 Steinmetz's Analog: The Phasor   *222*
9-5 Complex Numbers   *224*
9-6 Sinusoids and Exponentials   *230*

10 IMPEDANCE AND ADMITTANCE   234

10-1 Kirchhoff's Laws   *234*
10-2 Ohm's Law Extended: Impedance   *235*
10-3 Impedance and Admittance   *242*
10-4 Circuit Analysis with Phasors   *246*

11 CIRCUIT FUNCTIONS AND ANALYSIS TECHNIQUES   256

11-1 Generalized or Complex Exponentials   *257*
11-2 Generalized Impedance   *260*
11-3 Circuit Functions   *265*
11-4 Kirchhoff's Laws and Analysis Methods   *268*
11-5 Series and Parallel Impedances   *271*
11-6 Voltage- and Current-Divider Circuits   *273*
11-7 Source Transformations   *277*
11-8 Thévenin and Norton Equivalent Circuits   *278*
11-9 Superposition   *284*

12 AVERAGE POWER   291

12-1 Power and Energy   *291*
12-2 RMS Values   *295*
12-3 Maximum Power Transfer   *297*

13 THE FREQUENCY SPECTRUM   302

13-1 Harmonics   *302*
13-2 Fourier Series   *304*
13-3 Spectrum   *307*
13-4 Bandwidth   *311*
13-5 Signal Processing   *312*
13-6 Distortion and Shaping   *314*

14 FOURIER ANALYSIS AND THE FREQUENCY SPECTRUM   321

14-1 Circuits as Signal Processors   *321*
14-2 Fourier Coefficients   *326*
14-3 Signals with Even and Odd Symmetry   *329*
14-4 The Rate at which Fourier Coefficients Decrease   *331*
14-5 Spectrum Shaping   *333*
14-6 Spectrum of a Pulse Train   *335*
14-7 Continuous Spectra: The Fourier Integral   *339*

15 CAPACITOR–OP AMP CIRCUITS   349

15-1 Frequency Response   *349*
15-2 Simple *RC* and Op-Amp Circuits   *351*
15-3 Poles and Zeros of $T(s)$ *356*
15-4 Magnitude and Phase from Pole–Zero Plots   *358*
15-5 Bode Plots   *365*

16 CIRCUIT DESIGN   379

16-1 The Design Process   *379*
16-2 Op-Amp Modules   *380*
16-3 The Low-Pass Circuit   *386*
16-4 The Band-Pass Circuit   *388*
16-5 The Band-Elimination Circuit   *392*
16-6 The Notch Filter   *396*
16-7 The High-Pass Circuit   *397*

I 7   **SWITCHED-CAPACITOR CIRCUITS**   404

     17-1   The MOS Switch   *404*
     17-2   The Switched Capacitor   *409*
     17-3   Analog Operations   *411*
     17-4   Range of Circuit Element Sizes   *416*
     17-5   First-Order Filters   *417*

     INDEX   423

PREFACE

In undertaking the writing of this textbook, we asked ourselves the question: Now that the pressures of new developments in electrical engineering are such that only one course can ordinarily be devoted to an introduction to electric circuits, what should we teach? This book comprises our answer.

Our question is that one that educators are facing in many different parts of the curriculum. In considering the question, the first thing that will be recognized is that some topics must be eliminated. The obvious way to accomplish this objective is to rank order all topics and then eliminate those deemed less essential. This will involve judgment. In addition, the topics chosen should be given a design orientation in keeping with current trends in engineering education as well as accreditation requirements imposed by ABET (formerly known as ECPD). Our experience is that such a design orientation, in contrast to a systematic presentation of abstract concepts, makes the subject more interesting for students.

Of course, the amount of material to be covered and the depth of the coverage must be played one against the other. The compromise that becomes the real course must have some elements of range of coverage as well as depth of coverage. This means that some topics will receive short shrift. On the other hand, some violence will be done to those who insist that any topics covered be dealt with "in depth." Our experience is that true in-depth coverage comes only with successive exposure in later courses that utilize the foundation material of the first course. We believe that coverage and depth should be chosen to advance the student understanding to the point that the student can solve engineering problems.

In this treatment, we have reduced our list of essential topics to give packages which we describe as follows:

Part 1, the Kirchhoff laws. The first four chapters cover a statement and application of the two Kirchhoff laws. This knowledge is critical in equipping students for all remaining studies. These laws are described in terms of resistive circuits with both independent and dependent sources. Our treatment could have been much simpler had we chosen to eliminate dependent sources. Unfortunately, the proliferation of silicon devices which are described by models that contain controlled sources makes this impractical. Resistive circuits with controlled sources represent facts of life that cannot be ignored.

Part 2, an introduction to the time domain. The introduction of the capacitor and the inductor together with a switch that closes or opens introduces the time-domain behavior of circuits. The most important concept to be taught is that of the time constant, and so we have stressed first-order circuits. The idea of the time constant will remain with the student as he becomes a graduate engineer and practices his or her profession. A related concept is that of circuit "ringing" or oscillatory behavior, and this requires our consideration of second-order circuits. In describing oscillatory behavior of circuits, we make extensive use of the concept of rise time, know that this too is a design parameter for the engineer.

Part 3, the sinusoidal steady state and the concept of impedance. We recognize the well known paradox: the concepts are simple yet routinely create confusion to those studying them for the first time. One aspect of our solution is that studying of complex numbers in depth is a regular topic. Past generations have relegated complex numbers to an appendix trusting that the appendix would suffice as a reminder of studies in pre-engineering mathematics. We must recognize that the structure of mathematics has changed and that this background is no longer part of the student background. Another concept that should be stressed is that of average power in contrast to instantaneous power and so we have devoted one chapter to this subject.

Part 4, Fourier methods. The concept of the frequency spectrum is one which, like the time constant or rise time, will be with the engineer throughout his or her career. An introduction to this subject must come early and be repeated in later courses which contain applications.

Part 5, The operational amplifier or "op amp". The op amp is in the same class as the microprocessor—both have altered the way in which circuits are designed. Because this fact is often well recognized by students, the study of op amp circuits helps instill excitement for them. The coupling of such studies with Bode plots makes both clear and provides a simple entry to design. We have decided to introduce the op amp early so that its status as "just another circuit element" can develop naturally. The last three chapters emphasize design with op amps. It is worth noting that this material relates easily to laboratory experiments which the student may carry out in parallel. At low frequencies, op amp circuits behave in a near-ideal way and give excellent agreement between theory and measurements.

As the op amp has become a conventional circuit element, the switched-capacitor circuit is regarded as a practical replacement for the resistor in integrated-circuit applications.

The class notes from which this book developed have been used for a number of years at the University of Hawaii at Manoa. The enthusiastic response of our students has been a factor in working toward the completion of the writing project. We have benefited from the comments and other assistance of our colleagues at the University of Hawaii. Paul Weaver, Franklin Kuo and James Yonemoto have been involved directly in the teaching of the course. We are particularly grateful to Professor Yonemoto who devised many of the problems used as examples and also for homework assignment.

It has been our good fortune to work with Virginia Huebner as our editor and production supervisor. We are pleased to acknowledge her valuable assistance and to express our appreciation for the many ways in which she has helped us. Finally, it is a pleasure to acknowledge our indebtedness to our wives, Evelyn Van Valkenburg and Marva Kinariwala, for encouragement and understanding during the writing of the book, and to Evelyn Van Valkenburg for typing the final version of the manuscript.

M. E. VAN VALKENBURG

BHARAT KINARIWALA

Urbana, Illinois
Honolulu, Hawaii

LINEAR CIRCUITS

I

INTRODUCTION

This is a book about electric circuits. In everyday usage, a circuit is a path or a route, starting at a given point in space and eventually returning to the same point. The "circuit rider" of another era in American history, a judge or a preacher, left home and visited several settlements, and eventually returned home. By tradition, electrical engineers use the word *loop* or *closed loop* to have the meaning just given. Instead, an electric *circuit* consists of electric elements (the components) interconnected in some specific way. It is unfortunate that we also use the word *network* to mean the same thing as circuit. We will use circuit exclusively.

In studying electric circuits, we will have one of two objectives in mind. The first is *analysis:* Given a complete description of the circuit, determine all voltages and all currents. From this information, charge, energy, or power can be found. The second objective is *design:* Given specified voltages or currents in certain parts of the circuit, find a circuit that meets the specifications.

We begin our study with basic definitions and conventions and also the units to be used.

I-I CHARGE AND CURRENT

The conceptual scheme we use in describing an atom pictures a positively charged nucleus surrounded by a cloud of negatively charged electrons. In the neutral atom, the number of positive charges is equal to the number of negative charges. In conductors, electrons are able to break away from the nucleus of the atom and move through the conducting material. Thus, we picture a positively charged nucleus which is fixed, and negatively charged electrons in motion as charge carriers. Motion in semiconducting materials is more complicated. In effect, charge carriers both negative and positive are possible. In insulators, the number of charge carriers is extremely small and is assumed to be zero. But in conductors and semiconductors, there are positive and negative charges which are mobile.

Figure 1-1 shows part of a conducting wire in which a cross-sectional area S is identified. If charge carriers are transported through the surface S at a

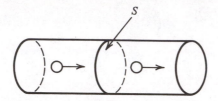

Figure 1-1. Representation of a wire of cross-sectional area S.

rate such that Δq of charge is transported in Δt of time, then the charge flow is said to constitute a current

$$i = \frac{\Delta q}{\Delta t} \tag{1-1}$$

or in the limit

$$i = \frac{dq}{dt} \tag{1-2}$$

In these equations, we have used i as the symbol for current, taken from the French word *intensité*. We have used a lowercase letter to indicate that i varies with time, $i(t)$; had the current been constant or time-invariant we would have indicated that fact by using a capital letter, I. This convention will be followed throughout the book.

The unit for charge is the *coulomb* (C), that for time is the *second* (s), and for current is the *ampere* (A) or coulomb per second (C/s). This example of the use of units illustrates the general practice. A unit is usually named for people who have made important contributions relating to electrical engineering. In this case, they are Charles A. Coulomb and André Marie Ampère. When we use their names for units, the units are never capitalized. More about units in the next section.

Now we return to Eq. 1-2 and consider the convention for the algebraic sign for the current, i. In Fig. 1-1, the charge can flow in either direction. We first must select a direction which will be designated as the *reference direction*. The reference direction selected is shown in Fig. 1-2 and is from left to right. The two possibilities for sign of charge are shown as (a) and (b) in the figure. Shall we select (a) or (b)? That choice has already been made for us and it was done by Benjamin Franklin, who selected (b). We know that he made the wrong choice, of course, but his convention has persisted. His choice is sometimes

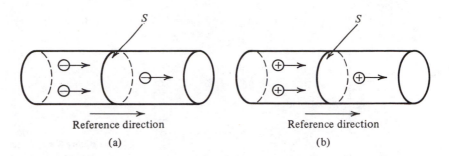

Reference direction   Reference direction

(a)   (b)

Figure 1-2. Flow of charge with respect to a reference direction.

called *conventional* current, in contrast with *electron* current. Unless stated otherwise, we will use conventional current. We will find that this creates no problems for us as long as we are consistent. When you later study semiconductor devices, the kind of current as related to the kind of carrier will assume importance.

Then if conventional current is directed in the reference direction that has been selected, it is given a positive algebraic sign; otherwise, it is given a negative sign. The relationship between direction and algebraic sign is illustrated in Fig. 1-3, which shows that changing sign is equivalent to changing reference direction.

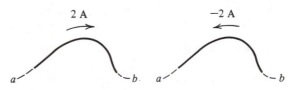

Figure 1-3. Relationship between reference direction and algebraic sign.

The relationship between reference direction and algebraic sign is important and, incidentally, often a point of difficulty for the beginning student. In circuit analysis, we tend to forget entirely about charge and deal exclusively in current. The current will always have a reference direction and the current will have an algebraic sign. We know that charge flow constitutes current. Does current flow? Logic dictates that it does not, and purists insist on this interpretation. But the fact that it is always shown on circuit diagrams with an arrow indicating reference direction has led to the widespread use of the term "current flow." We will follow this practice.

Another concept that often creates difficulty for the beginning student is that all circuit elements are *electrically neutral*, meaning that no net positive or negative charge will accumulate on the elements. Even if an element stores charge, the positive and negative charges will be equal, so that the net charge

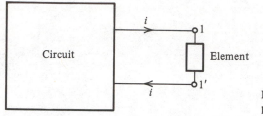

Figure 1-4. Specific element that is part of a circuit.

is zero. Consider the circuit element identified in Fig. 1-4, where one element has been selected for study and its terminals marked 1 and 1′. If positive current is flowing throughout the loop shown, then positive charge is entering terminal 1 at the rate of i coulombs/second. An equal amount of charge must flow out of terminal 1′ so that the element remains uncharged overall. The charge in and out of the element is related to the current by the equation found by integrating Eq. 1-2:

$$q_{total} = q(t) - q(t_0) = \int_{t_0}^{t} i(t)\, dt \tag{1-3}$$

Here t_0 is a reference time from which the charge q_{total} is to be found, and $q(t_0)$ is the charge at that reference time. The concept we have discussed is often called the principle of *conservation of charge*.

1-2   SI UNITS AND PREFIXES

The choice of units was discussed in Section 1-1 in connection with the description of charge and current. It is important that we use proper units in solving problems, so it is important that we discuss this issue early. Fortunately, the metric system of measurements is standard in electrical engineering. The International System of Units (abbreviated SI in all languages) has been adopted by all organizations interested in standards, including the Institute of Electrical and Electronics Engineers (IEEE) and the National Bureau of Standards. The system is based on the standard measurements of the meter, kilogram, second, ampere, and degree Kelvin. Quantities that will be used, the associated symbol, and the units used are shown in Table 1-1.

Because of the large range of applications that are the domain of electrical engineers, from tiny modules in large-scale integrated (LSI) circuits to huge rotating machines, it is essential that we use prefixes with the units just given in Table 1-1. In power systems, for example, capacitors approaching a value of 1 F may be used for power factor correction, while in LSI circuits, capacitances in the range of 1 femtofarad (fF) become significant. The prefixes that are commonly in use are shown in Table 1-2.

Table 1-1

Quantity	Symbol	Unit
Charge	Q or q	coulombs (C)
Energy	W or w	joules (J) = N-m
Power	P or p	watts (W) = J/s
Current	I or i	amperes (A) = C/s
Voltage	V or v	volts (V) = J/C
Electric field	$\mathcal{E}$	volts/meter (V/m)
Flux linkages	Ψ	webers (Wb) = V-s
Magnetic flux density	B	teslas (T) = Wb/m^2
Frequency	f	hertz (Hz) = cycles/second
Frequency	ω	radians/second = $2\pi f$
Temperature	T	kelvin (K)
Time	t	seconds (s)
Resistance	R	ohms (Ω) = V/A
Conductance	G	siemens (S) or mhos ($\mho$) = A/V
Inductance	L	henrys (H) = Wb/A
Capacitance	C	farads (F) = C/V
Elastance	S	(farad)$^{-1}$, $S = 1/C$
Reciprocal inductance	Γ	(henry)$^{-1}$, $\Gamma = 1/L$

Table 1-2

Prefix	Factor	Symbol
atto	10^{-18}	a
femto	10^{-15}	f
pico	10^{-12}	p
nano	10^{-9}	n
micro	10^{-6}	μ
milli	10^{-3}	m
centi	10^{-2}	c
deci	10^{-1}	d
deka	10	da
hecto	10^2	h
kilo	10^3	k
mega or meg	10^6	M
giga	10^9	G
tera	10^{12}	T

EXERCISES

1-2.1. The charge of the electron is known to be 1.6021×10^{-19} C. How many electrons does it take to make 1 C of charge?

Ans. 6.242×10^{18}

1-2.2. A dielectric substance is rubbed until the charge is 10^4 free electrons. Find this charge expressed in femtocoulombs.

Ans. 1.6021 fC

1-2.3. The charge entering a terminal of an element is given by the equation

$$q(t) = k_1 t^2 + k_2 t \quad \text{C}$$

If $i(0) = -5$ A and $i(1) = +5$ A, find k_1 and k_2.

Ans. 5; -5

1-2.4. The current entering the terminal of an element is given by the equation

$$i(t) = 10^{-6} t^2 - 10^{-7} t \quad \text{A}$$

Find the total charge flowing into the element between $t = 1$ and $t = 2$ s.

Ans. 2.183×10^{-6} C

1-3 ENERGY AND VOLTAGE

Two electrically neutral plates are shown in Fig. 1-5(a). We wish to separate charge from the plates so that the top plate is charged to $+Q$ and the bottom to $-Q$. We know that like charges have a force of repulsion and unlike charges attract each other. To overcome this force of attraction, an *electromagnetic force (EMF)* must be applied. This is accomplished by connecting to the plates a *source of energy*, shown in Fig. 1-5(b), to accomplish the work of separating the charges. When the charges are separated, it is said that a *potential difference* between the two plates exist, and the work or energy per unit charge utilized in this process is known as the *voltage* or sometimes the potential difference. In equation form, when w joules (J) of energy is supplied to q coulombs (C) of

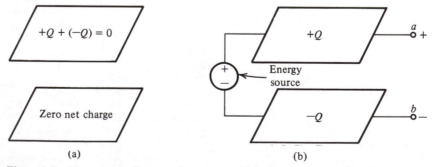

Figure 1-5. Two electrically neutral plates on which charges are placed.

charge, the voltage is

$$v = \frac{w}{q} \qquad \text{volts (V)} \qquad (1\text{-}4)$$

Now the upper plate, which is charged positively, is at a higher potential than the negative lower plate. If some element is connected between the two plates, as in Fig. 1-6, the motion of the positive charges of the upper plate will be toward the negative plate, from *a* to *b*, and this charge transfer will constitute a positive current, shown on the figure as *i* with the given reference direction. Had we

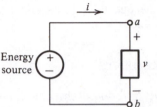

Figure 1-6.   Current and voltage reference directions for all elements.

reversed the *direction* of the *energy source* so that the upper plate was charged negatively and the lower positively, then for the same reference direction, the current would have been reversed.

From this discussion, we see that voltage and current have in common the need for a reference direction and the fact that the voltage may be positive or negative and so is an algebraic quantity. The reference direction for the voltage is provided by the $+$ and $-$ signs on either side of the element as shown in Fig. 1-6. A positive sign for *v* indicates that the polarity marks correctly describe the voltage, while a negative sign indicates that the polarity marks are opposite to the actual voltage.

For the simple circuit of Fig. 1-6 with one loop, it is easy to solve for the voltage *v* and determine the actual polarity. In a more complicated circuit, we mark each element with a set of polarity marks. A positive value for voltage means that the $+$ terminal has the higher voltage, while a negative value for the voltage implies that the $-$ terminal has the higher voltage. These values are always interpreted in terms of the general element shown in Fig. 1-7, across which the voltage is

$$v = (\text{voltage at } + \text{ terminal}) - (\text{voltage at } - \text{ terminal}) \qquad (1\text{-}5)$$

which is algebraic in the sense that it is either positive or negative (or zero).

Figure 1-7.   Reference direction for voltage.

EXERCISES

1-3.1. An electron in a vacuum device is observed to lose 1.6×10^{-20} J of energy in moving from point A to point B. Calculate the voltage of point B with respect to point A.

Ans. -0.1 V

1-3.2. An electron is emitted from point A in an electronic device and is observed at point B, 0.1 ns later. If the voltage between points A and B is 5 V, find the energy acquired by the electron.

Ans. 8×10^{-19} J

1-3.3. Show that the quantities $\frac{1}{2}CV^2$ and $\frac{1}{2}LI^2$ have the dimensions of energy. (*Hint:* Use Table 1-1.)

1-4 BATTERIES AND MEASUREMENTS

In Section 1-3 we introduced the concept of voltage in terms of an energy source. One of the most familiar of energy sources is the electric battery, depicted in Fig. 1-8 together with the circuit symbol (or model) we will use to represent it. Batteries are encountered in many ways: the energy source for a transistor

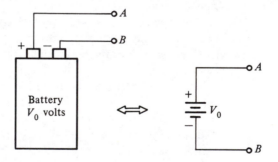

Figure 1-8. Physical construction of the electric battery and the symbol used to represent it.

radio, a critical part of every automobile. The magnitude of a battery voltage is fixed by the manner in which it is manufactured. Batteries have two terminals to which connections are made, one marked $+$ (A) and one $-$ (B). In terms of Fig. 1-8, we say that point A is V_0 volts positive with respect to point B, or that B is V_0 volts negative with respect to A. This important idea was introduced in the preceding section: voltage must be defined in terms of reference marks.

Measurements of voltages are made with a *voltmeter*. A voltmeter for measurement of constant (or dc) voltages from batteries is shown in Fig. 1-9. Connected to the voltmeter will be two leads, one lead red and one black for example, and one of the two leads will be indicated as the reference. The instrument will record the voltage as being positive or negative with respect to the

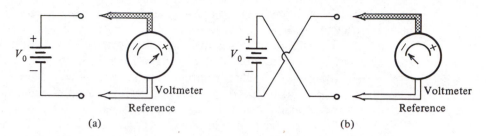

Figure 1-9. Measurement of battery voltage using a voltmeter.

reference, as illustrated by Fig. 1-9(a) and (b). In (a) the voltmeter measures the voltage of A with respect to B, which is $+V_0$. In (b) the voltmeter is connected so as to measure the voltage of B with respect to A, which is $-V_0$.

The battery terminals thus have values only relative with respect to each other. The terminals have no absolute values. For these reasons the battery terminals are called *floating* and so is the battery. Frequently, it is advisable to fix one of the terminals to a reference value, say 0 volts, and measure all other voltage values with respect to such a *datum* or *reference*. It has been generally agreed to regard earth as being at 0 volts, and the appropriate battery terminal is tied to the earth or ground by an electrical wire. The corresponding terminal is called *grounded*, and the ground or reference symbols are shown in Fig. 1-10(a). Any of these symbols may be used to indicate the reference. The refer-

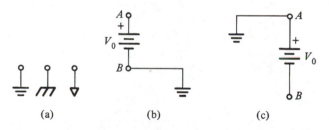

Figure 1-10. (a) Three symbols used to represent the voltage reference (or ground or datum), (b) the grounding of terminal B, and (c) terminal A is grounded.

ence may also be other than ground, such as chassis, housing, metal plate, and so on. The term "ground" or "datum" is still used in all such cases of a battery terminal connected to a reference. In some situations the battery is left floating, but some point in the rest of the circuit may be grounded. Voltages in the circuit are then assumed to be measured with respect to the ground unless otherwise stated.

In Fig. 1-10(b), terminal B is grounded, so A is now at $+V_0$ volts. Thus, A is given an absolute value since the reference is always assumed at 0 volts. In (c), A is grounded. Since B is $-V_0$ volts with respect to A, we now say that

B is at $-V_0$ volts. Voltages in a grounded circuit are generally defined with respect to the ground and stated as being a positive or negative number.

If a circuit element is connected to the battery as shown in Fig. 1-11, the current will flow external to the battery in the element in the direction of the arrow from the $+$ to the $-$ terminal. Inside the battery, the current will flow from $-$ to $+$. If the polarity of the battery is reversed, the direction of current will be reversed and it will flow in a direction opposite to the arrow. The manner in which this relates to measurements is shown in Fig. 1-12. There an *ammeter* is connected in the circuit in such a way that current in the direction of the arrow causes the meter to record a positive current, so current in the direction opposite to that of the arrow causes the meter to record a negative current.

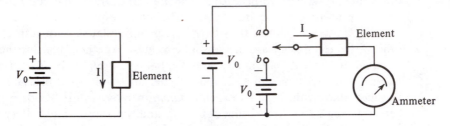

Figure 1-11. Reference direction for the current I.

Figure 1-12. Relationship of current measurements with an ammeter to reference directions.

Thus, both voltage and current may be described by a magnitude and a direction: $+$ and $-$ signs for voltage, an arrow for current. Reversing the $+$ and $-$ signs or the arrow direction is equivalent to changing the sign of the voltage or current. Conversely, changing the sign of the voltage or current is equivalent to changing reference directions. These statements are illustrated in terms of a number in Fig. 1-13.

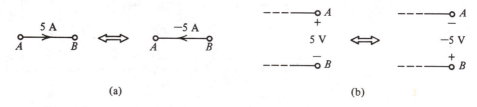

(a)   (b)

Figure 1-13. Relationship of reference direction and the sign of current or voltage.

Frequently, a composite voltage or current is made up of several components. The value of the composite is obtained by combining the components in the manner shown in Fig. 1-14(a) and (b). If any of the polarities is reversed, the corresponding component appears with a negative sign. In (b), all the currents are presumably entering at B and flowing toward A. The converse situation is the more usual one in analysis, as shown in (c). A current I_0 comes to a junction and splits into unknown currents. What are the values of I_1, I_2, and I_3? These

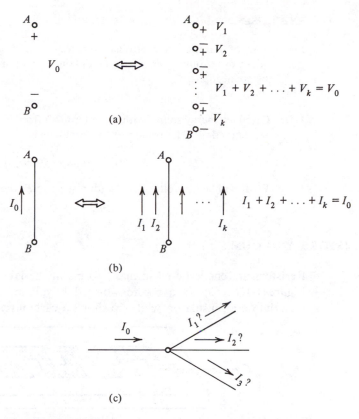

Figure 1-14. (a) The components of voltage sum to give an overall voltage, (b) the components of current sum to give the overall current, and (c) the converse of (b) in which a current subdivides into components that are to be determined.

and other questions will be answered in Chapter 2 by the use of Kirchhoff's laws.

Batteries connected in series have composite terminal voltages which depend upon the manner in which the battery polarities are arranged. So in Fig. 1-15(a), the terminal voltage V_{T1} is the sum $V_1 + V_2$, while in (b) of the figure, the terminal voltage V_{T2} is the difference $V_3 - V_4$.

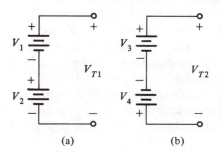

Figure 1-15. Overall voltage of two batteries can be either (a) the sum or (b) the difference of the individual voltages.

EXERCISES

1-4.1. A set of batteries have terminal voltages of 1, 3, 5, and 7 V. How many integer values of terminal voltage can be obtained by appropriate connections? Show the connections.

Ans.   16

1-4.2. Given an energy source with the A terminal from it marked $+$ and the B terminal marked $-$. The voltage at the terminals is

$$v(t) = 10 \cos \left(3t + \frac{\pi}{6}\right)$$

What will be the voltage of A with respect to B when $t = \pi/3$? $2(\pi/3)$?

Ans. -8.66 V, $+8.66$ V

I-5   RESISTORS AND OHM'S LAW

Resistors are encountered in many forms in the laboratory and in the home. Figure 1-16(a) shows a resistor formed by plating a thick film of resistive material on an insulator, while (b) shows a carbon resistor in cylindrical form.

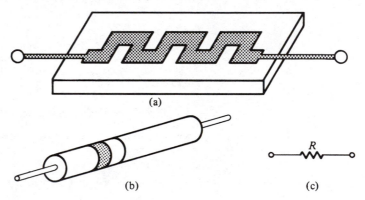

(a)

(b)   (c)

Figure 1-16. Parts (a) and (b) show forms in which resistors are constructed, and (c) shows the model used to represent all resistors.

A familiar resistor in the home is the heating element of a toaster or an oven. Such resistors are represented by the symbol shown in Fig. 1-16(c). When voltage and current references are added to the symbol, it is also called a *model* for a resistor. We distinguish the property of these materials by calling it *resistance*, as distinguished from resistor to designate the device or its model.

When a resistor is connected to a battery as in Fig. 1-17(a), the current I is positive in the direction of the arrow. To this figure we add the polarity marks of the voltage across the resistor. These are chosen to be identical to the reference marks of the battery since there is a direct connection at A and B. Concentrating on this resistor in the circuit in Fig. 1-17(b), we may interpret

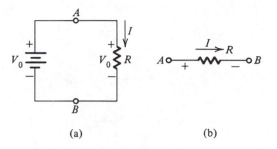

Figure 1-17. Conventions associated with the connection of a resistor to a battery.

these reference marks as telling us that voltage is positive with respect to the
given + and − marks and the current is positive with respect to the given arrow
direction for the resistor. The relationship between this specific voltage and
current is known as *Ohm's law*, which is

$$V_0 = RI \tag{1-6}$$

When the current I is in amperes and the voltage V_0 is in volts, then the resistance R is in *ohms* (designated by the Greek letter Ω). This equation applies
with the polarities and reference directions shown in the figure for V_0 and I.
If either reference direction is reversed, as in Fig. 1-18, then the corresponding
equation is not that just given, but

$$V_0 = -RI \tag{1-7}$$

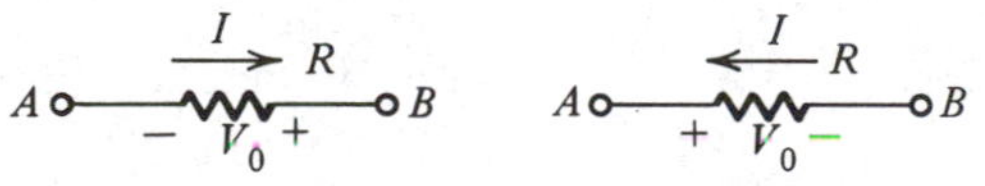

Figure 1-18. Effect of reversing one of
the reference directions.

Reference directions are important, and will be central to the discussion of the
next few sections.

A graphical interpretation of Ohm's law of Eq. 1-6 will give further insight.
Let the battery of Fig. 1-17(a) be adjustable in voltage so that V_0 can be adjusted
to various positive and negative values, as shown in Fig. 1-19(a). As we assume
different values for V_0, shown as V_{01}, V_{02}, V_{03}, and so on, the corresponding
values of current can be measured and V_0 and I plotted as shown in Fig. 1-19(b).
The result will be a straight line of slope $1/R$. If the value of R is changed and
the experiment repeated, the plots will constitute a family of straight lines,
shown in Fig. 1-19(c) for low and high values of R.

From studies of elementary algebra, we recall that the standard equation
of a straight line is

$$y = mx + b \tag{1-8}$$

where m is the slope, b is the y intercept, and x and y are the variables. Comparing this equation with Eq. 1-6, we see that $b = 0$ and the lines always go

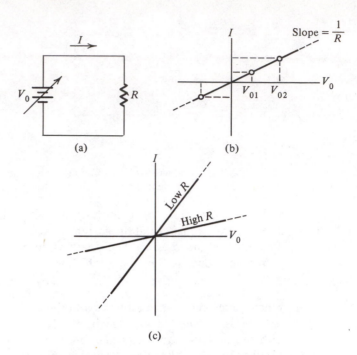

Figure 1-19. (a) A circuit in which the applied voltage is varied,
(b) a plot of the current for different values of applied voltage, and
(c) two plots showing how the plots of (b) change as the value of R
is varied.

through the origin. The value of the slope is determined by differentiating
Eq. 1-6, giving

$$\frac{dI}{dV_0} = \frac{1}{R} \qquad (1\text{-}9)$$

Because of the straight-line relationship of Fig. 1-19(b), the resistor is known
as a *linear* resistor. In the event that this straight-line relationship does not
exist, the resistor is said to be *nonlinear*. The I–V_0 characteristic of two classes
of nonlinear resistors are shown in Fig. 1-20. In this text we will be concerned
only with the linear case.

Equation 1-6 may be solved for the current in terms of the voltage, giving

$$I = \frac{1}{R}V_0 = GV_0 \qquad (1\text{-}10)$$

where the quantity $G = 1/R$ is known as the *conductance* in *mhos* (℧). From
Eq. 1-9, we see that the slope of the straight-line relationship between I and V_0,
shown in Fig. 1-19(b), is the conductance, G.

We digress here to discuss another word that will find frequent usage
throughout the book, the word *model*. When we have described the resistor as
simply made up of a certain amount of resistive material such as carbon, we

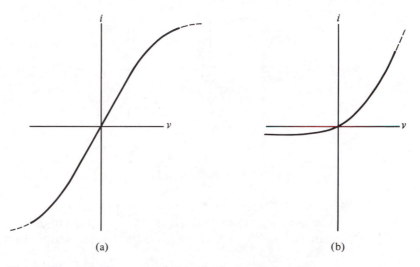

Figure 1-20.   Voltage–current relationships for nonlinear resistors.

have greatly simplified the picture. Under certain conditions, other effects will become important. In Chapters 6 and 7 we discuss inductive and capacitive effects, which may already be familiar to you from your high school course in physics. In such cases we say that we model the resistor with a combination of resistive and inductive or capacitive effects. Thus, the model of the device relates to the assumptions made in describing it. This model may be simply a word picture, it may be an equation such as $V = RI$, or it may be a schematic circuit diagram such as that used for the simple description of the resistor as in Fig. 1-17. Before any analysis is undertaken, it is important to state the model that is being employed to describe each device being used. This concept will be further clarified when, in Chapter 5, we describe various different models used to represent the operational amplifier.

EXERCISES

1-5.1. A resistor has the value of 30 kΩ and the current in it is measured to be 0.5 mA. Find (a) the conductance, (b) the terminal voltage, and (c) plot the I–V_0 characteristic.

Ans.   (a) 3.33×10^{-5}; (b) 15 V

1-6   REFERENCE DIRECTIONS

In Fig. 1-4 the arrow direction for the current I was chosen to be in the actual direction of the current. What do we do if we *do not know* the actual direction? You may suspect that the answer is this: *Assign any direction.* If the direction assigned is opposite to the actual direction of current, the answer will turn out to have a negative sign! To establish this result, consider the circuit shown

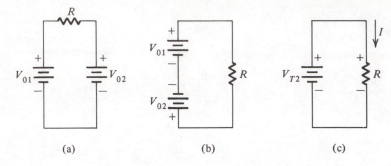

| (a) | (b) | (c) |

Figure 1-21. Circuit used in arriving at the result of Eq. 1-13.

in Fig. 1-21(a), which has two batteries connected in series. We may redraw our circuit as shown in Fig. 1-21(b). Now we found in discussing Fig. 1-15 that two batteries with opposing polarities were equivalent to a single battery of voltage

$$V_{T2} = V_{01} - V_{02} \tag{1-11}$$

so we may redraw the circuit as in Fig. 1-21(c). If we treat this circuit as we did that of Fig. 1-17(a), then

$$V_{T2} = RI \tag{1-12}$$

with the reference direction as assigned. Combining these equations and solving for I, we have

$$I = \frac{V_{T2}}{R} = \frac{V_{01} - V_{02}}{R} \tag{1-13}$$

From this, we see that I will be positive and so in the reference direction, provided that $V_{01} > V_{02}$. Otherwise, I will be negative and so in the opposite of the reference direction. Had we chosen the reference direction in Fig. 1-21(c) opposite to that actually chosen, we would have found a current identical to Eq. 1-13 except for a change in sign. Hence we reach this important conclusion: *Reference directions for both voltage and current may be chosen arbitrarily, and interpreted in terms of the directions so chosen; this process will always give consistent results.* If you do not assign reference directions, it will not be possible to interpret results.

The circuit of Fig. 1-22 illustrates this result. We cannot know in advance what the actual directions of currents in the three resistors will be. That does not matter, however. A first step in analysis is to *arbitrarily assign reference*

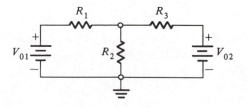

Figure 1-22. Circuit for which reference directions must be chosen.

directions for voltages and currents. It will simplify matters if voltages and currents for each resistor are chosen with the $+$ at the tail of the arrow as in Fig. 1-17(b), but even this is not necessary. These statements will be illustrated throughout our discussion and we will assign reference directions to all unknown voltages and currents quite arbitrarily in all our circuits.

1-7 CIRCUIT CONNECTIONS

Electric circuits or *networks* are constructed by interconnections of circuit elements—batteries and resistors so far. Electrical wires used to connect the elements will be assumed ideal, that is, resistanceless. By Ohm's law, there is no voltage across an ideal (i.e., zero-resistance) wire regardless of the current flowing through it.

Circuits in our figures will be drawn with clarity and convenience in mind. Frequently, long wires will result as a consequence. In practice, most wires are kept as short as possible to avoid undesirable effects due to their nonideal nature. Where two or more elements are connected together, the connections are indicated by dots when it is necessary to emphasize them, as in Fig. 1-23.

It is important to realize that identical circuits do not have to look the same on paper. As long as the connections are the same for the same elements, the circuits are the same. Satisfy yourself that all the circuits in Fig. 1-23 are identical. In each case: R_1 connects the battery at the $+$ terminal; R_1, R_2, and

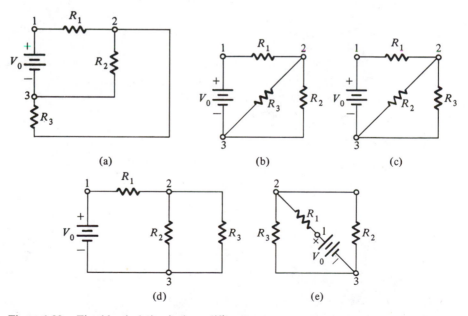

Figure 1-23. Five identical circuits have different appearances.

R_3 are connected together; and R_2 and R_3 connect to the battery at the $-$ terminal.

1-8 SHORT AND OPEN CIRCUITS

Two circuit connections that have special importance are the *short-circuit* and the *open circuit*. When two points are connected together by an ideal wire as in Fig. 1-24, they are *short-circuited*. A short circuit represents an ideal connection and no voltage exists across it. However, any amount of current can pass through it, depending upon the rest of the circuit. Thus, $V_{ab} \equiv 0$ *with an arbitrary value of* I_{ab} *represents a short circuit (sc).*

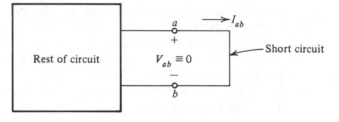

Figure 1-24. If the resistance between a and b is zero, these terminals are said to be short-circuited.

Two points are *open-circuited* if there is no circuit element or a direct connection between them, as shown in Fig. 1-25. Since there is no connecting path between them, $I_{ab} \equiv 0$, whereas V_{ab} is determined by the rest of the circuit. Thus, *two points with arbitrary voltage across them and zero current between them represent an open circuit (oc).*

Frequently, we make use of such voltage–current relationships as above to identify the presence of a short or open circuit. We can also identify these situations in terms of a resistance between two points, as in Fig. 1-26. If $R = 0$, then for arbitrary I_{ab},

$$V_{ab} = RI_{ab} \equiv 0 \tag{1-14}$$

Thus, $R = 0$ *represents a short circuit.* If $R = \infty$, then for arbitrary V_{ab},

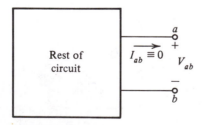

Figure 1-25. If the current between a and b is zero, the terminals a and b are said to be open-circuited.

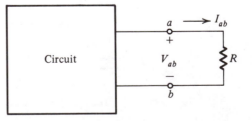

Figure 1-26. The circuits of Figs. 1-24 and 1-25 can be considered as limiting cases of the circuit shown for which $R = 0$ or $R = \infty$.

$$I_{ab} = \frac{V_{ab}}{R} \equiv 0 \qquad (1\text{-}15)$$

An *open circuit is thus represented by* $R = \infty$.

The terms "zero resistance" and "short circuit" may thus be used interchangeably, just as may the terms "infinite resistance" and "open circuit."

EXERCISES

1-8.1. In the circuit of Fig. 1-27, when the switch, Sw, is closed, what is the value of (a) I_1 and (b) I_2 in terms of I?

Ans. (a) 0; (b) *I*.

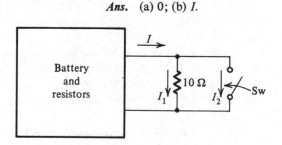

Figure 1-27. Circuit for Exercise 1-8.1.

1-8.2. In Fig. 1-28, the voltages across two of the resistors are given. Determine the value of the voltage V_3.

Ans. 10 V

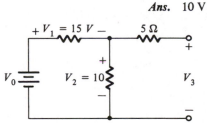

Figure 1-28. Circuit for Exercise 1-8.2.

1-8.3. The circuit shown in Fig. 1-29 is made up of three resistors and two switches connected as shown. Determine the values of the voltages V_2, V_3, V_4, V_5, and

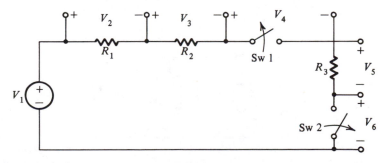

Figure 1-29. Circuit for Exercise 1-8.3.

V_6 if $V_1 = 10$ V if (a) the switch Sw$_1$ is open and Sw$_2$ is closed; (b) both switches are open.

Ans.   (a) $V_4 = 10$.

I-9   SERIES AND PARALLEL CONNECTIONS

Two types of connections are encountered quite extensively. They form either complete circuits or parts of circuits. *Series* connection of circuit elements (i.e., batteries and resistors so far) is distinguished by the property that the *same current* must flow through each of them. *Parallel* connection is distinguished by the property that the *same voltage* must exist across each of the circuit elements. These connections are shown in Fig. 1-30. In (a) all the current flowing out of any circuit element must flow into the next one, since it can escape nowhere else. In (b) the same voltage exists across each circuit element, since they are all connected together by ideal resistanceless wires. Each of the elements shown is either a battery or a resistor or any other element to be discussed in later chapters.

Such series and parallel combinations may also occur as parts of larger,

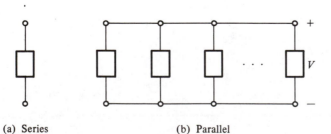

(a) Series   (b) Parallel

Figure 1-30.   Series and parallel connections of elements.

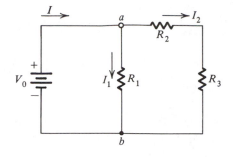

Figure 1-31.   Circuit with both series and parallel connections.

more complicated circuits as shown in Fig. 1-31. The battery V_0 and the resistor R_1 are in parallel since the same voltage V_0 exists across them. Resistors R_2 and R_3 are in series since the same current I_2 must flow through them. Observe that the battery current I splits at a into I_1 and I_2. I_2 flows through R_2 and R_3 and then joins I_1 at b and the sum returns to the battery.

Many properties of series and parallel combinations will be derived in later sections. For now, let us make use of them to derive some obvious results. Consider the series combination of Fig. 1-32(a). We have a battery V_0 in series

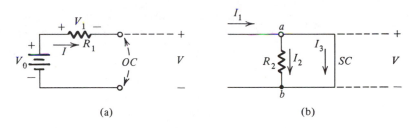

(a)   (b)

Figure 1-32.   Special cases in which (a) $V_1 = 0$ and (b) $I_2 = 0$.

with R_1 in series with an open circuit. The same current I flows through all of them. But the current in the open circuit must be zero, as seen in Section 1-8, so that $I \equiv 0$. By Ohm's law, the voltage across R_1 is $V_1 = R_1 I \equiv 0$. The composite voltage across V_0 and R_1 combination is (by Fig. 1-14) $V_0 - V_1 = V_0$. Thus, the voltage across the open circuit, V, is equal to V_0.

Next consider the parallel combination in Fig. 1-32(b). Here we have R_2 in parallel with a short circuit. The voltage V across the combination must be the same. Since the voltage across a short circuit is always zero, $V \equiv 0$. This implies by Ohm's law that $I_2 = V/R_2 \equiv 0$. The current I_1 coming in at a will not split but all of it will go into the short circuit, so that $I_3 = I_1$.

In some ways we are anticipating here the consequences of Kirchhoff's laws, to be discussed in Chapter 2. But these items should form such an intuitive part of our circuit thinking that we have presented them first. For example, it should be possible by a mere glance at the circuit of Fig. 1-33 to see that the voltage across R_1 is just V_0 (since R_2 carries no current and the voltage across

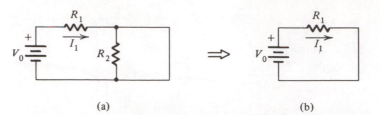

(a)

(b)

Figure 1-33. · Since there is no current in R_2, it may be removed from the circuit.

it is zero). In effect, the circuit of part (b) is all that is necessary to determine the current, which by Ohm's law is V_0/R_1.

I-10 **SUMMARY**

1. Current is the time rate of flow of charge, $i = dq/dt$. Conventional current is the flow of positive charges. $q_T = q(t) - q(t_0)$, or

$$q_T = \int_{t_0}^{t} i(t)\, dt$$

2. Voltage between two points is defined as the work (energy) per unit of charge to move the charge from one point to the other: $v = w/q$.

3. Reference directions are assigned for current and voltage, an arrow for current, $+$ and $-$ signs for voltage. An algebraic sign is used to indicate whether the current or voltage are in the reference directions, a positive sign indicating in the reference direction, and a negative sign indicating not in the reference direction.

4. Ohm's law relates voltage and current in a resistor. With reference directions shown in Fig. 1-34, then

$$V = RI \qquad \text{or} \qquad I = \left(\frac{1}{R}\right)V = GV$$

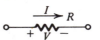

Figure 1-34. Reference directions for the resistor.

5. Resistors connected in the manner shown in Fig. 1-35(a) are said to be in *series;* those connected as in Fig. 1-35(b) are said to be in

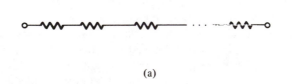

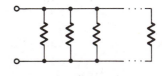

(a)

(b)

Figure 1-35. (a) Series and (b) parallel resistors.

parallel. A circuit with a combination of series and parallel connections is called a *series–parallel* connected circuit.

6. Two special conditions in circuits are shown as a short circuit and an open circuit. A short circuit is characterized by zero voltage by any current, whereas an open circuit has zero current by any voltage (see Fig. 1-36).

SC	OC
Zero voltage	Zero current
Any current	Any voltage

(a)   (b)

Figure 1-36. (a) Short circuit and (b) open circuit.

7. The SI system of units is used exclusively in electrical engineering for electrical quantities. These are shown in Table 1-1. The units and the system of prefixes of Table 1-2 should be memorized!

PROBLEMS

1-1. The electrical engineer makes use of an extensive range of values for basic quantities in describing large power systems as well as electronic measurement systems. It is important to be able to convert units using Table 1-2. Convert each of the following:

(a) 3 A to microamperes
(b) 0.05 μA to nanoamperes
(c) 0.04 μs to picoseconds
(d) 0.5 ps to nanoseconds
(e) 0.09 V to millivolts
(f) 5 mV to microvolts.

1-2. The current entering a terminal is given by the equation

$$i = e^{-10^5 t} \text{ mA}$$

Find the total charge flowing into the element between $t = 1$ μs and $t = 2$ μs.

1-3. The current entering a terminal is given by the equation

$$i = 10 \cos \left(1000t + \frac{\pi}{6} \right) \text{ mA}$$

Find the charge flowing into the element between $t = 0$ and $t = 2\pi/3$ ms.

1-4. The figure shows the charge entering the positive terminal of a circuit element as a function of time. (a) Find the total charge that entered the element between 1 s and 5 s. (b) Find the current entering the element at $t = 3$ s.

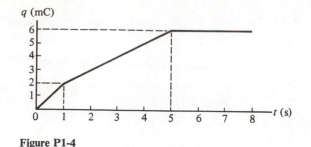

Figure P1-4

1-5. In the figure of Prob. 1-4, replace q in mC by i in mA, where i is the current entering the positive terminal of an element. (a) Find the charge that has entered the terminal between $t = 1$ s and $t = 5$ s. (b) Find the rate at which the charge is entering at $t = 3$ s.

1-6. The current into the positive terminal of a 6-V battery is a constant 3 A for 1 h and is then reduced to 2 A for an additional hour. (a) What is the total charge delivered to the battery? (b) How much energy is supplied to the battery?

1-7. The battery of Prob. 1-4 is replaced by a battery that has a terminal voltage that varies linearly from 2 to 6 V over a 2 h period that the charging current is 3 A. (a) What is the total charge delivered to the battery? (b) How much energy is supplied to the battery?

1-8. A resistor can be manufactured as a film of resistive material on a plate of insulator. Final adjustment or "trimming" of the value of the resistance can be accomplished by burning away selected portions of the resistive film with a laser. The figure shows the pattern of the resistive film for such a thin-film resistor. It includes a "top hat" section, portions of which are removed during laser trimming of the resistance to a final value. The laser trim of the top hat may involve a straight cut or an L-shaped cut as shown in the figures (center and right). Discuss the relative merits of both kinds of cuts for trimming.

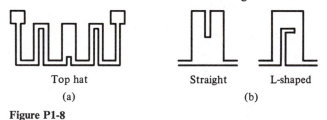

Top hat   Straight   L-shaped

(a)   (b)

Figure P1-8

1-9. The figure shows a ladder circuit in which two ammeters are located. If ammeter A_1 reads the current to be I_1, what will ammeter A_2 read for the current I_2? Why?

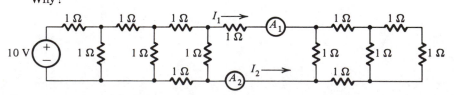

Figure P1-9

1-10. For the two simple circuits shown in the figure, find the currents labeled *I*.

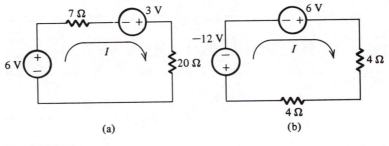

(a)   (b)

Figure P1-10

2

THE KIRCHHOFF LAWS

In this chapter we embark on a study of two important laws given by Kirchhoff in 1847. These laws will prove to be the foundation laws for the analysis of circuits. What do we mean by *analysis*? A circuit is completely analyzed when all voltages and all currents in the elements of the circuit are determined. Such complete analysis is not often required; we may be interested in only one or two voltages. In any case, we always make use of one or both of the Kirchhoff laws as a starting point.

2-1 KIRCHHOFF'S VOLTAGE LAW

We introduce this law by first introducing the concept of *summing voltages around a closed loop*. This we do in the following steps:

1. Define node.
2. Define closed loop.
3. Distinguish between a voltage rise and a voltage drop.
4. Describe the process of summing voltages.

A *node* is an electrical junction in a circuit at which two or more circuit elements are connected together. The circuit shown in Fig. 2-1(a) contains a battery and two resistors. The three elements are connected together by connecting wires which are assumed to have zero resistance. Points *a*, *b*, and *c* in

26

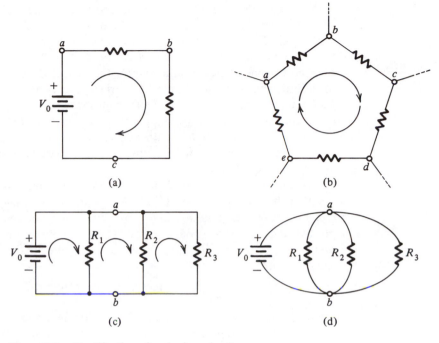

Figure 2-1.   Identification of nodes in a circuit.

this circuit are nodes. In Fig. 2-1(b), nodes are identified at *a, b, c, d,* and *e.* In Fig. 2-1(c), nodes are at *a* and *b.*

   All points connected together by resistanceless wires represent the same electrical junction or node. Thus, in Fig. 2-1(a) the node *c* could have been placed anywhere along the bottom line, since the entire line represents that node. In Fig. 2-1(c), nodes *a* and *b* could be placed anywhere on the top and the bottom lines, respectively, or simply combined as in Fig. 2-1(d). The points *a* and *b* are used to indicate the nodes, but the entire top line is the same node and so is the bottom line. Thus, *any two points short-circuited represent the same node.* In Fig. 2-1, closed loops are identified simply as *abca* in (a) and *abcdea* in (b). In (c), the loops are very simple as shown. Circular lines with direction-indicating arrows are drawn within the closed loops to both identify and give reference directions to the loops. Note that we consider only loops that close on themselves without crossing; there are certainly other "figure 8"-shaped loops in a more complicated circuit.

   We have selected the arrow direction of the loops in Fig. 2-1 to indicate the order in which we anticipate *summing the voltages* across the circuit elements in the loop. In other words, the arrow direction tells us the order in which we sum the voltages around a closed loop as we apply KVL. Thus, in Fig. 2-1(a), the direction that has been selected implies that we measure the voltage of *b* with respect to *a*, then of *c* with respect to *b*, and finally of *a* with respect to *c*. We then sum these voltages around the closed loop *abca*. Had we selected the arrow in the opposite direction, the summation procedure would have been reversed. Next consider the circuit shown in Fig. 2-1(b), which is more complicated or general than that shown in Fig. 2-1(a). Here dashed lines are used to indicate that there is more of the circuit that is not shown. The process of summation is exactly the same as that illustrated for (a) of the figure. Record (measure if you will) the voltages around the loop starting with *b* with respect to *a*, of *c* with respect to *b*, *d* with respect to *c*, *e* with respect to *d*, and finally *a* with respect to *e*. If you add up all these voltages, you have summed the voltages around the closed loop *abcdea*.

In Chapter 4 we consider the case of several loops, as in Fig. 2-1(c).

Kirchhoff's voltage law (KVL) states that *the algebraic sum of voltages around a closed loop must equal zero.* Since some of these voltages may be positive and some may be negative, such a summation is termed *algebraic summation,* to distinguish from others discussed below. KVL may be stated in equation form as

$$\sum_{\text{loop}} \text{voltages across elements} = 0 \qquad\qquad (2\text{-}1)$$

where the summation is around a closed loop in a specified direction.

In doing the summation we have just discussed, there is a need to sort out the voltages with respect to "algebraic sign" or polarity. It is convenient to define the terms "voltage rise" and "voltage drop" in order to do this sorting. In Fig. 2-2(a), the voltage of *b* is larger than that of *a* by the amount of the battery voltage, and this is spoken of as a *voltage rise* in the direction shown.

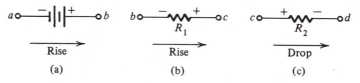

Figure 2-2.   Differentiation between voltage rise and voltage drop.

Such a voltage will appear with a positive sign in Eq. 2-1. The same polarity marks apply to the voltage across the resistor R_1 in Fig. 2-2(b); hence this, too, is a voltage rise. On the other hand, the reversed polarity of the resistor R_2 in Fig. 2-2(c) implies that the voltage at *d* is less than that at *c*, and this is a *voltage drop.* A voltage drop will appear with a negative sign in the summation of Eq. 2-1. In other words, a voltage drop is a negative voltage rise, and vice versa.

All the voltages appearing in Eq. 2-1 are either voltage rises, in which case they appear with positive sign, or they are voltage drops and appear in the summation with negative sign. So Eq. 2-1 may be written as

$$\sum \text{voltage rises} - \sum \text{voltage drops} = 0 \qquad (2\text{-}2)$$

or

$$\sum \text{voltage rises} = \sum \text{voltage drops} \qquad (2\text{-}3)$$

Equation 2-3 is the alternate form of KVL, which states that *the sum of voltage rises is equal to the sum of voltage drops* around a closed loop. It is also possible to write Eq. 2-2 in a different form by giving voltage drops positive signs:

$$\sum \text{voltage drops} - \sum \text{voltage rises} = 0 \qquad (2\text{-}4)$$

To summarize, we may apply KVL to any closed loop by either equating the sum of voltage drops to the sum of voltage rises as in Eq. 2-3 or by setting the algebraic summation of voltages to zero as in Eq. 2-1. In Eq. 2-1 we may assign positive sign to voltage rises, as in Eq. 2-2, or to voltage drops, as in Eq. 2-4.

The Kirchhoff voltage law is justified in terms of the conservation of energy. If there was voltage left over after we had traversed a closed loop, then voltage would have been generated. It is sometimes advantageous to think of voltage as being analogous to elevation in the mapping of a mountain. Thus, if we start at a given point on the mountain and record the change in elevation as various points on the mountain are visited, then when we return to the point of origin we can be sure that the summation of elevation increases would just equal the summation of elevation descreases—for otherwise we would not be back to our point of origin.

A circuit must first be prepared by assigning unknown currents and voltages for all the elements in the circuit and by taking care to assign the same current to all the circuit elements in series. Loops are identified and assigned directions. After this, the KVL is applied around each loop. Ohm's law may then be used, if necessary, to write down the voltage rises or drops across the resistors in terms of the (unknown) currents through them. It is then possible to solve for these currents if desired.

We will concentrate for now on circuits that consist of single loops. In a single loop all the elements are in series and therefore the same current, called the *loop current*, flows through all of them, as can be seen in Figs. 2-3 and 2-4. It is usually convenient to assign the same direction to both the loop and the current in the loop. The steps involved for *single-loop circuits* are:

1. Identify and assign the loop, the loop direction, the loop current, and the voltage polarities across all elements.
2. Apply the KVL to the loop and use Ohm's law, if necessary, for each resistor.
3. Solve the equation (if desired).

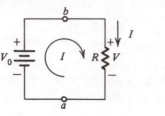
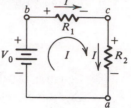

Figure 2-3. Description of a circuit path called a loop.

Figure 2-4. Circuit analyzed in Eq. 2-3.

A few examples will illustrate the use of KVL. In the circuit of Fig. 2-3, the closed loop *aba* is shown in the figure. The loop current *I* and the polarities of the voltages are assumed as shown. There is one voltage rise, V_0, in the direction of the loop and one voltage drop across the resistor, *V*. Then using Eq. 2-2,

$$V_0 = V \tag{2-5}$$

which by Ohm's law is

$$V_0 = RI \tag{2-6}$$

If we use Eq. 2-2, then

$$V_0 - V = 0 \tag{2-7}$$

which by Ohm's law becomes

$$V_0 - RI = 0 \tag{2-8}$$

The same equation is obtained in either case and may be solved for *I* if V_0 and *R* are known.

For most circuits, it is easy to use Ohm's law directly in the application of KVL to write down the appropriate voltage rises and drops across resistors. Thus, Eq. 2-6 or Eq. 2-8 is written directly without writing the preceding step in each case. This is the policy we will follow when there is no chance of confusion. We will also use the loop direction arrow to indicate the loop current direction, as in Fig. 2-4, when there is only a single-loop current.

The circuit of Fig. 2-4 consists of two resistors in series with a battery. The closed loop, the current, and the voltage polarities are shown in the figure. The KVL is

$$V_0 = R_1 I + R_2 I \tag{2-9}$$

Here if V_0, R_1, and R_2 are known, we may solve for the current, *I*.

A graphical interpretation of Eq. 2-9 may be made as shown in Fig. 2-5. We assume that the elements in the circuit are distributed in such a way that the voltage along the length of the element varies linearly. Point *a* is the reference or datum, so the voltage is zero. In moving from point *a* to point *b*, the voltage increases by the amount of the battery voltage V_0. From *b* to *c* there is a voltage

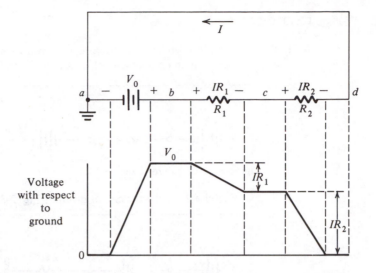

Figure 2-5. Graphical description of the voltage variation at different points around a loop.

drop by the amount IR_1, and in going from c to d, there is an additional drop of amount IR_2. Point d is connected to point a, so they have the same voltage or zero value. As we climb the hills on this figure and then descend in two steps, we know that we must return at d to the same voltage we started at—at point a. This is a statement of the Kirchhoff voltage law.

This simplicity of single-loop circuits is not to be found when we consider more complicated circuits, such as that of Fig. 2-6. The figure identifies a closed loop that does not include R_2. Summing voltages around this loop, *abcda*, and equating voltage rises to voltage drops, we have

$$V_1 = R_1 I_1 + R_3 I_2 + V_2 \qquad (2\text{-}10)$$

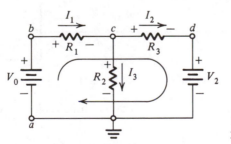

Figure 2-6. Circuit analyzed in Eq. 2-10.

Clearly, we will need additional information to solve the equation if both I_1 and I_2 are unknowns. We postpone a more complete analysis until additional concepts are introduced.

EXERCISES

2-1.1. For the circuit given in Fig. 2-7, prepare an analysis similar to that given in Fig. 2-5, determining numerical values for the voltages.

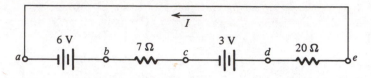

Figure 2-7. Circuit for Exercise 2-1.1.

2-1.2. Repeat Exercise 2.1-1 for the circuit given in Fig. 2-8.

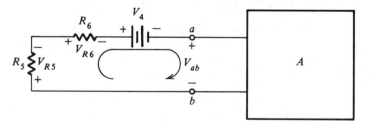

Figure 2-8. Circuit for Exercise 2-1.2.

2-2 SOME KVL APPLICATIONS

The KVL statement is quite general and can be applied to any situation. The loop voltages may be across any entities: for example, the circuit elements discussed so far, any other elements, groups of elements or subcircuits, open circuits, and so on. Consider the circuit in Fig. 2-9, where only part of the

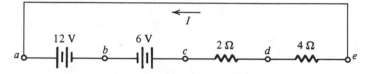

Figure 2-9. Circuit analyzed in Eq. 2-11.

circuit is specified and the rest is some subcircuit A. Applying KVL to the loop shown,

$$V_{ab} + V_{RS} + V_{R6} + V_4 = 0 \qquad (2\text{-}11)$$

or

$$V_{ab} = -V_{RS} - V_{R6} - V_4 \qquad (2\text{-}12)$$

Equation 2-11 is just a statement giving us the composite voltage across the combination R_5, R_6, V_4. Thus, the discussion in Section 1-4 about composite voltages is a consequence of KVL. Of course, the subcircuit A can be any combination of elements, open circuit, short circuit, and so on.

    Naturally, if the entire loop is specified, we can obtain the voltage between any two nodes in the loop in two alternative ways. Suppose that we wish to determine the voltage V_{ab} with the polarities as shown in the circuit of Fig. 2-10. We may use the approach described above and go along the dashed loop as shown to obtain

$$V_{ab} = -V_{R1} + V_1 - V_{R2} \tag{2-13}$$

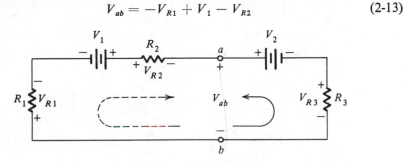

Figure 2-10.   Circuit analyzed in Eqs. 2-13 and 2-14.

Or we may use the solid loop and obtain

$$V_{ab} = V_{R3} + V_2 \tag{2-14}$$

The two values of V_{ab} must be the same and if we equate the two right-hand sides of Eqs. 2-13 and 2-14, we have

$$V_{R3} + V_2 = -V_{R1} + V_1 - V_{R2} \tag{2-15}$$

Equation 2-15 is clearly valid since it is a statement of KVL applied to the entire loop.

EXAMPLE 2-1

In Fig. 2-11, given $I = 2$ A, find V.

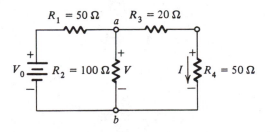

Figure 2-11.   Circuit for Example 2-3.

Solution

Going along the path R_3, R_4 from a to b,

$$V = V_{R3} + V_{R4} = R_3 I + R_4 I \tag{2-16}$$

$$= 20 \times 2 + 50 \times 2 = 140 \text{ V} \tag{2-17}$$

Voltages across circuit elements are *node-to-node* voltages. If the circuit is grounded, the voltages of the nodes are usually given with respect to ground. Such voltages, defined for the various nodes, are called *node-to-datum* voltages to distinguish them from node-to-node voltages. We frequently need to find other voltages from these node-to-datum voltages. For example, suppose that the node-to-datum voltages V_1, V_2, and V_3 are known in Fig. 2-12. Then

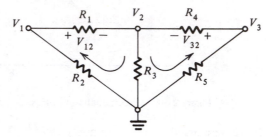

Figure 2-12. Circuit analyzed in Eqs. 2-18 and 2-19.

by the discussion above, it follows that

$$V_{12} = V_1 - V_2 \tag{2-18}$$

and

$$V_{32} = V_3 - V_2 \tag{2-19}$$

Rearranging the order of elements in series does not change the current in the circuit. Consider the circuit of Fig. 2-13(a), where elements in series are

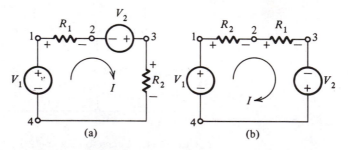

(a)   (b)

Figure 2-13. Circuits described by Eq. 2-20.

rearranged as in (b). The KVL for both circuits gives

$$V_1 + V_2 = V_{R1} + V_{R2} \tag{2-20}$$

or

$$V_1 + V_2 = R_1 I + R_2 I \tag{2-21}$$

so

$$I = \frac{V_1 + V_2}{R_1 + R_2} \tag{2-22}$$

The currents are the same in both cases as long as the values of V_1, V_2, R_1, and R_2 are the same in the two cases. Further, the KVL equations are the same and so are the voltages (drops, rises) across the elements. The circuits are *different*, however. Also, the voltages of nodes 2 and 3 with respect to 4 are quite different in the two cases. For example, the voltage of node 2 with respect to 4 is $V_1 - R_1 I$ in (a) and $V_1 - R_2 I$ in (b).

Similarly, an interchange of elements in parallel does not change the voltage across them. In fact, even the circuit does not change in this case, as can be easily verified.

EXERCISES

2-2.1. In the circuit shown in Fig. 2-14, the node marked e is grounded and the remaining node-to-ground voltages are recorded as follows: $V_a = 13$ V, $V_d = 21$ V,

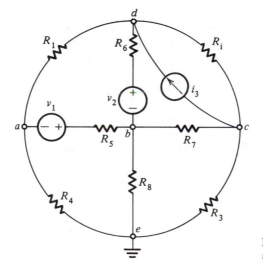

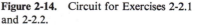

Figure 2-14. Circuit for Exercises 2-2.1 and 2-2.2.

$V_c = 9$ V, and $V_b = 12$ V. Determine all the node-to-node voltages in the circuit.

> *Ans.* $V_{ab} = 1$ V, $V_{bd} = -9$ V, $V_{cd} = -12$ V,
> $V_{ad} = -8$ V, $V_{bc} = 3$ V

2-2.2. In the circuit shown in Fig. 2-14, the node-to-ground voltages are those given in Exercise 2-2.1, with node e grounded. That ground is removed, and node d is grounded. What will the remaining node-to-ground voltages be after this change is made?

Ans. $V_a = -8$ V, $V_c = -12$ V, $V_b = -9$ V,
$V_e = -21$ V

2-3 THE VOLTAGE-DIVIDER CIRCUIT

The voltage-divider circuit is one that we will frequently use in the studies to follow. The circuit, shown in Fig. 2-15, consists of two resistors, R_1 and R_2, connected in series and driven by a battery. It is assumed that no other resistor is connected to the node marked 2; this is indicated by stating that $I_2 = 0$.

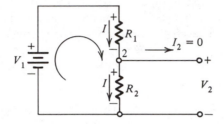

Figure 2-15. Voltage-divider circuit.

We are interested in finding V_2 as a function of V_1. We are sometimes interested in the fraction of V_1 that appears as V_2, so that we frequently solve for the ratio V_2/V_1. This is known as a *gain* or a voltage-ratio *transfer function*, designated T in later chapters. Summing the voltages around the loop, we have

$$V_1 = R_1 I + R_2 I \qquad (2\text{-}23)$$

and in addition, observe that

$$V_2 = R_2 I \qquad (2\text{-}24)$$

Solving the first of these equations for I, and substituting into the second, we obtain

$$\frac{V_2}{V_1} = \frac{R_2}{R_1 + R_2} \qquad (2\text{-}25)$$

This equation indicates that the voltage V_2 is a fraction $R_2/(R_1 + R_2)$ of the voltage V_1, and hence the circuit may be used when it is necessary to reduce a voltage. Writing Eq. 2-25 as

$$\frac{V_2}{V_1} = \frac{1}{1 + R_1/R_2} \qquad (2\text{-}26)$$

we plot V_2/V_1 as a function of R_1/R_2 in Fig. 2-16. Note that when R_1 is zero, no division takes place, so that $V_2 = V_1$. The larger the value of R_1 compared to R_2, the smaller the voltage V_2 will be.

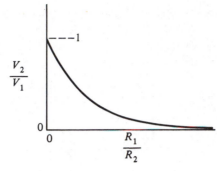

Figure 2-16. Equation 2-26 plotted, showing the behavior of a voltage-divider circuit.

EXAMPLE 2-2

The input to an amplifier is required to be $\frac{1}{10}$ of that available from a given source. Design a circuit to give the required voltage to the amplifier.

Solution

From Eq. 2-26 we see that

$$\frac{V_2}{V_1} = \frac{1}{10} = \frac{1}{1 + R_1/R_2} \tag{2-27}$$

requires that $R_1/R_2 = 9$. If we select $R_2 = 1000 \ \Omega = 1 \ \text{k}\Omega$, it is necessary that $R_1 = 9 \ \text{k}\Omega$. Hence, a circuit that meets the requirement is that shown in Fig. 2-17. Clearly, there will be other considerations that will determine whether R_2 should be 1 kΩ or 1 MΩ or some other value.

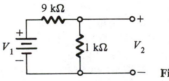

Figure 2-17. Circuit for Example 2-2.

There are many occasions when it is desirable to have the (output) voltages with respect to ground (i.e., node-to-datum voltages). All that is needed is to ground the negative terminal of the battery in Fig. 2-15 as shown in Fig. 2-18.

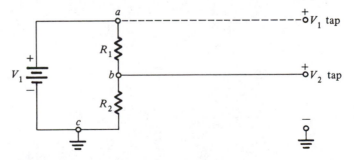

Figure 2-18. Generalization of the voltage-divider circuit to introduce the concept of a tap.

Then V_2 is obtained at node b. The second wire extending from c is not always shown, since ground is available everywhere and the voltage at b is called a *tap*. It is also possible to have V_1 available as a tap, as shown. Other taps are possible by using several resistors, as shown in Fig. 2-19. The tap

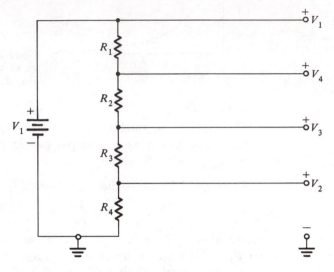

Figure 2-19. Voltage-divider circuit with four taps.

values in this case will, of course, depend on the values of R_1, R_2, R_3, and R_4. Appropriate design equations are obtained by using KVL as before.

An interesting case occurs when both positive and negative tap values are desired (with respect to ground). If we ground the node b in Fig. 2-18 (instead of c), then node a will be positive and node c will be negative with respect to ground, as in Fig. 2-20. The values are given by

$$I = \frac{V_1}{R_1 + R_2} \qquad (2\text{-}28)$$

$$V_2 = -R_2 I = \frac{-R_2}{R_1 + R_2} V_1 \qquad (2\text{-}29)$$

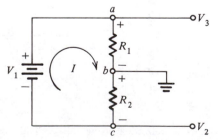

Figure 2-20. Shifting of the datum or ground node in a voltage-divider circuit.

$$V_3 = R_1 I = \frac{R_1}{R_1 + R_2} V_1 \tag{2-30}$$

For example, if $V_1 = 100$ V and $R_1 = R_2$, then $V_2 = -50$ V and $V_3 = +50$ V.

EXERCISES

2-3.1. Design a resistor circuit to provide the following voltages on taps as in Fig. 2-19: 2, 4, and 9 V. The battery has the value $V_0 = 10$ V. Assume that the taps are unloaded and that no part of the battery is grounded (it is said to be floating). The voltages required are all as measured with respect to ground. Assume that $R_4 = 2\,\Omega$.

Ans. $R_1 = 1\,\Omega$, $R_2 = 5\,\Omega$, $R_3 = 2\,\Omega$, $R_4 = 2\,\Omega$

2-3.2. Repeat Exercise 2-3.1 with the specifications changed to the following: Tap voltages: 18, 5, -7, and -18 V; 36-V battery to be used and assume that $R_2 = 5\,\Omega$.

Ans. $R_1 = 13\,\Omega$, $R_2 = 5\,\Omega$, $R_3 = 7\,\Omega$, $R_4 = 11\,\Omega$

2-4 KIRCHHOFF'S CURRENT LAW

Kirchhoff's current law (KCL) relates to summing currents at a node and, as we shall see, it may be used to determine the node voltages in a circuit. KCL simply states that *the sum of the currents directed into any node must equal the sum of the currents coming out of that same node*. It is a consequence of conservation of charge, for if the two sums were not equal, charge would pile up at the node and charge conservation would not hold. KCL is written for any node, say a, as

$$\sum_{\text{node } a} \text{currents in} = \sum_{\text{node } a} \text{currents out} \tag{2-31}$$

Alternatively, KCL may be written in terms of the *algebraic* sum of the currents at a node. If the incoming currents are assumed to have positive directions, then the outgoing currents have negative directions, and vice versa. In any case, KCL then states that *the algebraic sum of the currents at a node is zero*. Thus,

$$\sum_{\text{node } a} \text{algebraic currents} = 0 \tag{2-32}$$

or if incoming currents are given positive directions, then

$$\sum_a \text{currents in} - \sum_a \text{currents out} = 0 \tag{2-33}$$

and if outgoing currents are given positive directions, then

$$\sum_a \text{currents out} - \sum_a \text{currents in} = 0 \tag{2-34}$$

Of course, Eqs. 2-32 to 2-34 all represent the same equation as Eq. 2-31 and can be obtained by simple manipulations.

The KCL we have just written in several equivalent forms applies at each node of a circuit. Part of a circuit is shown in Fig. 2-21, showing the resistors connected to node 1. For each of the four resistors, a current is identified along

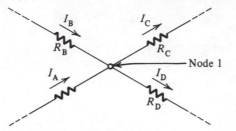

Figure 2-21. Circuit used to illustrate KCL and to arrive at Eq. 2-37.

with an assigned reference direction. Summing currents directed into the node and equating these to currents directed out of the node, we have

$$I_A + I_B = I_C + I_D \tag{2-35}$$

which is in the form of Eq. 2-31. Rearranging this equation to be like Eq. 2-33, we have

$$I_A + I_B - I_C - I_D = 0 \tag{2-36}$$

or in the form of Eq. 2-34,

$$I_C + I_D - I_A - I_B = 0 \tag{2-37}$$

We note that Eq. 2-37 may be obtained from Eq. 2-36 by multiplying by -1.

KCL gives us a complete and precise statement about the currents at a node. We anticipated some intuitive aspects of KCL regarding composite currents in section 1-4. Similarly, the fact that the same current flows through a series combination of elements is a consequence of KCL, as will be seen below.

Another interpretation of the KCL is often useful in visualizing a problem. Figure 2-22 shows a resistive ladder circuit with two currents, I_1 and I_2, identified. Two ammeters, A_1 and A_2, are provided to measure I_1 and I_2. If we extend the circuit as shown in Fig. 2-22(b) and then cut it with a plane surface as shown, then at the surface the KCL must apply: Any current entering the surface must leave that surface. Then at the surface, $I_1 + I_2 = 0$ or $I_2 = -I_1$. Thus, the ammeters would read the same magnitude, but opposite sign.

In analyzing a circuit, KCL enables us to determine all the unknown node voltages. From these, it is easy to determine any desired voltage or current in the circuit. The steps involved in the use of KCL require first the identification of all nodes and designating any one of them as the reference node (or datum or ground). Node voltages with respect to the reference are then assigned to all nodes. Current directions are also assigned for all elements. KCL is then written for each node (with an unknown node voltage to be determined).

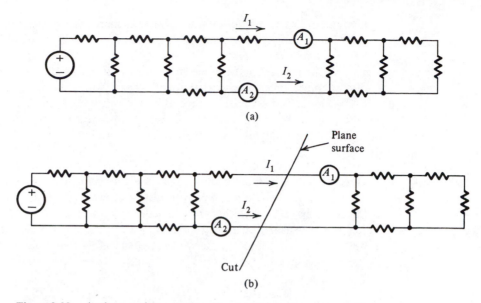

(a)

(b)

Figure 2-22. A plane surface separates the circuit into two parts; KCL applies where the surface cuts the circuit.

Ohm's law is then used, if necessary, to write down the currents at the node in terms of node voltages. The resulting equations may then be solved if desired.

In this chapter we concentrate on those circuits that have only a single unknown node voltage. The steps are:

1. Assign a reference node, the node-to-datum voltage for the unknown node, and current directions for all elements.
2. Write KCL for the unknown node and use Ohm's law for the currents through resistors, if necessary.
3. Solve the equations, if desired.

The manner in which we make use of KCL can be seen by considering the simple two-resistor circuit of Fig. 2-23. In that circuit the battery has the value V_1, and we desire to determine a voltage V_2, which is the voltage measured across the resistor R_2. To avoid the confusion that sometimes comes from an

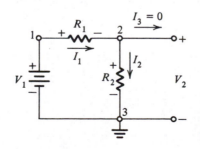

Figure 2-23. Circuit used to illustrate the concept of Fig. 2-22.

analysis of this type of circuit, we note that there is no current in the wire which is provided for an indication of the manner in which V_2 would be measured; hence the indication $I_3 = 0$.

We first identify the three nodes and select one of them, say node 3, as our reference. We observe that node 1 has a known voltage V_1 (with respect to 3). The unknown node voltage at node 2 is the same as V_2, the voltage to be determined. Assumed currents I_1 and I_2 and the voltage polarities are also shown. Then applying KCL at node 2, we write

$$I_1 = I_2 \tag{2-38}$$

Observe that this merely shows that the current in the series combination of R_1 and R_2 must be the same. Note that R_1 and R_2 are in series, since $I_3 = 0$, implying that no other element is connected at node 2.

We next use Ohm's law to express the currents in Eq. 2-38 in terms of the node voltages V_1 and V_2. As we have seen before, the voltage across R_1 with the desired polarity is $V_1 - V_2$ and that across R_2 is simply V_2. Then Eq. 2-38 becomes

$$\frac{V_1 - V_2}{R_1} = \frac{V_2}{R_2} \tag{2-39}$$

Solving this equation for V_2, we find that

$$V_2 = \frac{R_2}{R_1 + R_2} V_1 \tag{2-40}$$

The same result was obtained in Eq. 2-25, which was determined by the application of KVL to the same circuit. Hence, we have shown an alternative way in which we can derive the voltage-divider equation. This result will turn out to be general: *We may use either KCL or KVL to analyze a given circuit* and obtain the same solution. Later we will find that one choice may result in more effort than the other.

2-5 THE CURRENT-DIVIDER CIRCUIT

The circuit shown in Fig. 2-24 is made up of two resistors connected in parallel with the battery of voltage V. A current I flows from the battery. Applying KCL at node 1, with the other node chosen as the reference, we have

$$I = I_1 + I_2 \tag{2-41}$$

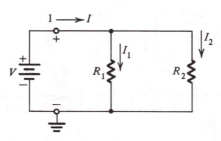

Figure 2-24. Current-divider circuit.

or the current I divides into two parts, I_1 and I_2. Given R_1 and R_2, what rule determines this division? The answer will be analogous to the result for the voltage-divider circuit of Fig. 2-25. Since $I_1 = V/R_1$ and $I_2 = V/R_2$, Eq. 2-41 is

$$I = \frac{V}{R_1} + \frac{V}{R_2} \tag{2-42}$$

or

$$V = I\frac{R_1 R_2}{R_1 + R_2} \tag{2-43}$$

Then

$$I_1 = \frac{V}{R_1} = \frac{R_2}{R_1 + R_2}I \tag{2-44}$$

and

$$I_2 = \frac{V}{R_2} = \frac{R_1}{R_1 + R_2}I \tag{2-45}$$

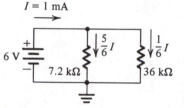

Figure 2-25. Circuit analyzed in Example 2-3.

Comparing the operation of the current divider with the result previously found for the voltage divider (Eq. 2-40), we see that the voltage division is directly related to the resistor across which the voltage is measured, whereas for the current divider, the division is related to the "other" resistor. In any case, observe that the current is split in half if $R_1 = R_2$.

EXAMPLE 2-3

Design a circuit such that the current from a 6-V battery is 1 mA and one resistor of the current divider circuit carries $\frac{1}{6}$ of the current.

Solution

From Eq. 2-43, we obtain

$$6 = 1 \times 10^{-3}\frac{R_1 R_2}{R_1 + R_2} \quad \text{or} \quad \frac{R_1 R_2}{R_1 + R_2} = 6000 \tag{2-46}$$

Let us select R_2 to carry $\frac{1}{6}$ of the total current, so that from Eq. 2-45 we require $I_2 = \frac{1}{6}I$ or

$$\frac{1}{6} = \frac{R_1}{R_1 + R_2} = \frac{1}{1 + R_2/R_1} \tag{2-47}$$

That is, we need $R_2/R_1 = 5$. Substituting $R_2 = 5R_1$ in Eq. 2-46, we find that

$$R_1 = 7.2 \text{ k}\Omega \quad \text{and} \quad R_2 = 36 \text{ k}\Omega \tag{2-48}$$

The complete circuit design is shown in Fig. 2-25.

The results obtained for the current divider and the voltage divider are quite general. They describe the manner in which the current or the voltage splits for a parallel or a series combination of resistors, respectively, anywhere in the circuit. Figure 2-26 shows a current divider in (a) and a voltage divider in (b) connected to arbitrary circuits. KCL at node 1 for the circuit in (a) will

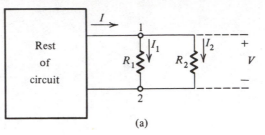

(a)

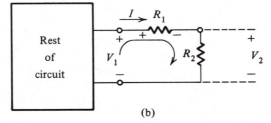

(b)

Figure 2-26.   Generalization of the current divider.

give us exactly the same equations as before, (Eqs. 2-41 through 2-43), and Eqs. 2-44 and 2-45 will follow. Similarly, KVL applied to the circuit in (b) will give us Eq. 2-23, from which will follow Eq. 2-25 as before.

More complicated current dividers such as that shown in Fig. 2-27 may be analyzed in a manner similar to the above. But more convenient methods are developed in Chapter 3.

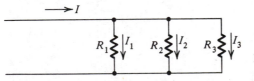

Figure 2-27.   Example of a more general current divider.

EXERCISES

2-5.1. Figure 2-28 shows a dc source of 100 V connected to a varying load R_1, the smallest value of which is 1 Ω. We would like to measure the current in the load, I_1, but the only ammeter available measures in the range 0 to 1 mA. This am-

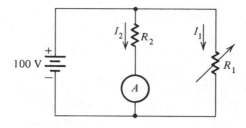

Figure 2-28. Circuit for Exercise 2.5-1.

meter is in series with a resistor R_2. We are required to design a current-dividing circuit so that the ammeter measures the current in the load.

$$\textit{Ans.} \quad R_1 = 1\,\Omega,\ R_2 = 0.1\ \text{M}\Omega$$

2-5.2. Repeat Exercise 2-5.1 under the condition that the only ammeter available has a range of 0 to 1 A.

$$\textit{Ans.} \quad R_1 = 1\,\Omega,\ R_2 = 100\ \Omega$$

2-6 ENERGY AND POWER IN RESISTOR CIRCUITS

Power (P) is defined as the time rate at which energy, w, is transferred,

$$P = \frac{dw}{dt} \tag{2-49}$$

As applied to a resistor, power is the time rate at which energy from the battery is being transferred to the resistor and dissipated in the form of heat. Since, in general,

$$\frac{dw}{dt} = \frac{dw}{dq} \times \frac{dq}{dt} = VI \tag{2-50}$$

we have the basic equation

$$P = VI \tag{2-51}$$

The unit for power may be determined from Eq. 2-50:

$$\frac{\text{joules}}{\text{coulomb}} \times \frac{\text{coulombs}}{\text{second}} = \frac{\text{joules}}{\text{second}} = \text{watts (W)} \tag{2-52}$$

In resistor circuits driven by batteries, V and I in Eq. 2-50 are constants, so the energy in the time interval from 0 to t is

$$w = \int_0^t VI\,d\tau = VIt \qquad \text{joules (J)} \tag{2-53}$$

which is shown in Fig. 2-29. The energy delivered to the resistor in the time interval from t_1 to t_2 is

$$w = VI(t_2 - t_1) \qquad \text{J} \tag{2-54}$$

The unit joules is equivalent to the electrical unit *watt-seconds*. One watt-second of energy is equivalent to 1 watt of power absorbed over 1 second. A more convenient unit is the *watt-hour* (Wh), which is obtained when time is measured in hours.

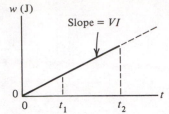

Figure 2-29. Plot illustrating Eq. 2-54.

As energy is being *absorbed by* the resistor, it is being *supplied* or *delivered* by the battery. In one case, energy is increasing; in the other, it is decreasing. Since dw/dt may thus be either positive or negative, so from Eq. 2-49, P may be positive or negative. How is the sign of P manifest in terms of circuits? The answer is that it is implied by the voltage and current reference directions.

Consider again the simple circuit shown in Fig. 2-30(a), with a resistor connected to the terminals of a battery. In (b) and (c) of the figure the battery

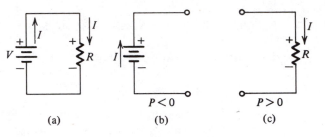

Figure 2-30. Conventions implying positive or negative energy.

and resistor are shown separately. The power P is positive for the resistor, negative for the battery. This is implied by the reference directions for voltage and current: If the $+$ sign is at the tail of the arrow, then $P > 0$; if the $-$ sign is at the tail of the arrow, then $P < 0$. This result for the battery and resistor holds for a general device, indicated by a box in Fig. 2-31. It illustrates the general rule about the sign of P as determined by voltage and current references. A positive value of P implies that the power is absorbed by the device, whereas a negative value of P implies that the power is supplied by the device.

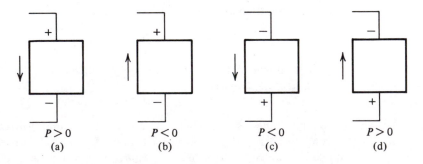

Figure 2-31. Postive and negative power conventions.

From the equation $P = VI$, and Ohm's law, $V = RI$, two other forms of equations for P_R, the power absorbed by a resistor, may be written:

$$P_R = RI^2 \quad \text{W} \tag{2-55}$$

or

$$P_R = \frac{V^2}{R} \quad \text{W} \tag{2-56}$$

where V and I for the resistor are assumed to be as shown in Fig. 2-31(a). Both forms find frequent use in the analysis of resistive circuits.

EXAMPLE 2-4

The voltage and current for the 5-Ω resistor shown in Fig. 2-32 are measured and found to be $V = 10$ V and $I = -2$ A. Determine the power and the energy dissipated by the resistor in 1 h.

Figure 2-32. Circuit for Example 2-4.

Solution

The reference directions indicate that P must be negative, so

$$P = -VI \tag{2-57}$$

Since $V = 10$ and $I = -2$, P is equal to $+20$ W. If the voltage and current remain constant for 1 h, then from Eq. 2-54,

$$w = 20 \text{ Wh} \tag{2-58}$$

A more common unit is the kilowatt-hour (kWh), and in terms of this unit

$$w = 0.02 \text{ kWh} \tag{2-59}$$

EXAMPLE 2-5

Part of a circuit is shown in Fig. 2-33. Find the power dissipated in each resistor if $V_1 = 20$ V.

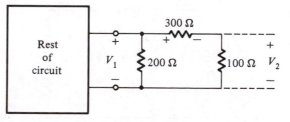

Figure 2-33. Circuit for Example 2-5.

Solution

V_1 is across the 200-Ω resistor, so by Eq. 2-56, power in the 200-Ω resistor is

$$P_1 = \frac{V_1^2}{200} = \frac{400}{200} = 2 \text{ W} \qquad (2\text{-}60)$$

The voltage V_1 is also across the series combination of 300 Ω and 100 Ω. So we can find V_2 by the voltage-divider rule of Eq. 2-25,

$$V_2 = \frac{100}{300 + 100} = \frac{100}{400} \cdot 20 = 5 \text{ V} \qquad (2\text{-}61)$$

V_2 appears across 100 Ω and the power absorbed by that resistor is

$$P_2 = \frac{V_2^2}{100} = \frac{25}{100} = 0.25 \text{ W} \qquad (2\text{-}62)$$

Finally, the voltage across the 300-Ω resistor is $V_1 - V_2 = 15$ V, and the power dissipated in that resistor is

$$P_3 = \frac{15^2}{300} = \frac{225}{300} = 0.75 \text{ W} \qquad (2\text{-}63)$$

In Fig. 2-30(a) the battery supplies the power that is absorbed by the resistor. The voltage and the current for both elements have the same magnitude. The *magnitude* of power in both cases is thus also the same and *conservation of energy* holds. This is true of all circuits consisting of resistors and batteries. Resistors are called *passive* circuit elements. The conservation principle holds true for passive circuits (consisting of passive elements) and batteries, or other similar devices supplying power and energy. Later we discuss *active* elements, which behave differently from passive ones.

EXERCISES

2-6.1. In the circuit given in Fig. 2-34, it is known that $V_0 = 10$ V, $R_1 = 10$ Ω, and $R_2 = 10$ Ω. If $R_3 = 5$ Ω, what is the power supplied by the battery?

Ans. 4 W

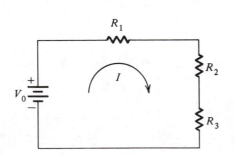

Figure 2-34. Circuit for Exercise 2-6.1.

2-6.2. The following information is given for the circuit shown in Fig. 2-35: $I = 15$ A, $R_1 = 3\,\Omega$, $R_3 = 4\,\Omega$, and $R_4 = 12\,\Omega$. If the voltage from a to b with the polarity as shown in the figure is 30 V, what is the power dissipated in R_2?

Ans. 75 W

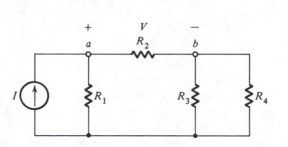

Figure 2-35. Circuit for Exercise 2-6.2.

2-7 GENERALIZATION

Thus far we have considered only simple circuits made up of resistors and batteries. We will, of course, consider more complicated circuits in later chapters. In anticipation, we use this opportunity to generalize to some extent.

In general, we use capital letters to designate quantities that do not vary with time. Thus, I or I_0 designate time-invariant (i.e., constant) currents and V or V_0 designate time-invariant voltages. If a quantity does vary with time, it will be represented by a lowercase letter, as $i(t)$ or $v(t)$. Then the symbol we have used for a battery will go through the transition suggested by Fig. 2-36.

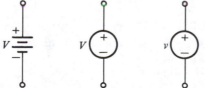

Figure 2-36. Symbols used to model an independent voltage source.

A circle with internal polarity marks will represent an *independent voltage source*, with V implying time-invariant voltage and v or $v(t)$ implying time dependence. The word *independent* is used to indicate that the source is not dependent on external conditions, such as other voltages or currents, or the circuits to which these independent sources are connected.

Another kind of widely used source is one producing current not influenced by external conditions, the *independent current source*. Its symbol is that shown in Fig. 2-37(a). If the voltage or current of a source *depends* in turn upon some other voltage or current, it is called a *dependent* or more often a *controlled source*. Such sources are common in electronics, for example, and are indicated by the diamond-shaped symbols of Fig. 2-37(b) and (c). The source may

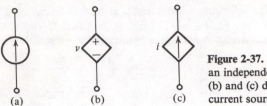

Figure 2-37. (a) Symbol used to model an independent current source, and (b) and (c) dependent voltage and current sources, respectively.

be either a voltage or a current source, and it may be controlled by another voltage or another current.

Kirchhoff's laws are, of course, valid for all types of circuits containing any and all types of elements. For resistors, Ohm's law is valid for constant as well as time-dependent voltages and currents. The same is true of the relations for power and energy. A few examples will illustrate.

EXAMPLE 2-6

In Fig. 2-38, assume that v_1, R_1, and R_2 are known. Find v_2.

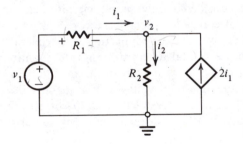

Figure 2-38.   Circuit for Example 2-6.

Solution

Here we have a current source of value $2i_1$ controlled by the current i_1. We can write KCL at node 2,

$$i_1 + 2i_1 = i_2 \tag{2-64}$$

The voltage across R_1 is $v_1 - v_2$, so $i_1 = (v_1 - v_2)/R_1$. i_2 is simply v_2/R_2. So

$$\frac{3(v_1 - v_2)}{R_1} = \frac{v_2}{R_2} \tag{2-65}$$

or

$$v_2\left(\frac{1}{R_2} + \frac{3}{R_1}\right) = \frac{3}{R_1}v_1 \tag{2-66}$$

and

$$v_2 = \frac{3R_2}{3R_2 + R_1}v_1 \tag{2-67}$$

EXAMPLE 2-7

Find the current i in the circuit of Fig. 2-39.

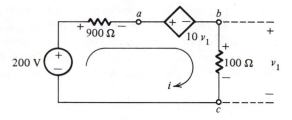

Figure 2-39. Circuit for Example 2-7.

Solution

By KVL,

$$200 = 900i + 10v_1 + v_1 \tag{2-68}$$
$$= 900i + 11(100i) \tag{2-69}$$
$$= 2000i \tag{2-70}$$

or

$$i = 0.1 \text{ A} \tag{2-71}$$

EXAMPLE 2-8

Find v in Fig. 2-40. Assume that i_1, i_2, R_1 and R_2 are known.

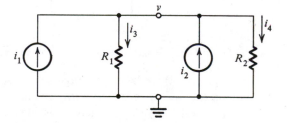

Figure 2-40. Circuit for Example 2-8.

Solution

We simply write the KCL at the single node whose voltage is desired,

$$i_1 + i_2 = \frac{v}{R_1} + \frac{v}{R_2} \tag{2-72}$$

or

$$v = \frac{R_1 R_2}{R_1 + R_2}(i_1 + i_2) \tag{2-73}$$

The controlled sources are *active* elements, and circuits containing such elements do not obey the conservation principle (more on this in Chapter 3). A second feature of these elements is that they cannot be defined completely in terms of only *two terminals*, as are resistors and independent sources. The controlled sources must always be defined together with the controlling voltages or currents. For example, in Fig. 2-39 the terminals (or nodes) *a*, *b*, and *c* are required to define the controlled source. In general, the controlling signal may be anywhere in the circuit, so *four terminals* are required to define controlled sources.

Some other terms of interest are *branch* and *branch current*. A *branch* is an element (or part of an element) placed between two nodes of a circuit. The current through the branch is called the *branch current*. The voltage across the branch is the node-to-node voltage across the element and is sometimes referred to as the *branch voltage*.

There will be many occasions when the *independent sources* are *set to zero* values or *turned off*. Notice they are not disconnected from the rest of the circuit. They are still connected, but the value of the source voltage or the source current is zero. The situation is as shown in Fig. 2-41.

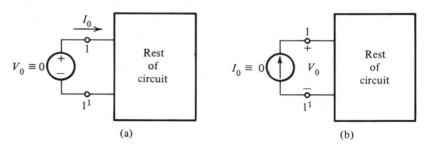

(a) (b)

Figure 2-41. Setting independent sources to zero.

In Fig. 2-41(a) we have a voltage source set at zero. Across 1–1' we have the voltage identically zero. The current value depends on the rest of the circuit and is quite arbitrary. This situation is identical with that of a short circuit and we see that a zero voltage source is equivalent to a short circuit.

Similarly, in Fig. 2-41(b) we have a zero-valued current source. Here the current is zero regardless of the voltage and the situation represents an open circuit. A zero current source is thus equivalent to an open circuit.

We can reach the same conclusions from the corresponding circuit equations. For the circuits of Fig. 2-42, we write KVL to give

$$R_1 I + R_2 I = 0 \tag{2-74}$$

for both the circuits, thus demonstrating that *a zero voltage source is equivalent to a short circuit*. Similarly, we can demonstrate that a zero current source is equivalent to an open circuit.

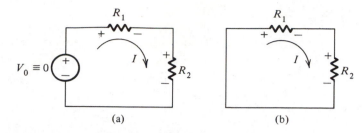

(a)   (b)

Figure 2-42. Circuit used to determine Eq. 2-74.

A final note concerns the controlled sources. They are models of certain circuit elements such as transistors and do not represent actual sources. They are not independent of the rest of the circuit, but depend on some voltage or current in the circuit. They cannot therefore be turned off or set to zero. *Only the independent sources may be set to zero.*

EXERCISES

2-7.1. For the circuit given in Fig. 2-43, it is known that $R_1 = 6\,\Omega$, $R_2 = 2\,\Omega$, $V_0 = 16$ V, and $I_0 = 4$ A. Determine the values of I_1, I_2, and V_A.

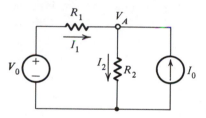

Figure 2-43. Circuit for Exercise 2-7.1.

Ans. 1 A, 5 A, 10 V

2-7.2. The circuit given in Fig. 2-44 contains one dependent and one independent source. Show that

$$v_2(t) = \frac{1}{1 + \dfrac{R_1}{R_2}\left(\dfrac{1}{1 + \alpha}\right)} v_1(t)$$

How do you explain the result you obtain if $\alpha = -1$?

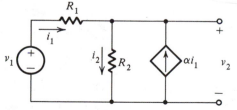

Figure 2-44. Circuit for Exercise 2-7.2.

PROBLEMS

2-1. For the circuit shown in the figure, determine I and V.

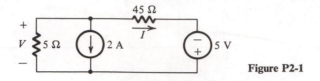

Figure P2-1

2-2. For the circuit given in the figure, find the voltage V.

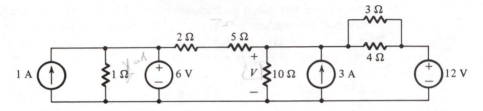

Figure P2-2

2-3. For the circuit described in the figure, find the current I.

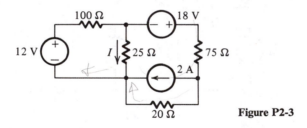

Figure P2-3

2-4. For the circuit shown in the figure, find the value of I_1 that will cause V_2 to have a value of 6.5 V.

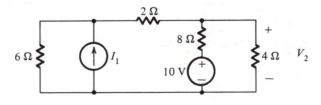

Figure P2-4

2-5. Solve for V_0, I_1, and I_2 as identified in the circuit shown in the figure. Make use of node equations.

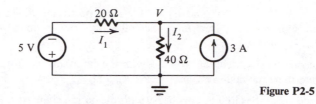

Figure P2-5

2-6. For the circuit given in the figure, find the following voltages and currents: V_1, I_1, I_2, I_3, and I_4.

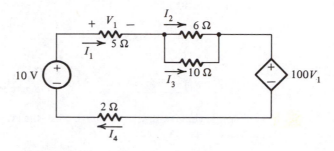

Figure P2-6

2-7. Find the value of the voltage V_1 for the circuit of the figure.

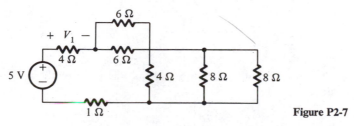

Figure P2-7

2-8. Find the value of the current I_1 for the circuit given in the figure.

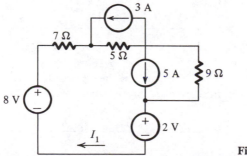

Figure P2-8

2-9. Consider the circuit given in Fig. 2-14 with node e indicated as the datum or ground connection. With respect to this datum, it is found that $V_a = 13$ V,

$V_b = 12$ V, $V_c = 9$ V, and $V_d = 21$ V for a particular choice of resistor values and source values. Determine the following node-to-node voltages: V_{ab}, V_{bd}, V_{cd}, V_{ad}, and V_{bc}.

2-10. Reconsider the circuit of Prob. 2-9 with the datum moved from node e to node d. Determine the node voltages V_a, V_b, V_c, and V_e.

2-11. Reconsider the circuit of Prob. 2-9 with the datum changed from node e to node b. What will be the voltages at the nodes, V_a, V_c, V_d, and V_e?

2-12. Find the value of R in the circuit given such that the power dissipated by the 2-Ω resistor is 40.5 W.

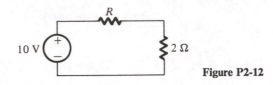

Figure P2-12

2-13. What is the value of the node voltage V_1 in the given circuit if the current I_2 has the value -48 mA?

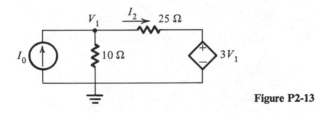

Figure P2-13

2-14. Use the current-divider relationship to find the value of the voltage V_1 in the circuit given.

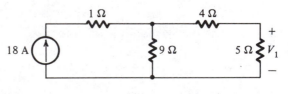

Figure P2-14

2-15. Suppose that 4.5 W is dissipated by the 8-Ω resistor R_2 in the circuit given in the figure. What is the current I_1 through the resistor R_1?

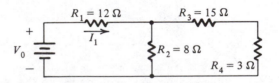

Figure P2-15

2-16. Design a resistive circuit to supply given voltages on taps. Assume that there is no load attached to the taps and that the power supply is floating (i.e., neither side is grounded). All the required voltages are to be measured with respect to ground (the datum). Select the resistor values so that the total power dissipated by the circuit is equal to the specified value.

(a) 24-V supply; 9-, 15-, and 24-V taps; 3 W dissipated.

(b) 36-V supply; 18-, 5-, −7-, and −18-V taps; 8 W dissipated.

2-17. A speaker is connected to an amplifier by 40 ft of 24-gauge wire which is known to have a resistance of 25 Ω/1000 ft. As a result of the use of such small wire for connection, the amplifier rated at 25 W delivers only 16 W. Assuming that the source voltage is capable of delivering full speaker power, what is the value of the speaker resistance, the voltage of the speaker?

2-18. Find the value of the resistor R in the figure given such that the voltage across the resistor is $V_R = 2$ V. Find the power in the controlled source, specifying whether the power is being absorbed or delivered by the source.

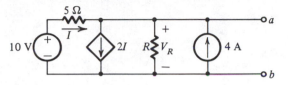

Figure P2-18

3

CIRCUIT SIMPLIFICATION TECHNIQUES

In this chapter we describe several techniques that make it possible to simplify a circuit by combining some of the elements into *equivalent* elements. Our objective is to *simplify* when possible before beginning analysis. We do this in anticipation of analysis methods to be given in Chapter 4.

3-1 SERIES AND PARALLEL CIRCUITS

Figure 3-1(a) shows a circuit of three *resistors connected in series* with a voltage source. The same current passes through each of the three resistors. Let this current be i; then from KVL,

$$v = R_1 i + R_2 i + R_3 i \tag{3-1}$$

$$= (R_1 + R_2 + R_3)i \tag{3-2}$$

$$= R_{eq}i \tag{3-3}$$

where

$$R_{eq} = R_1 + R_2 + R_3 \tag{3-4}$$

Equation 3-3 represents the circuit shown in Fig. 3-1(b), in which the three series resistors are now replaced by the equivalent resistance, R_{eq}, whose value is the sum of the individual resistances. Notice that *the relation between the voltage and the current at the nodes or terminals 1–1' are unchanged*. This is the basic concept of *equivalent circuits*. In terms of measurements or calculations

58

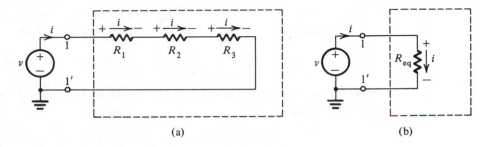

(a)   (b)

Figure 3-1. Circuit containing three resistors in series, and the equivalent resistor.

made at these nodes, the two circuits of the figure could not be distinguished. Should it be desired to make measurements at an *internal node* of Fig. 3-1(a), then the two circuits are not equivalent. If the internal nodes are of no interest, then the simplified circuit of (b) can replace that of (a). Such internal nodes are sometimes called *suppressed* nodes because they do not appear in the simplified equivalent circuit.

Note that if there had been N resistors in series, the result would have been in the form of Eq. 3-4 but with more terms added,

$$R_{eq} = R_1 + R_2 + R_3 + \ldots + R_N \tag{3-5}$$

In like manner, we may consider the connection of three *resistors in parallel*, as in Fig. 3-2(a). The circuit is actually a two-node network and at node 1 we write

$$i = i_1 + i_2 + i_3 \tag{3-6}$$

Ohm's law is used to express the currents in terms of voltage,

$$i = \frac{v}{R_1} + \frac{v}{R_2} + \frac{v}{R_3} \tag{3-7}$$

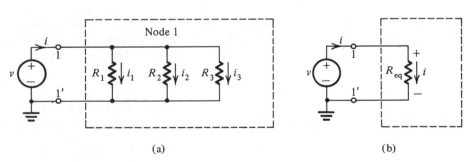

(a)   (b)

Figure 3-2. Circuit containing three parallel resistors and the equivalent resistor.

Again, we let $i/v = 1/R_{eq}$ [Fig. 3-2(b)] and attach the same meaning to equivalent resistance that was used for the series-connected resistors:

$$\frac{1}{R_{eq}} = \frac{1}{R_1} + \frac{1}{R_2} + \frac{1}{R_3} \tag{3-8}$$

Had there been N resistors in parallel rather than three, Eq. 3-8 would have additional terms,

$$\frac{1}{R_{eq}} = \frac{1}{R_1} + \frac{1}{R_2} + \frac{1}{R_3} + \cdots + \frac{1}{R_N} \tag{3-9}$$

The case $N = 2$ is of sufficient importance in analysis that a special notation has been introduced for it, and its form is simple enough to memorize:

$$R_{eq} = R_1 \| R_2 = \frac{R_1 R_2}{R_1 + R_2} \tag{3-10}$$

which is remembered easily as the product over the sum of the two resistances.

Series voltage sources and *parallel current sources* may also be routinely combined to form an equivalent source. The rules by which these combinations are accomplished are simple and are summarized in Fig. 3-3. These are

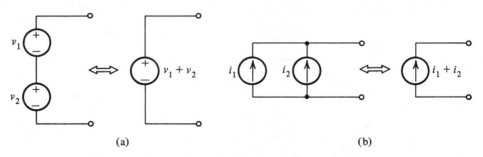

(a)   (b)

Figure 3-3. (a) Voltage sources in series add, and (b) current sources in parallel add.

simple consequences of Kirchhoff's laws. Obviously, voltage sources can be placed in parallel only if they have the same voltages, and current sources can be placed in series only if they have the same currents.

EXAMPLE 3-1

In the circuit of Fig. 3-4(a), we wish to find the current through the two sources, and we decide to solve this problem by circuit simplification.

Solution

The equivalent resistances for the two and the three resistors in parallel are found using Eqs. 3-10 and 3-8, and the result is shown in Fig. 3-4(b). We may then use Eq. 3-5 for the series connection to reduce the network

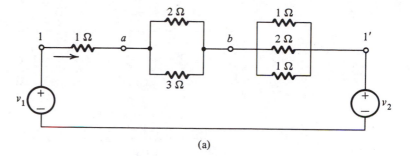

(a)

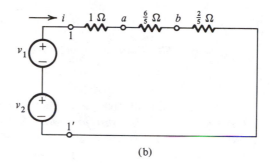

(b)

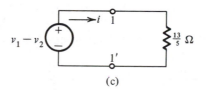

(c)

Figure 3-4. Circuit for Example 3-1.

into the equivalent or simplified form of Fig. 3-4(c). The current is now obtained by Ohm's law,

$$i = \frac{v_1 - v_2}{\frac{13}{5}} \tag{3-11}$$

Note that the circuit in (c) cannot be used directly to find the current in the 3-Ω resistor, for example. But it is now possible to return to the original circuit of (a) with the value of i given by Eq. 3-11. We can then use the current-divider equations of Chapter 2 to determine the currents in the various resistors, if desired.

EXAMPLE 3-2

The circuit shown in Fig. 3-5(a) has a parallel structure and, in addition, a controlled source. Simplify this network prior to completing an analysis.

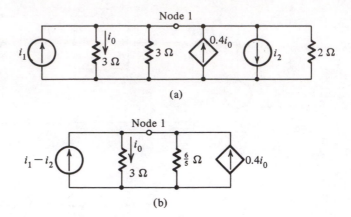

(a)

(b)

Figure 3-5.   Circuit for Example 3-2.

Solution

Simplification of this circuit is different from that of Example 3-1. *The current-controlled current source is controlled by the current in the 3-Ω resistor, which must therefore remain explicitly in the circuit.* Independent and controlled sources must *not* be combined. We can, however, combine two parallel resistors and two independent current sources. The result is that shown in Fig. 3-5(b). We cannot simplify the circuit any further at this point. Analysis must be performed on this circuit to determine whatever voltage or current we are interested in.

EXERCISES

3-1.1. In the circuit shown in Fig. 3-6, find the resistance from a to b if (a) the switch is open, and (b) if the switch is closed. Explain why the two are equal.

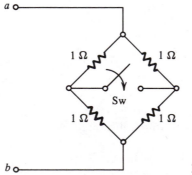

Figure 3-6.   Circuit for Exercise 3-3.1.

Ans.   1, 1

3-1.2. For the circuit shown in Fig. 3-7, find the equivalent circuit in which one voltage source is in series with one resistor.

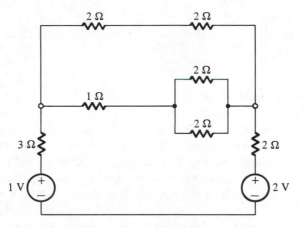

Figure 3-7.   Circuit for Exercise 3-3.2.

Ans. $V = 1\,\text{V}, R = \frac{19}{3}$

3-1.3. For the circuit given in Fig. 3-8, determine the equivalent circuit having one current source in parallel with one resistor.

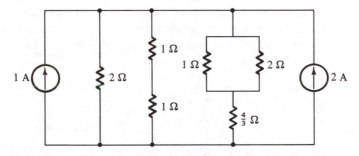

Figure 3-8.   Circuit for Exercise 3-3.3.

Ans. $I = 3\,\text{A}, R = \frac{2}{3}$

3-2 SOURCE TRANSFORMATION

In this section we present a procedure by which a voltage source can be converted into an equivalent current source, provided only that the voltage source has a resistor in series with it. Further, the transformation will work in reverse: a current source can be converted into an equivalent voltage source, provided only that the current source has a resistor in parallel with it. The meaning that

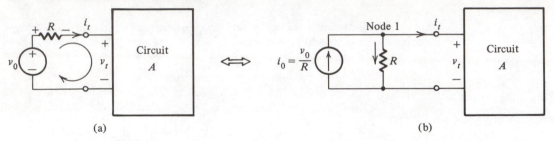

Figure 3-9. Source transformations.

we attach to *equivalent* in this case is illustrated in Fig. 3-9. With the source transformation, voltages and currents within circuit A are not affected by the transformation. We recognize, of course, that this is a fictional change in type of source, and that we are speaking of a mathematical equivalence in terms of equations in the circuits. It will prove to be a powerful tool in analysis.

To derive the source transformation equations, we begin with Fig. 3-9(a). Writing KVL around the loop shown, we have

$$v_0 = R i_t + v_t \tag{3-12}$$

where v_t and i_t are the voltage and the current at the terminals of the remainder of the circuit A. We require that v_t and i_t not be changed, a necessary condition if source transformation is not to affect the remaining circuit. In (b) of the figure, we show the rest of the circuit A at whose terminals a current source and a resistor R in parallel are connected. The terminal voltage and the current must, of course, retain the original values v_t and i_t. We now write KCL at node 1 of this circuit to give

$$i_0 = i_t + \frac{v_t}{R} \tag{3-13}$$

or

$$R i_0 = R i_t + v_t \tag{3-14}$$

Comparing this equation with Eq. 3-12, we see that for the two equations to be identical,

$$v_0 = R i_0 \tag{3-15}$$

or

$$i_0 = \frac{v_0}{R} \tag{3-16}$$

We are thus able to *transform the voltage source v_0 with the series resistance R to a current source*, whose value is given by Eq. 3-16, *with a parallel resistance R*. Conversely, we can *transform a current source–parallel resistance combination into a voltage source–series resistance combination* using Eq. 3-15. From the last two equations we see that the transformation has no meaning if $R = 0$ or $R = \infty$. Otherwise, the *source transformation* preserves the voltage and the current at the terminals of the remainder of the circuit. The operation does, however, suppress the node at the $+$ side of the voltage source in Fig. 3-9(a).

EXAMPLE 3-3

Figure 3-10(a) shows a resistor circuit with both voltage and current sources. For this network, we are to find the voltage at node 2 with respect to the datum.

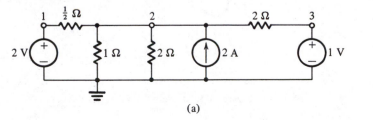

(a)

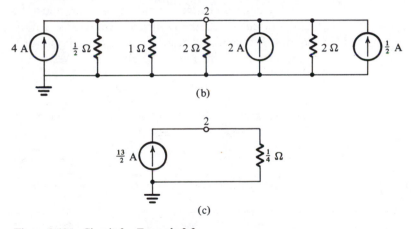

(b)

(c)

Figure 3-10. Circuit for Example 3-3.

Solution

After we use the source transformation on the two voltage sources, we obtain the parallel circuit of Fig. 3-10(b). In this circuit all of the current sources may be combined, as well as all resistors. The result of this combination is the circuit shown in Fig. 3-10(c). From this we determine the voltage at node 2,

$$v^2 = Ri = \tfrac{13}{2} \times \tfrac{1}{4} = \tfrac{13}{8} \text{ V} \tag{3-17}$$

Source transformation can be applied to *controlled* sources as well. The controlling signal, however, must *not* be tampered with in any way since the operation of the controlled source depends on it. An example will illustrate.

EXAMPLE 3-4

Find the voltage v_2 at node 2 in the circuit of Fig. 3-11.

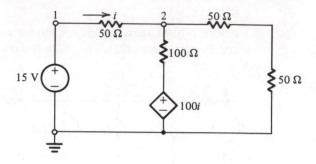

Figure 3-11. Circuit for Example 3-4.

Solution

We cannot change the circuit between nodes 1 and 2 since the controlling current i, for the controlled source, is in the resistor between these nodes. We can, however, apply circuit simplification anywhere else. The steps are shown in Fig. 3-12. Then, by KVL applied to the circuit in Fig. 3-12(c),

$$15 = 50i + 50i + 50i \tag{3-18}$$

The current $i = \frac{15}{150} = 0.1$ A and the voltage v_2 is $15 - 50i = 10$ V.

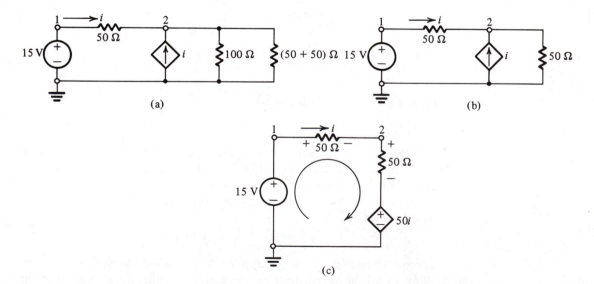

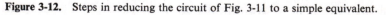
Figure 3-12. Steps in reducing the circuit of Fig. 3-11 to a simple equivalent.

EXERCISES

3-2.1. Make use of source transformations in determining values of V and I for the circuit shown in Fig. 3-13.

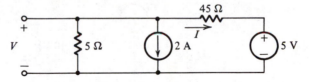

Figure 3-13. Circuit for Exercise 3-2.1.

Ans. $\ I = -0.1$ A, $V = -9.5$ V

3-2.2. Make use of source transformations to suppress nodes and so determine the value of v_1 and then the value of v_2 for the circuit shown in Fig. 3-14.

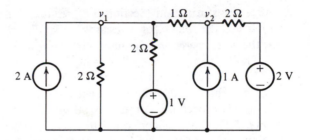

Figure 3-14. Circuit for Exercise 3-2.2.

Ans. $\ v_1 = 2.875$ V, $v_2 = 3.25$ V

3-2.3. For the circuit shown in Fig. 3-15, make use of source transformations to obtain an equation for the voltage, V.

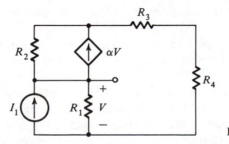

Figure 3-15. Circuit for Exercise 3-2.3.

Ans. $V = \dfrac{I_1 R_1 (R_2 + R_3 + R_4)}{R_2 + R_3 + R_4 + (1 + \alpha R_2)R_1}$ V

3-2.4. Find the current I_1 to give $v_2 = \frac{13}{2}$ V in the circuit given in Fig. 3-16.

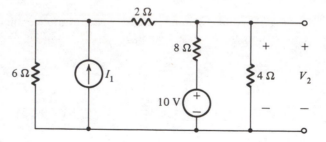

Figure 3-16.   Circuit for Exercise 3-2.4.

Ans. $I_1 = \frac{8}{3}$ A

3-3   SOURCE SHIFTING

Source transformation (Section 3-2) required that there be a resistor in series with a voltage source, or a resistor in parallel with a current source. If this is not the case, *source shifting* may often be used to advantage.

The voltage source v_1 shown in Fig. 3-17(a) has no resistor in series with it. This will not present any special problems of analysis, for the voltage at node a is then fixed to be v_1. But if circuit simplification is required, then we follow this procedure. An additional voltage source v_1 is added to the circuit in Fig. 3-17(b) and is connected to node b, which is actually identical to node a. Now since nodes a and b are maintained at the same voltage by the sources, the wire connecting them may be cut, giving the circuit of Fig. 3-17(c). Here the two voltage sources now have resistors in series, and source transformation may be done if desired.

In Fig. 3-17(a) and (c) two closed loops are shown. If KVL is applied around these loops, the same voltage equations will result. This line of reasoning may be followed in showing that the two circuits are equivalent.

The voltage source shift we have just demonstrated may be thought of as *pushing a voltage source through a node*. In fact, this is a good way to remember the procedure. Figure 3-18 demonstrates "pushing through a node" for two branches. The method may be used for any number of branches connected to the node.

A similar procedure may be employed when there is no resistor in parallel with a current source. Figure 3-19(a) shows a circuit in which the current source i_0 is connected between nodes a and c. There is no way in which the source i_0 may be converted to a voltage source as one step in circuit simplification. In Fig. 3-19(b) one source has been changed to two sources, and connected in such a way that the currents into the three nodes, a, b, and c, are

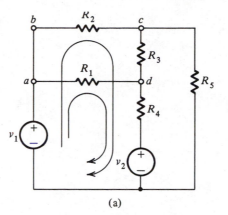

(a)

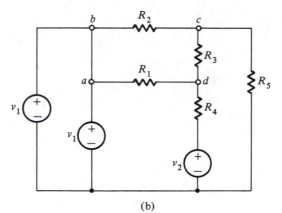

(b)

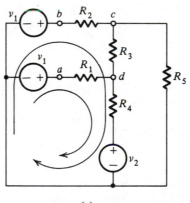

(c)

Figure 3-17. Steps involved in voltage source shifting: in (b) v_1 is broken into two sources, and in (c) each is "pushed through the node" to obtain an equivalent circuit.

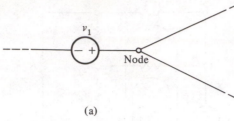

(a)   (b)

Figure 3-18. "Pushing a source through a node."

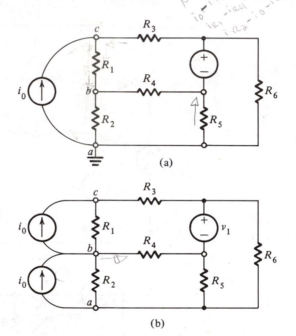

Figure 3-19. The two circuits are equivalent since the currents into nodes *a*, *b*, and *c* are the same.

the same in the two cases. At node *b*, the source current in equals the source current out. Now source transformation may be applied directly to the circuit of Fig. 3-19(b), resulting in circuit simplification.

The three techniques we have described—combining resistors or sources, source transformation, and source shifting—may be used in combination to accomplish circuit simplification. The sequence in which they are used is seldom unique and may require some ingenuity in choice of steps.

EXAMPLE 3-5

Let us return to the circuit of Fig. 3-17(c) and require that the voltage at node *c* be found through circuit simplification.

Solution

The steps in circuit simplification are shown in Fig. 3-20. In this problem, only source transformations are needed. In going from Fig. 3-17(c) to Fig. 3-20(a), the v_1, R_1 combination is converted to a current source in order that it may be combined with the current source obtained by converting v_2 and R_4. This is accomplished in the circuit of Fig. 3-20(b). Next, the current source and parallel resistor are changed back to a voltage source in order that the resulting series resistor may be combined with R_3, giving the result shown in Fig. 3-20(c). Next, both voltage sources are converted to current sources, giving the circuit of Fig. 3-20(d). In this form, the

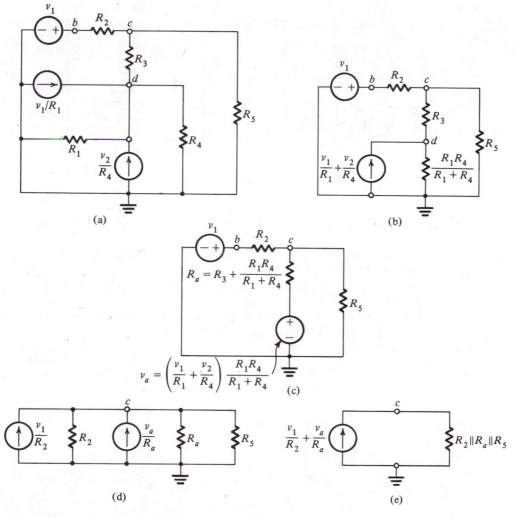

Figure 3-20. Steps in circuit simplification for Example 3-5.

current sources are combined and the parallel resistors are similarly combined, giving the simple circuit of Fig. 3-20(e). From the circuit of (e), the voltage at node c is found by multiplying current and resistance. Note that the circuit simplification has resulted in all nodes being suppressed except node c. Clearly, a different strategy would have to be followed if a different node voltage had been required.

EXERCISES

3-3.1. For the circuit shown in Fig. 3-21(a), it is given that $I_1 = 5$ A, $I_2 = -2$ A, $R_1 = 12 \, \Omega$, and $R_2 = 5 \, \Omega$. Make use of the source-shifting procedure to deter-

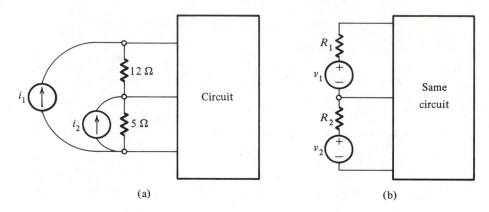

(a)   (b)

Figure 3-21. Circuit for Exercise 3-3.1.

mine the values for v_1 and v_2 in Fig. 3-21(b) if it is required that the two circuits be equivalent with respect to the rest of the circuit.

$$\textit{Ans.} \quad v_1 = 60 \text{ V}, v_2 = 15 \text{ V}$$

3-3.2. Making use of source-shifting procedures, simplify the circuit shown in Fig. 3-22 in such a way that the voltage v_x is determined.

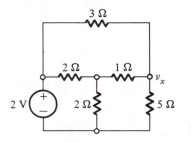

Figure 3-22. Circuit for Exercise 3-3.2.

$$\textit{Ans.} \quad v_x = \tfrac{35}{31} \text{ V}$$

The circuit in Fig. 3-23 consists of two parts, N and L, connected together at the two nodes a and b. Part of the circuit, L, is often called a *load*. The load may be a circuit of some complexity but more often is a simple circuit, such

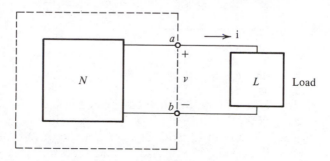

Figure 3-23. Circuit representation for derivation of the Thévenin equivalent circuit.

as a resistor, an open circuit, and so on. The rest of the circuit, marked N, is to be subjected to circuit simplification. The circuit N may contain resistors, independent sources, and controlled sources. The objective is to replace N by *an equivalent circuit*, shown in Fig. 3-24, consisting of an equivalent voltage source, v_{eq}, in series with an equivalent resistance, R_{eq}. The voltage and the current at the terminals a and b in the two circuits must, of course, be identical regardless of what the load L consists of. In fact, *the equivalence must hold true for all possible loads*. Let us first write KVL for the circuit in Fig. 3-24 for an arbitrary load,

$$v_{eq} = R_{eq}i + v \qquad (3\text{-}19)$$

We will derive the values of v_{eq} and R_{eq} by considering two very simple loads, an open circuit and a short circuit.

Now, let us consider the voltage and current at terminals a–b when the load L is *temporarily removed and replaced by an open circuit (oc)*, as shown in

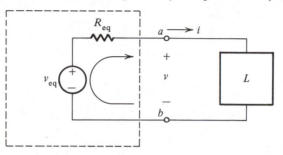

Figure 3-24. Circuit used in writing Eq. 3-19.

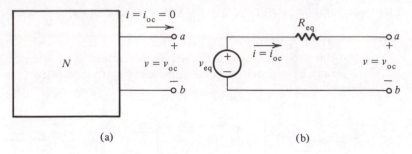

(a)   (b)

Figure 3-25. Circuit used in writing Eq. 3-20.

Fig. 3-25 for both the original circuit and the equivalent. In the original circuit of Fig. 3-25(a) the voltage with an open circuit at a–b will be some value, say v_{oc}. The current i_{oc} will, of course, be zero. We require that the equivalent circuit of Fig. 3-25(b) also exhibit the same values of voltage and current, $v = v_{oc}$ and $i = i_{oc} = 0$. Substituting in Eq. 3-19 yields

$$v_{eq} = R_{eq}i_{oc} + v_{oc} \qquad (3\text{-}20)$$

$$= v_{oc} \qquad (3\text{-}21)$$

We thus see that v_{eq} is simply the value of the open-circuit voltage at a–b in the original circuit of Fig. 3-23 when L is replaced by an open circuit.

Next, let us consider the case when the load L in the original circuit of Fig. 3-23 is *removed and replaced by a short circuit (sc)*. In this case, as shown in Fig. 3-26(a), the voltage v_{sc} will, of course, be zero and the current will have some value, say i_{sc}. These same values must be present in the equivalent circuit

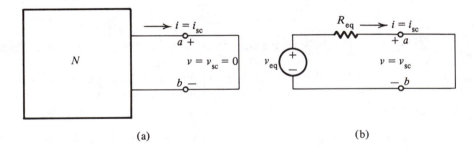

(a)   (b)

Figure 3-26. Circuit used in writing Eq. 3-22.

of Fig. 3-26(b). Substituting $v = v_{sc} = 0$ and $i = i_{sc}$ in Eq. 3-19, we obtain

$$v_{eq} = R_{eq}i_{sc} + v_{sc} \qquad (3\text{-}22)$$

$$= R_{eq}i_{sc} \qquad (3\text{-}23)$$

or

$$R_{eq} = \frac{v_{eq}}{i_{sc}} \qquad (3\text{-}24)$$

Substituting Eq. 3-21 into Eq. 3-24 yields

$$R_{eq} = \frac{v_{oc}}{i_{sc}} \tag{3-25}$$

Thus, R_{eq} is obtained by a simple ratio of the open-circuit voltage and the short-circuit current obtained from the original circuit by, respectively, open-circuiting and short-circuiting the terminals.

The equivalent circuit of Fig. 3-24 is thus obtained from Eqs. 3-21 and 3-25, and it is known as the *Thévenin equivalent circuit*. The R_{eq} is known as the Thévenin equivalent resistance and v_{eq} as the Thévenin equivalent source. Thévenin equivalent circuit is an important concept and we summarize the steps:

1. Remove the actual load leaving only the circuit N with terminals *a-b*.
2. Find the voltage v_{oc} when the terminals of N are open-circuited and the current i_{sc} when the terminals of N are short-circuited.
3. The Thévenin equivalent of N is obtained by setting $v_{eq} = v_{oc}$ and $R_{eq} = v_{oc}/i_{sc}$.

Sometimes it is preferable to represent the circuit N by an equivalent circuit, as shown in Fig. 3-27. This is obtained by a simple source transfor-

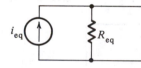

Figure 3-27. Norton equivalent circuit.

mation of the Thévenin equivalent. Thus,

$$i_{eq} = \frac{v_{eq}}{R_{eq}} \tag{3-26}$$

where v_{eq} and R_{eq} are given by Eqs. 3-21 and 3-25. Substituting their values into Eq. 3-26, we obtain

$$i_{eq} = \frac{v_{oc}}{v_{oc}/i_{sc}} \tag{3-27}$$

$$i_{eq} = i_{sc} \tag{3-28}$$

The equivalent circuit of Fig. 3-27 is thus given by Eqs. 3-25 and 3-28, and it is known as the *Norton equivalent circuit*.

The Thévenin equivalent resistance R_{eq} may be alternatively derived as follows. The circuit N and its equivalent must still remain equivalent if all the *independent* sources are set to zero or turned off, as shown in Fig. 3-28(a) with the load L again removed temporarily. Observe that all voltage sources must be replaced by short circuits and all current sources by open circuits, as discussed in Chapter 2, for both the circuit N and the Thévenin equivalent.

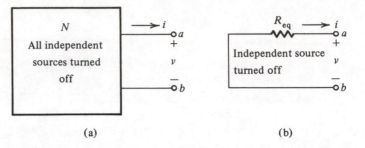

Figure 3-28. Circuit used in the alternative derivation of the Thévenin equivalent circuit.

Either from Eq. 3-19 or directly by observation in Fig. 3-28(b), we write, by Ohm's law,

$$R_{eq} = \frac{v}{-i}\bigg|_{v_{eq} \equiv 0} \tag{3-29}$$

The same v and i must exist for the circuit N in Fig. 3-28(a) so that

$$R_{eq} = \frac{v}{-i}\bigg|_{\text{indep. sources in } N \equiv 0} \tag{3-30}$$

Interpretation of Eq. 3-30 is easier if we change the direction of current and let $i_1 = -i$, as in Fig. 3-29. Now we see that Eq. 3-30 merely represents

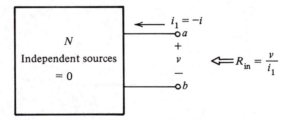

Figure 3-29.   Circuit used in deriving Eq. 3-31.

the equivalent resistance between terminals a and b of the circuit N with all independent sources turned off. Such a resistance is also called the *input resistance*, R_{in}. It may be obtained by a mere *inspection* of the circuit and the use of circuit simplification techniques, *or from the ratio*

$$R_{in} = \frac{v}{i_1} \tag{3-31}$$

Equation 3-31 should be used whenever it is not easy to obtain R_{in} by inspection (e.g., when controlled sources are present in N). The procedure then is to *assume a voltage v at the terminals and find i_1 in terms of v.* Then use Eq. 3-31 to find R_{in}. No matter how R_{in} is obtained,

$$R_{eq} = R_{in} \tag{3-32}$$

Equation 3-32 may be used in place of Eq. 3-25 whenever desired. It is usually more convenient when there are no controlled sources present. In that case, R_{in} is usually obtained by inspection and circuit simplification. We need only determine v_{oc} then to obtain the Thévenin equivalent. When controlled sources are present, it is usually easier to use Eq. 3-25. Some examples will illustrate these concepts.

EXAMPLE 3-6

We wish to determine both the Thévenin and Norton equivalent circuits for the voltage divider studied in Chapter 2 and shown again in Fig. 3-30(a).

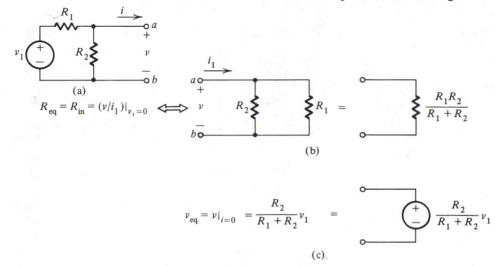

$$R_{eq} = R_{in} = (v/i_1)|_{v_1=0}$$

$$v_{eq} = v|_{i=0} = \frac{R_2}{R_1 + R_2} v_1$$

(c)

Figure 3-30. Circuit for Example 3-6, showing steps in calculations.

Solution

The steps in finding R_{eq} and v_{eq} are shown in Fig. 3-30(b) and (c), where we have used Eq. 3-32 to determine R_{eq}. The results are combined to give the Thévenin equivalent in Fig. 3-31(a). The Norton equivalent of Fig. 3-31(b) may be obtained by the usual source transformation.

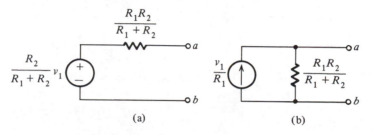

Figure 3-31. Resulting circuit for Example 3-6.

For the simple circuit considered in this case, we could have used repeated source transformations to arrive at the Thévenin equivalent. By first using voltage-to-current source transformation and combining R_1 and R_2 in parallel, we have the Norton circuit of Fig. 3-31(b). It is then easy to obtain the Thévenin equivalent of Fig. 3-31(a).

EXAMPLE 3-7

The circuit of Fig. 3-32 contains both an independent source and a current-controlled current source. We wish to find the Thévenin equivalent circuit.

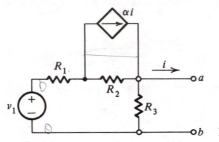

Figure 3-32. Circuit for Example 3-7.

Solution

We first observe that v_{eq} may be determined directly from Fig. 3-32 since the controlled current source becomes an open circuit when $i = 0$. The circuit remaining is a voltage-divider circuit and analysis follows Example 3-6; thus by Eq. 3-21,

$$v_{eq} = v_{oc} = \frac{R_3}{R_1 + R_2 + R_3} v_1 \tag{3-33}$$

Now R_{eq} may be determined in two different ways, given by Eqs. 3-25 and 3-32. In this example we use both ways to derive R_{eq}. Let us first use Eq. 3-25, for which we need to determine i_{sc} obtained when the terminals *a-b* are short-circuited as in Fig. 3-33(a). We do not show R_3 since it is across a short circuit. Observe that $i_{sc} = i$ in this case. The controlled current source is transformed to a controlled voltage source in (b). Now, by KVL,

$$v_1 = R_1 i + R_2 i - R_2 \alpha i \tag{3-34}$$

or

$$i = \frac{v_1}{R_1 + R_2(1 - \alpha)} \tag{3-35}$$

Since $i_{sc} = i$ in this case, we substitute Eqs. 3-33 and 3-35 into Eq. 3-25 to obtain

$$R_{eq} = \frac{v_{oc}}{i_{sc}} = \frac{R_3[R_1 + R_2(1 - \alpha)]}{R_1 + R_2 + R_3} \tag{3-36}$$

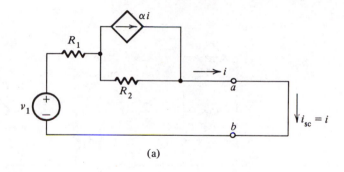

(a)

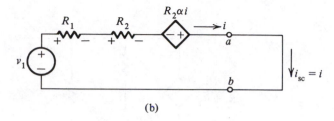

(b)

Figure 3-33. Part of the solution of Example 3-7.

The alternative way is to use Eq. 3-32. Here R_{eq} is the input resistance with $v_1 = 0$, so that we may determine R_{eq} from the simplified circuit of Fig. 3-34. By routine repeated source transformation with the individual steps as shown, we arrive at the circuit of (e), from which KVL gives

$$\frac{R_a}{R_1 + R_2}\alpha i R_2 = R_a i + v \tag{3-37}$$

From this equation we solve for $R_{in} = v/-i$ and obtain

$$R_{eq} = R_{in} = R_a\left(1 - \alpha\frac{R_2}{R_1 + R_2}\right) \tag{3-38}$$

where

$$R_a = \frac{(R_1 + R_2)R_3}{R_1 + R_2 + R_3} \tag{3-39}$$

Substituting the value of R_a into Eq. 3-38 yields

$$R_{eq} = \frac{R_3[R_1 + R_2(1 - \alpha)]}{R_1 + R_2 + R_3} \tag{3-40}$$

which is the same as before.

The Thévenin equivalent is defined by Eqs. 3-33 and 3-36.

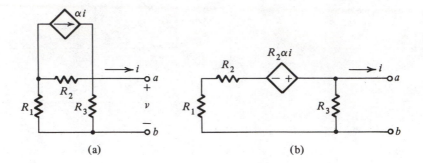

(a)   (b)

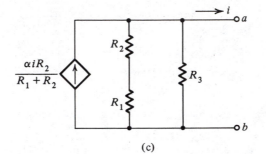

(c)

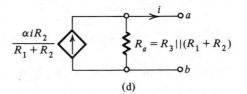

(d)

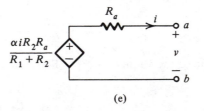

(e)

Figure 3-34. Steps in part of the solution of Example 3-7.

EXAMPLE 3-8

In the circuit of Fig. 3-35, find the load current i_L.

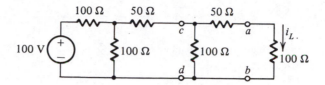

Figure 3-35. Circuit for Example 3-8.

Solution

If we can find the Thévenin equivalent for the circuit to the left of the terminals marked $a-b$ in the figure, the rest will be easy. However, it is hard to find the desired equivalent directly. We will do it in two steps. First, we find the Thévenin equivalent of the circuit to the left of the terminals $c-d$. This circuit and its equivalent are shown in Fig. 3-36(a). Figure

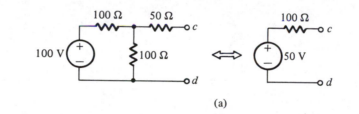

(a)

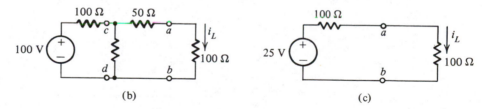

(b)   (c)

Figure 3-36. Steps in the solution of Example 3-8.

3-36(b) shows the original circuit, in which we have substituted the equivalent circuit of part (a). In Fig. 3-36(c) the Thévenin equivalent at $a-b$ is shown and the current $i_L = \frac{25}{200} = 0.125$ A.

Whenever it is easier to find i_{sc} rather than v_{oc}, the Norton equivalent may be used.

EXAMPLE 3-9

Find the load current i_L in the circuit of Fig. 3-37.

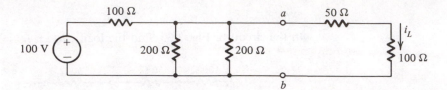

Figure 3-37.   Circuit of Example 3-8.

Solution

It is easy to determine the Norton equivalent for the circuit to the left of terminals *a–b*. When these terminals are short-circuited as in Fig. 3-38(a), $i_{sc} = 100/100 = 1$ A. The input resistance at *a–b* with the voltage source turned off is $200 \,\|\, 200 \,\|\, 100 = 50 \,\Omega$, as in Fig. 3-38(b). The Norton equivalent, together with the rest of the circuit, is shown in Fig. 3-38(c). The current i_L is obtained by the current-divider action as

$$i_L = \frac{50}{50 + (50 + 100)} \times 1 = 0.25 \text{ A} \qquad (3\text{-}41)$$

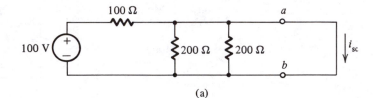

(a)

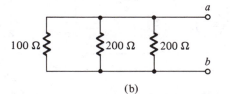

(b)

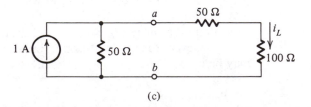

(c)

Figure 3-38.   Steps in the solution of Example 3-8.

It should be clear that the Thévenin equivalent without any independent source in the circuit is just the Thévenin resistance. If there is no independent source in the circuit, $v_{oc} = 0$, and the R_{eq} is obtained directly by determining the input resistance, R_{in}. This and several other equivalents are summarized in Fig. 3-39.

Practical independent sources (such as a *real* battery) are not usually

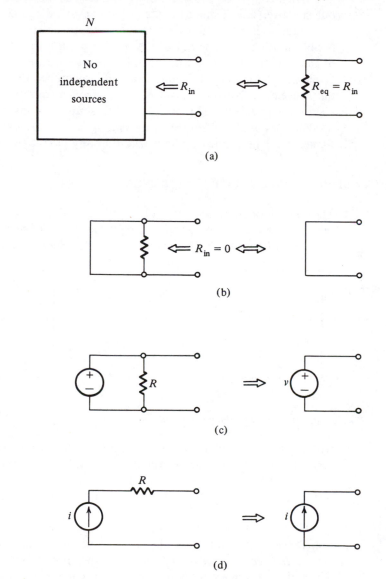

(a)

(b)

(c)

(d)

Figure 3-39. Thévenin–Norton equivalent circuits.

representable by ideal sources. Their representations involve ideal sources as part of complex circuits. Such circuits are usually too cumbersome to handle and consequently, the Thévenin or the Norton equivalent is used to represent the source circuit or network. In future, the sources shown in our circuits are presumed to be such equivalent source networks.

When controlled sources are present, it is at times possible for R_{in} to turn out negative. This is due to the fact that controlled sources are *active* elements and the models we use to represent them suppress certain aspects. The apparent nonphysical values arise out of the specific models we use for the applications of interest in this text. More on this and other aspects of active elements is given in Chapter 4. It is important to remember that only when we use the Thévenin or Norton equivalent are controlled sources incorporated into the resulting equivalent resistance. All other simplifications do not eliminate the presence of controlled sources; and as long as they are present, the controlling signals must be explicitly retained during any circuit simplification.

3-5 EQUIVALENT VERSUS INPUT RESISTANCE

The concept of equivalent circuit is a replacement of the existing circuit between two given terminals by another circuit that leaves the terminal voltage and current unchanged. This is shown in Fig. 3-40. Here, N_1 and N_2 are equivalent,

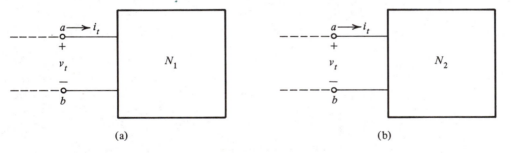

(a)   (b)

Figure 3-40. Two circuits that are potentially equivalent.

with the terminal voltage v_t and current i_t remaining unaltered in the two cases regardless of what the rest of the circuit is at the dashed ends. Equivalence at the terminals *a–b* implies that circuits N_1 and N_2 are *indistinguishable at these terminals*. Internal to N_1 and N_2, these circuits may be very different, as we have seen.

If circuit N_1 is resistive and does not contain any independent sources, the Thévenin equivalent N_2 is just a resistance. This (Thévenin) *equivalent resistance is equal in value to the ratio of the terminal voltage and current* v_t/i_t. The ratio v_t/i_t for any resistive circuit, not containing independent sources, is a constant known as the *input resistance*. Conceptually, the input resistance is the value of the ratio of the terminal (or input) voltage and current of a circuit. The equivalent resistance is the *replacement* of the circuit by a resistance that has the same value as the input resistance.

EXERCISES

3-5.1. For the circuit shown in Fig. 3-41, find the Thévenin equivalent circuit at terminals *a–b* with R_L disconnected.

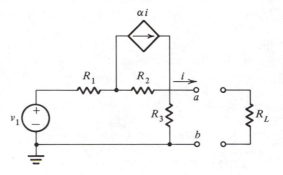

Figure 3-41. Circuit for Exercise 3-5.1.

$$\textbf{\textit{Ans.}} \quad v_{oc} = \frac{R_3}{R_1 + R_2 + R_3}v_1,$$

$$R = \frac{R_1R_3 + (1 - \alpha)R_2R_3}{R_1 + R_2 + R_3}$$

3-5.2. The circuit shown in Fig. 3-42 contains a controlled voltage source that is controlled by the difference of the independent voltage v_1 and a node voltage v_a. For the numerical values given, determine the Thévenin equivalent circuit and the current in resistor, R_L.

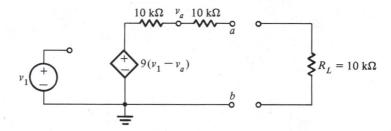

Figure 3-42. Circuit for Exercise 3-5.2.

$$\textbf{\textit{Ans.}} \quad v = 0.9v_1, \ R = 11 \ k\Omega, \ i_L = \tfrac{0.9}{21} \text{ mA}$$

3-5.3. For the circuit shown in Fig. 3-43, determine the Thévenin equivalent circuit.

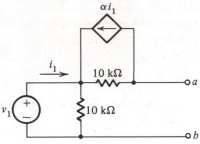

Figure 3-43. Circuit for Exercise 3-5.3.

Ans. $v = (1 - \alpha)v_1$, $R = (1 + \alpha)\, k\Omega$

3-5.4. For the circuit shown in Fig. 3-44, determine the Thévenin equivalent circuit. Note that the voltage-controlled voltage source has the value $v = kv_1$ with $k = 1$.

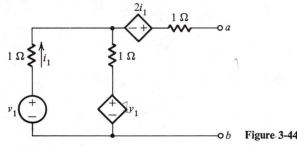

Figure 3-44. Circuit for Exercise 3-5.4.

Ans. $v = v_1$, $R = 0.5\, \Omega$

PROBLEMS

3-1. Use source transformation to determine the value of the current I_1 in the circuit given in the figure.

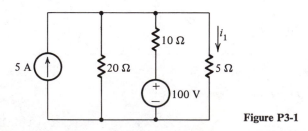

Figure P3-1

3-2. Simplify the circuit given in the figure using source transformation and other techniques and determine the value of V_1.

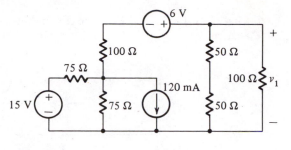

Figure P3-2

3-3. Find the currents I_1, I_2, and I_3 and the voltage V_1 by node analysis for the circuit shown in the figure.

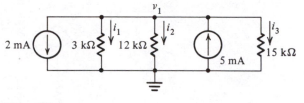

Figure P3-3

3-4. Find the currents I_1, I_2, and I_3 and the voltages V_1 and V_2 by nodal analysis for the circuits shown in the figure.

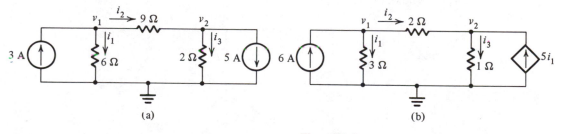

(a)   (b)

Figure P3-4

3-5. Find V_1, V_2, I_1, I_2, and I_3 for the circuit given in the figure, making use of loop analysis.

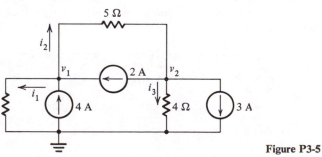

Figure P3-5

3-6. Find I_1, I_2, V_1, V_2, and V_3 of the circuit shown in the figure, making use of loop analysis.

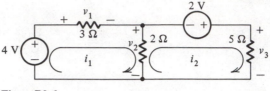

Figure P3-6

3-7. Find I_1, I_2, V_4, V_5, and V_6 of the circuit shown in the figure, making use of loop analysis.

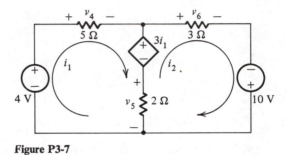

Figure P3-7

3-8. Find expressions for I_1 and I_2 in terms of V, I, R_1, R_2, and R_3 for the circuit shown in the figure.

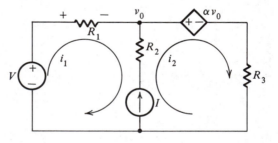

Figure P3-8

3-9. Determine the open-circuit voltage and the short-circuit current at the terminal pair a–b in the circuit shown in the figure. Use these results to obtain the Thévenin equivalent circuit.

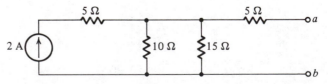

Figure P3-9

3-10. Find the Norton equivalent circuit of the portion of the circuit shown in the figure within the dashed lines. Find the current I_1 using the Norton equivalent of the circuit.

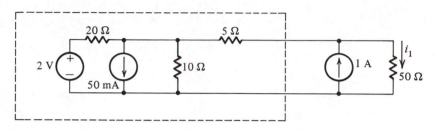

Figure P3-10

3-11. Find the Thévenin equivalent circuit with respect to the terminal pair *a–b* for the circuit in the figure.

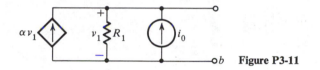

Figure P3-11

3-12. Find the Thévenin and Norton equivalent circuits for the terminal pair *a–b* for the circuit shown in the figure.

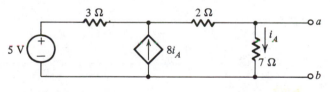

Figure P3-12

3-13. Solve for I_1, I_2, I_3, V_1, and V_2 for the circuit given in the figure, making use of the principle of superposition.

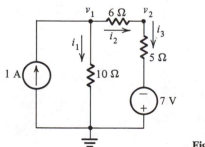

Figure P3-13

3-14. Find the Thévenin equivalent of the ladder circuit shown in the figure.

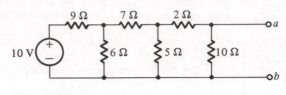

Figure P3-14

3-15. Find the Thévenin equivalent of the circuit shown in the figure with respect to terminal pair *a–b*.

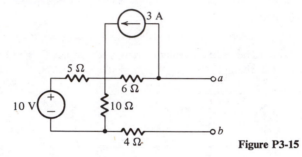

Figure P3-15

3-16. Find the Thévenin equivalent resistance at terminals *a–b* of the circuit given in the figure.

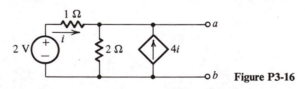

Figure P3-16

3-17. Replace the portion of the circuit within the dashed lines by its Thévenin equivalent, and then find the voltage V_1.

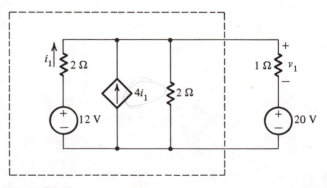

Figure P3-17

3-18. The dashed line of the network shown in the figure encloses a simplified equivalent circuit for a transistor. The controlled current source has the value αI_e. Find the Thévenin equivalent of the circuit at terminals a–b.

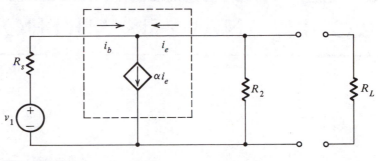

Figure P3-18

3-19. For the circuit shown in the figure, find the Thévenin equivalent circuit with respect to terminals a–b.

3-20. For the circuit given in the figure, show that the Thévenin equivalent with respect to terminals a–b is a 12-Ω resistor and a voltage source of 5.4 V.

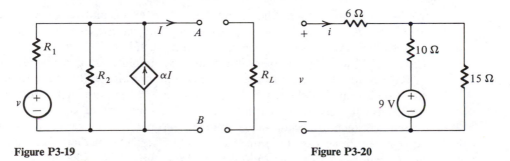

Figure P3-19   Figure P3-20

3-21. Consider the circuit shown in the figure. For this circuit determine the voltage ratio v_0/v_s and the resistance v_s/i_s.

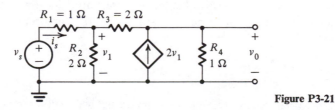

Figure P3-21

4

CIRCUIT ANALYSIS METHODS

We have seen in Chapter 2 how Kirchhoff's laws (KVL and KCL) are used to determine *the current flowing in a single loop* or *the single-node voltage* in a simple circuit where it is the only unknown node voltage. We have also seen in Chapter 3 how many complex circuits may be simplified so that a single KVL or KCL suffices to determine the desired current or voltage. But there are many situations when simplification is not practical. There are other, infrequent situations when *complete analysis* is required and we must determine all branch currents, branch or node-to-node voltages, and perhaps several node-to-datum voltages. In either case the same methods apply. These methods are based on the Kirchhoff's laws (KCL and KVL) and on Ohm's law, as before.

In this chapter we present these methods of analysis. When all simplifications fail or become impractical, one can always depend on these methods. The first is based on KCL and determines all the *node-to-datum voltages* in a given circuit and is known as *node analysis*. The second method, based on KVL, determines all *loop currents* (to be defined later) and is known as *loop analysis*. Both methods are just more general versions of the simple methods developed in Chapter 2.

4-1 NODE ANALYSIS

Remember that when we say node analysis, we really mean node voltage analysis. Our objective is to determine the voltages of all nodes with respect to the datum or reference from which node-to-node voltages can be found using KVL.

Let us review the steps involved in the application of KCL, as discussed in Chapter 2. We must first identify a reference node, the (unknown) node voltages, and the reference directions for all branch currents. We then write KCL at *each node* (whose voltage is not known) and use Ohm's law to express the branch currents in terms of the node-to-datum voltages and the resistances. The resulting equations can then be solved. This is node analysis, and we saw how it is used when there is only one unknown node voltage. Now we have more complex situations.

In our circuits the node-to-datum voltages will frequently be indicated at the nodes without any polarity marks, as in Fig. 4-1. It should always be under-

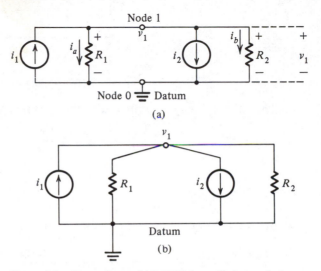

Figure 4-1.   Example to which KCL is applied at node 1.

stood that these are node voltage values with respect to the datum. Such voltages are also called *node voltages*. Let us now apply the node-analysis concept to several circuits.

The simplest case is that of two nodes, with one designated as the reference. Such a circuit is shown in Fig. 4-1(a), with two resistors and two sources connected between node 0 (reference) and node 1, and the node voltage designated as v_1. We sometimes draw circuits in equivalent forms, as shown in Fig. 4-1(b). Figure 4-1(a) has all reference directions shown, and in terms of these we now use KCL to equate currents in to currents out:

$$i_1 = i_a + i_2 + i_b \qquad (4\text{-}1)$$

Ohm's law is used to express i_a and i_b in terms of v_1; as $i_a = v_1/R_1$ and $i_b = v_1/R_2$,

$$i_1 = \frac{v_1}{R_1} + \frac{v_1}{R_2} + i_2 \qquad (4\text{-}2)$$

From this equation we solve for the voltage v_1,

$$v_1 = \frac{R_1 R_2}{R_1 + R_2}(i_1 - i_2) \tag{4-3}$$

The form of this equation reminds us that in this case we might have used the techniques of Chapter 3 to simplify the circuit before beginning analysis. Thus, the two sources combine to give $i_{eq} = i_1 - i_2$, and the two resistors in parallel combine to give $R_{eq} = R_1 \| R_2 = R_1 R_2/(R_1 + R_2)$, as shown in Fig. 4-2.

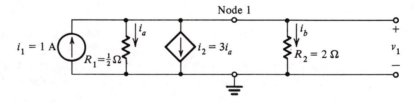

Figure 4-2. Circuit represented by Eq. 4-3.

Then Eq. 4-3 may be written directly from this figure using KCL

$$i_{eq} = \frac{v_1}{R_{eq}} \tag{4-4}$$

or

$$v_1 = R_{eq} i_{eq} \tag{4-5}$$

which is Eq. 4-3.

A different situation in node analysis is seen in examining Fig. 4-3, in that one of the current sources is a controlled source. Again, we apply KCL at

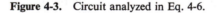

Figure 4-3. Circuit analyzed in Eq. 4-6.

node 1, summing the currents out and equating these to the one current directed in:

$$i_1 = i_a + i_2 + i_b \tag{4-6}$$

Using the numerical values given yields

$$i_1 = 1$$

$$i_a = \frac{v_1}{R_1} = 2v_1$$

$$i_2 = 3i_a = 6v_1 \tag{4-7}$$

$$i_b = \frac{v_1}{R_1} = \frac{1}{2}v_1$$

and Eq. 4-6 becomes

$$1 = 2v_1 + 6v_1 + \tfrac{1}{2}v_1 \tag{4-8}$$

from which

$$v_1 = \tfrac{2}{17} \text{ V} \tag{4-9}$$

Note that having found v_1, the four branch currents of the circuit may be found using Eq. 4-7.

If a circuit has three nodes and one is selected as the reference, there remain two nodes for which node-to-datum voltages may be determined. Such a circuit is shown in Fig. 4-4, with two current sources and reference directions selected. The node-to-datum voltages are indicated at the nodes. Analysis of

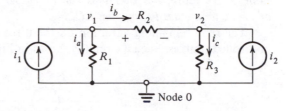

Figure 4-4.   Two-node circuit analyzed in Eq. 4-10.

this circuit requires that *KCL be applied at each of the two nodes.* At node 1 the KCL gives

$$i_1 = i_a + i_b \tag{4-10}$$

The voltage across R_1 is v_1, and the voltage across R_2 with the polarity marks as shown is $v_1 - v_2$. Then by Ohm's law, $i_a = v_1/R_1$ and $i_b = (v_1 - v_2)/R_2$. Substituting in Eq. 4-10, we obtain

$$i_1 = \frac{v_1}{R_1} + \frac{v_1 - v_2}{R_2} \tag{4-11}$$

Similarly, at node 2,

$$i_2 + \frac{1}{R_2}(v_1 - v_2) = \frac{1}{R_3}v_2 \tag{4-12}$$

We next tidy up a bit by arranging these equations in standard form:

$$\left(\frac{1}{R_1} + \frac{1}{R_2}\right)v_1 - \frac{1}{R_2}v_2 = i_1 \tag{4-13}$$

and

$$-\frac{1}{R_2}v_1 + \left(\frac{1}{R_2} + \frac{1}{R_3}\right)v_2 = i_2 \tag{4-14}$$

These are two *simultaneous algebraic equations* in the two variables v_1 and v_2. Let $R_1 = 1\,\Omega$, $R_2 = \tfrac{1}{2}\,\Omega$, and $R_3 = \tfrac{1}{3}\,\Omega$, with $i_1 = 1$ A and $i_2 = 2$ A, for example. Then these equations become

$$3v_1 - 2v_2 = 1 \tag{4-15}$$

$$-2v_1 + 5v_2 = 2 \tag{4-16}$$

Multiplying the first equation by 2, the second by 3, and adding, we eliminate the variable v_1 and find that $v_2 = \frac{8}{11}$ V. Substituting this value into Eq. 4-15 gives $v_1 = \frac{27}{33}$ V, and the two node-to-datum voltages are determined. Having obtained these values, we can find the currents in all branches of the circuit.

Return to Eq. 4-12 and note that its form comes from summing currents in before equating these to currents out. We made use of the branch current, i_b, which is directed from node 1 to node 2. We might have introduced a new current when writing KCL for node 2, calling it i_d and directing it from node 2 to node 1. Then $i_d = (v_2 - v_1)/R_2$, but this term would then appear on the right-hand side of Eq. 4-12. In other words, using i_d gives exactly the same result as using i_c! Reference directions can indeed be arbitrarily chosen at each node. Once more, *the directions can be chosen differently as we proceed from one node to another in applying KCL.*

The only remaining variation on the three-node case is one incorporating controlled sources. Such a circuit is shown in Fig. 4-5, which is too complicated

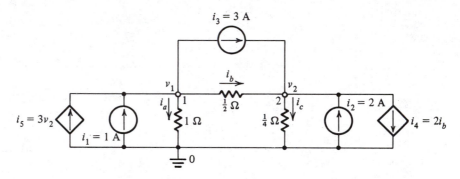

Figure 4-5.   Three-node circuit analyzed in Eqs. 4-17 and 4-18.

to be practical but has been contrived to show a principle. The KCL applied to node 1 gives

$$i_1 + i_5 = i_3 + i_a + i_b \tag{4-17}$$

and at node 2

$$i_3 + i_2 + i_b = i_4 + i_c \tag{4-18}$$

We use Ohm's law to express the currents in the resistors in terms of node voltages. If the numerical values given on Fig. 4-5 are substituted into these equations, Ohm's law used where necessary, and the terms put in a standard form, we have

$$3v_1 - 5v_2 = -2 \tag{4-19}$$

and

$$2v_1 + 2v_2 = 5 \tag{4-20}$$

Solution of these equations gives

$$v_1 = \tfrac{21}{16} \text{ V} \quad \text{and} \quad v_2 = \tfrac{19}{16} \text{ V} \tag{4-21}$$

From these values all branch currents can be found, if required.

In the studies to follow, we seldom encounter circuits with more than three nodes. But the procedures we have established apply in general. With four nodes, for example, we are required to apply KCL at each of the three nodes other than the reference or datum, giving three simultaneous equations in three unknowns—the node-to-datum voltages. We may solve for these voltages using any available method, such as systematic elimination, determinants, the Gauss elimination method, and so on. The steps to be followed will be illustrated in terms of the circuit shown in Fig. 4-6.

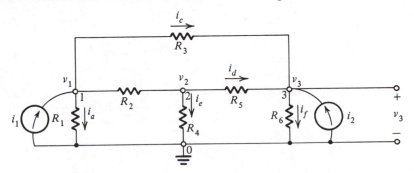

Figure 4-6. Circuit used to illustrate the general method of node analysis.

1. Simplify the circuit, if possible, by combining resistors in series or parallel or by combining current sources in parallel. This is not possible for our circuit.
2. Assign a reference node and the reference directions, and assign a current and a voltage name for each branch and node, respectively.
3. Apply KCL at each node except for the reference node; apply Ohm's law to the branch currents.
4. Simplify algebraically; put in a standard form.
5. Solve the simultaneous equations for the unknown node voltages, v_1, v_2, and v_3.
6. Using these voltages, find any branch currents required.

EXERCISES

4-1.1. In the circuit shown in Fig. 4-7, it is given that $i_1 = 1$ A and $i_2 = 2$ A. Determine the voltages at the nodes with respect to the datum, v_1 and v_2.

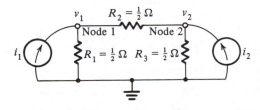

Figure 4-7. Circuit for Exercises 4-1.1 and 4-1.2.

Ans. $v_1 = \frac{2}{3}$ V, $v_2 = \frac{5}{6}$ V

4-1.2. In the circuit of Fig. 4-7, the current sources i_1 and i_2 may be adjusted. (a) What values of i_1 and i_2 will cause v_2 to equal 0? (b) What will be the value of v_1 then?

$$\textbf{Ans.} \quad \text{(a) } i_1 = -2i_2; \text{(b) } v_1 = \tfrac{1}{4}i_1$$

4-1.3. Find the branch current in the circuit of Fig. 4-8 using KCL and node analysis.

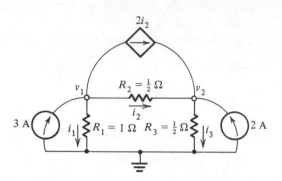

Figure 4-8. Circuit for Exercise 4-1.3.

$$\textbf{Ans.} \quad i_1 = 1.8 \text{ A}, \; i_2 = 0.4 \text{ A}, \; i_3 = 3.2 \text{ A}$$

4-2 LOOP ANALYSIS

Loop analysis is accomplished by the use of KVL and is in reality loop current analysis. We saw in Chapter 2 how a single-loop circuit has the same current through all the elements. This current common to all the elements in the loop is called a loop current. Soon, we will define the concept of loop current when there are several loops present in a circuit. But first, let us review the steps for the single-loop circuit.

We identify the loop, assign a direction to it, and assign an unknown but identical current through all the elements in the loop. The loop current and the loop, as well as their directions, are identified with a common symbol, as in Fig. 4-9, for convenience. Assign the polarities for voltages across the branches. Apply KVL around the loop and use Ohm's law to express the branch voltages in terms of the unknown loop current and the resistances. This is the basis of the loop analysis and we will generalize the method to

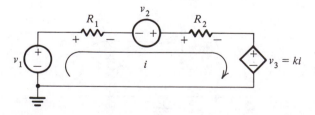

Figure 4-9. Single-loop circuit described by Eq. 4-22.

circuits with more than one loop. We first illustrate the single-loop case with a simple circuit.

The circuit of Fig. 4-9 has elements arranged in a single loop: two resistors, two independent voltage sources, and one current-controlled voltage source. All reference signs are indicated and the current that is common to all elements is shown in a loop directed in a clockwise direction. Equating voltage rises to voltage drops and using Ohm's law to express the drops across R_1 and R_2,

$$v_1 + v_2 = R_1 i + R_2 i + v_3 \qquad (4\text{-}22)$$

Substituting $v_3 = ki$, we may solve this equation for current,

$$i = \frac{v_1 + v_2}{R_1 + R_2 + k} \qquad (4\text{-}23)$$

This current may be multiplied by either R_1 or R_2 to find the voltages across the resistors, or by k to find the voltage of the controlled source. This procedure is called loop analysis of circuits.

We next describe *loop currents*, which were invented many years ago in an attempt to simplify analysis. In Fig. 4-9, the current i is a loop current which just happens to be identical to all the branch currents. Consider the circuit shown in Fig. 4-10. The three branch currents are labeled i_a, i_b, and i_c. Also

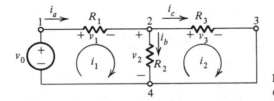

Figure 4-10. Two-loop circuit described by Eqs. 4-29.

shown on the circuit is an *alternate system of currents*—the two loop currents like these of Fig. 4-9, labeled i_1 and i_2. This implies that the *loop current* i_1 flows through all the elements in the loop 1241 in the direction specified, as shown. Similarly, the *loop current* i_2 flows through all the elements in the loop 2342. Thus, the current i_1 flows through the source, R_1 and R_2. The current i_2 flows through R_3 and R_2. Remember that this is an *alternate system* to the branch current designation. Therefore, we must be able to express one set in terms of the other.

We need to express the branch currents in terms of the loop currents, since Ohm's law is always expressed in terms of the branch current. The composite current in any branch, resulting from several loop currents flowing through the same branch, gives us the desired branch current. Thus, from the figure we observe that

$$i_a = i_1$$
$$i_b = i_1 - i_2 \qquad (4\text{-}24)$$

and

$$i_c = i_2$$

Thus, if the two loop currents are known, the three branch currents may be determined. The result of this simple example is true in general.

The circuits we use in our study have outlines like those shown in Fig. 4-11, which are sometimes described as *window panes*. Such circuits are said to be planar, in that they may be drawn without the wires crossing one another.

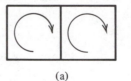

(a)

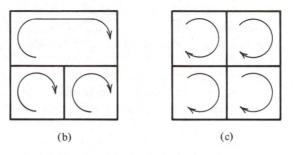

(b)   (c)

Figure 4-11. Generalization of choice of loop directions.

Circuits that are sufficiently complicated that they are nonplanar are also so complicated that analysis is done by digital computer programs. But, again, all the circuits we study will be planar and so represented in window-pane form. In each window pane of Fig. 4-11 is shown a loop current, each loop directed in a clockwise direction. Actually, the direction chosen is arbitrary, but for simplicity we follow the practice of drawing all loops clockwise.

Returning to the circuit of Fig. 4-10 with loop currents as shown, we apply KVL around loop 1 to obtain

$$v_0 = v_1 + v_2 \tag{4-25}$$

The current through R_1 is just i_1 and the composite current through R_2, in the direction of the voltage drop, is $(i_1 - i_2)$. Then, by Ohm's law, $v_1 = R_1 i_1$ and $v_2 = R_2(i_1 - i_2)$, so that

$$v_0 = R_1 i_1 + R_2(i_1 - i_2) \tag{4-26}$$

Similarly, around loop 2,

$$v_2 = v_3 \tag{4-27}$$

or

$$R_2(i_1 - i_2) = R_3 i_2 \tag{4-28}$$

Rearranging these two equations in standard form, we obtain

$$(R_1 + R_2)i_1 - R_2 i_2 = v_0$$
$$-R_2 i_1 + (R_2 + R_3)i_2 = 0$$

(4-29)

We have a set of simultaneous algebraic equations, which may be solved using any of the techniques mentioned in Section 4-1.

Again, it is possible to *reassign* the polarity marks of the branch voltages when we proceed from one loop to the next. For example, when we consider loop 2 above, we could have reversed the polarity marks for the voltage across R_2 so that it is a *voltage drop in the direction of* i_2. Conveniently, then, all the voltages in loop 2 are drops in the direction of i_2. Then the drop across R_2 is $R_2(i_2 - i_1)$, since now the composite current, in the direction of the voltage drop across R_2, is $(i_2 - i_1)$. By KVL around loop 2, then,

$$0 = R_2(i_2 - i_1) + R_3 i_2$$

(4-30)

which is of course the same as Eq. 4-28, as it should be. In summary, we can assign polarities arbitrarily as we move from loop to loop.

A two-loop circuit of added complexity is shown in Fig. 4-12. The rules

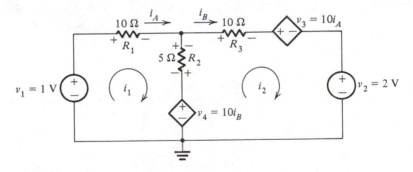

Figure 4-12. Two-loop circuit described by Eqs. 4-35.

described previously apply directly. Around loop 1,

$$v_1 = R_1 i_1 + R_2(i_1 - i_2) + v_4$$

(4-31)

$$= R_1 i_1 + R_2(i_1 - i_2) + 10 i_2$$

(4-32)

and around loop 2,

$$v_4 = R_2(i_2 - i_1) + R_3 i_2 + v_3 + v_2$$

(4-33)

or

$$10 i_2 = R_2(i_2 - i_1) + R_3 i_2 + 10 i_1 + v_2$$

(4-34)

Substituting the numbers given on Fig. 4-12, we arrive at the simple set of equations

$$15 i_1 + 5 i_2 = 1$$
$$5 i_1 + 5 i_2 = -2$$

(4-35)

which may be solved to give $i_1 = \frac{3}{10}$ A, and $i_2 = -\frac{7}{10}$ A. From these values, any branch voltage in the circuit may be found.

EXERCISES

4-2.1. Use loop analysis for the circuit given in Fig. 4-13 to determine the branch currents and the voltage at node A.

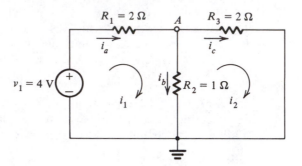

Figure 4-13.   Circuit for Exercise 4-2.1.

> *Ans.* $i_a = \frac{3}{2}$ A, $i_b = 1$ A, $i_c = \frac{1}{2}$ A, $v_A = 1$ V

4-2.2. The circuit shown in Fig. 4-14 contains a voltage-controlled voltage source. Use loop analysis to find the loop current, i, and determine the value of k that will cause $i = 0$ if $v_1 = 10$ and $v_2 = 2$.

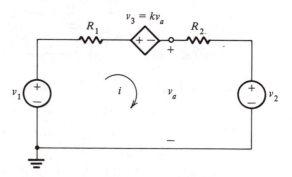

Figure 4-14.   Circuit for Exercise 4-2.2.

> *Ans.* $i = \dfrac{v_1 - (1 + k)v_2}{R_1 + (1 + k)R_2}$, $k = 4$

4-2.3. Use loop analysis to determine the branch currents in the circuit shown in Fig. 4-15.

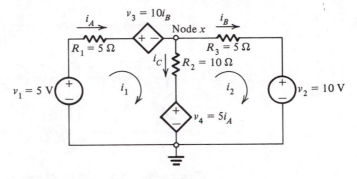

Figure 4-15.   Circuit for Exercises 4-2.3 and 4-2.4.

Ans. $i_A = \frac{1}{4}$ A, $i_B = -\frac{5}{12}$ A, $i_C = \frac{2}{3}$ A

4-2.4. For the circuit shown in Fig. 4-15, find the voltage at node x using either KCL or KVL.

Ans. $v_x = \frac{95}{12}$ V

4-3   CIRCUITS WITH MIXED SOURCES

We have specified that in node analysis the KCL is written for every node whose voltage is unknown. Similarly, in loop analysis, the KVL is written for every loop whose loop current is unknown. The circuits we considered, however, were such that for node analysis, all the node voltages (except the reference) were unknown, and for loop analysis, all the loop currents were unknown. This is always the case when node analysis is applied to circuits containing only (independent) current sources and loop analysis is applied to circuits containing only (independent) voltage sources.

When a circuit to be analyzed has both types of sources, it is frequently easier to write KCL for each node (except for reference) or KVL for each loop, anyway. We may need to assign unknown voltage or current variables for the source branches. After the KCL or KVL are written, we must eliminate any known variables from these equations. Alternatively, circuit simplification may be employed. We thus have the following choices:

1. Use the Kirchhoff laws assigning voltage or currents as needed and eliminate any known quantity from the equations so written.
2. Use source transformation or source shifting to condition the circuit so that it contains only the desired kind of sources.

We illustrate these choices by means of an example.

Figure 4-16 is a resistive circuit with mixed sources. There are three nodes, plus the node 0 as the reference, but the voltage at node 1 is fixed by the voltage

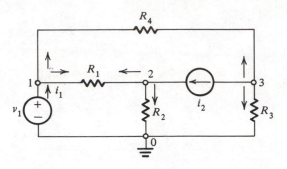

Figure 4-16. Circuit with mixed sources.

of the source and so is known. Evidently, only equations at nodes 2 and 3 will be needed to solve for v_2 and v_3. Assign the current in source v_1 to be i_1 as shown, and write the KCL at each of the three nodes. There results:

Node 1: $$\frac{1}{R_1}(v_1 - v_2) + \frac{1}{R_4}(v_1 - v_3) = i_1 \tag{4-36}$$

Node 2: $$\frac{1}{R_1}(v_2 - v_1) + \frac{1}{R_2}v_2 = i_2 \tag{4-37}$$

Node 3: $$\frac{1}{R_4}(v_3 - v_1) + \frac{1}{R_3}v_3 = -i_2 \tag{4-38}$$

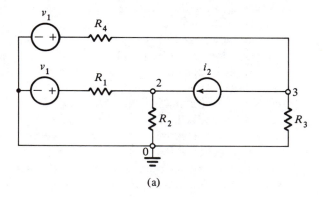

(a)

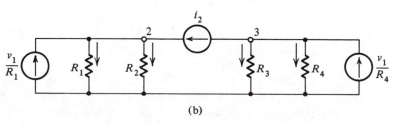

(b)

Figure 4-17. Steps in an alternative solution of the circuit with mixed sources.

Now the first equation is not needed, unless the value of i_1 is required, for the voltage at node 1 is fixed. The voltage v_1 in the second and third equations is not a variable but a known quantity. Hence, the three equations can be written in a form to determine v_2 and v_3:

$$\left(\frac{1}{R_1} + \frac{1}{R_2}\right) v_2 = i_2 + \frac{1}{R_1} v_1 \tag{4-39}$$

and

$$\left(\frac{1}{R_3} + \frac{1}{R_4}\right) v_3 = -i_2 + \frac{1}{R_4} v_1 \tag{4-40}$$

As our alternative approach, we first push the source v_1 through the node, giving the circuit shown in Fig. 4-17(a), and then the voltage source and series resistor combinations are converted to current sources. The result is shown in Fig. 4-17(b). Observe that this operation sequence suppressed node 1. Applying KCL to nodes 2 and 3 of Fig. 4-17(b) gives Eqs. 4-39 and 4-40 directly.

EXERCISES

4-3.1. The circuit shown in Fig. 4-18 contains both voltage and current sources. Analyze the circuit to determine the value of the voltage v_a.

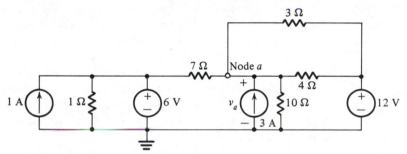

Figure 4-18. Circuit for Exercise 4-3.1.

Ans. $v_a = 13.14$ V

4-3.2. Given the circuit shown in Fig. 4-19, which contains both voltage and current sources. Analyze the circuit to determine the value of the current i_a.

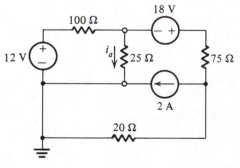

Figure 4-19. Circuit for Exercise 4-3.2.

Ans. $i_a = -0.324$ A

The circuit of Fig. 4-20 is known as a *simple ladder*. Ladder networks are encountered so often in engineering analysis that special consideration is justified. In the figure, resistors with odd-number subscripts are known as

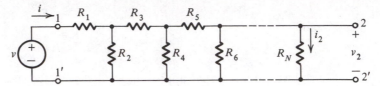

Figure 4-20. General form of a ladder circuit.

series arms, those with even-number subscripts as shunt or parallel arms. The individual resistors might have resulted from simplification of a series or parallel circuit into an equivalent resistor. If R_1, R_2, or any other resistor is replaced by a series–parallel combination of resistors, the structure is still known as a ladder, but not as a simple ladder. We will describe procedures for (1) determining the resistance of the ladder at terminals 1–1', R_{in}, and (2) determining the voltage, v, or current, i, given the current, i_2, or voltage, v_2. The voltage or current at terminals 1–1' are known as the *input* quantities and the ones at 2–2' are known as the *output* quantities. More on these terms later.

In the ladder of Fig. 4-20, assume that the ladder is truncated or chopped off after R_6. Call R_6 the *far end* of the ladder, and work toward the *near end* of the ladder as follows. R_5 and R_6 are in series and so may be added together. The combination of R_5 and R_6 is in parallel with R_4, so the reciprocals of these quantities may be added. Proceeding in this manner from the far end to the near end, there results

$$R_{in} = R_1 + \cfrac{1}{\cfrac{1}{R_2} + \cfrac{1}{R_3 + \cfrac{1}{\cfrac{1}{R_4} + \cfrac{1}{R_5 + \cfrac{1}{\cfrac{1}{R_6}}}}}} \tag{4-41}$$

where this equation is read from the bottom toward the top. The form of the equation is a *continued fraction*. The continued fraction may be evaluated by simply putting numbers in it and working from the bottom up. You should verify that if all R's have the value 1, then $R_{in} = \frac{13}{8}\ \Omega$. Note that there is a pattern in the continued fraction—we use R's for series arms and $1/R$'s for shunt arms.

Since we use $1/R$'s repeatedly in Eq. 4-41, it is convenient to define the reciprocal of resistance,

$$G = \frac{1}{R} \tag{4-42}$$

which is called the *conductance*, and the unit is siemens. This unit was initially named mhos (℧), ohm spelled backward. The continued fraction of Eq. 4-41 may now be written

$$R_{in} = R_1 + \cfrac{1}{G_2 + \cfrac{1}{R_3 + \cfrac{1}{G_4 + \cfrac{1}{R_5 + \cfrac{1}{G_6}}}}} \qquad (4\text{-}43)$$

which is much neater to write.

EXAMPLE 4-1

Find the input resistance, R_{in}, at terminals 1–1′, assuming that terminals 2–2′ are open, in Fig. 4-21.

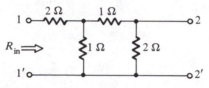

Figure 4-21.   Circuit for Example 4-1.

Solution

We follow the pattern of Eq. 4-43 and write

$$R_{in} = 2 + \cfrac{1}{1 + \cfrac{1}{1 + \cfrac{1}{\cfrac{1}{2}}}} = \frac{11}{4}\ \Omega \qquad (4\text{-}44)$$

The procedure for determining the input given the output will be illustrated by means of the ladder network of Fig. 4-22. The output voltage is designated as v_b and currents and voltages in the branches have the same subscript as that of the resistor. The objective of this analysis is to begin at the output end of the ladder and write every voltage and current in terms of v_b, working step by step toward the input. Each step involves writing Ohm's law or one of the Kirchhoff laws.

First observe that $v_4 = v_b$, and since the output terminals are assumed to be open, $i_3 = i_4$. Follow these steps with the circuit of Fig. 4-22.

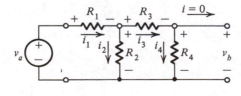

Figure 4-22.   Method for determining input resistance of the circuit of Fig. 4-21.

Ohm's law: $\quad i_4 = \dfrac{v_4}{R_4} = \dfrac{v_b}{R_4} = i_3$ $\hfill$ (4-45)

Ohm's law: $\quad v_3 = R_3 i_3 = \dfrac{R_3 v_b}{R_4}$ $\hfill$ (4-46)

KVL: $\quad v_2 = v_3 + v_4 = \left(1 + \dfrac{R_3}{R_4}\right) v_b$ $\hfill$ (4-47)

Ohm's law: $\quad i_2 = \dfrac{v_2}{R_2} = \dfrac{1}{R_2}\left(1 + \dfrac{R_3}{R_4}\right) v_b$ $\hfill$ (4-48)

KCL: $\quad i_1 = i_2 + i_3$ $\hfill$ (4-49)

$\qquad = \dfrac{1}{R_2}\left(1 + \dfrac{R_3}{R_4}\right) v_b + \dfrac{1}{R_4} v_b$ $\hfill$ (4-50)

Ohm's law: $\quad v_1 = R_1 i_1 = \left[\dfrac{R_1}{R_2}\left(1 + \dfrac{R_3}{R_4}\right) + \dfrac{R_1}{R_4}\right] v_b$ $\hfill$ (4-51)

KVL: $\quad v_a = v_1 + v_2$ $\hfill$ (4-52)

$\qquad = \left[\dfrac{R_1}{R_2}\left(1 + \dfrac{R_3}{R_4}\right) + \dfrac{R_1}{R_4} + 1 + \dfrac{R_3}{R_4}\right] v_b$ $\hfill$ (4-53)

We have worked our way from the output end of the circuit to the input end and have determined v_a. We did so by successively applying KVL and KCL— at each step updating the equation to be a function of v_b by using earlier results. This resulted in every branch voltage and branch current being written in terms of v_b. The procedure is continued until the input is found, which then will be a function of v_b. Often, we will be interested in the ratio v_b/v_a, in which case it is convenient to let $v_b = 1$ V and then carry out the derivation. Because of this practice, the method is sometimes called the *unit-output method*.

The method we have described applies if the output is a current, and clearly it applies if the input is a current source rather than a voltage source. The method works only for ladder networks. For other circuits, the rhythm

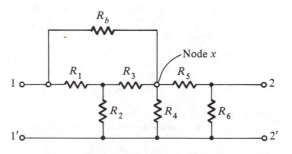

Figure 4-23. Bridged-T circuit for which the unit-output method does not apply.

of moving backward one node at a time is interrupted. Part of the circuit of Fig. 4-23 is shown as a *bridged-T*. When in working our way from output to input, we encounter node x, the method fails, because the bridging of a node involves more than one unknown node voltage at a time. Even so, the method is important simply because so many of the circuits you will use as an engineer are ladder circuits.

EXAMPLE 4-2

In the ladder circuit of Fig. 4-24, determine i_4/i_0 by assuming the output and then determining the input.

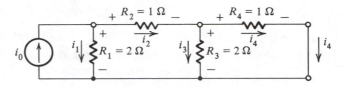

Figure 4-24.   Circuit for Example 4-2.

Solution

The resistors in the circuit are all assigned numerical values. Let current directed from $+$ to $-$ have the same subscript as the resistor and the voltage. We follow the procedure outlined step by step:

Identity:	$v_3 = v_4 = 1 \cdot i_4$	(4-54)
Ohm's law:	$i_3 = \dfrac{v_3}{R_3} = \dfrac{i_4}{2}$	(4-55)
KCL:	$i_2 = i_3 + i_4 = \dfrac{i_4}{2} + i_4 = \tfrac{3}{2}i_4$	(4-56)
Ohm's law:	$v_2 = R_2 i_2 = \tfrac{3}{2}i_4$	(4-57)
KVL:	$v_1 = v_2 + v_3 = \tfrac{5}{2}i_4$	(4-58)
Ohm's law:	$i_1 = \dfrac{v_1}{R_1} = \tfrac{5}{4}i_4$	(4-59)
KCL:	$i_0 = i_1 + i_2 = \tfrac{11}{4}i_4$	(4-60)

Finally,

$$\frac{i_4}{i_0} = \tfrac{4}{11} \tag{4-61}$$

We see that for a current input to the ladder, $\tfrac{4}{11}$ of this current will reach the output.

EXERCISES

4-4.1. For the ladder circuit shown in Fig. 4-25, determine the Thévenin equivalent circuit at terminals a–b.

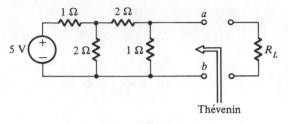

Figure 4-25. Circuit for Exercise 4-4.1.

$$\textit{Ans.} \quad v = \tfrac{10}{11} \text{ V}, \; R = \tfrac{8}{11} \, \Omega$$

4-4.2. In Fig. 4-26, if $R_L = 1 \, \Omega$, determine the current in the load, making use of Thévenin's equivalent-circuit concepts.

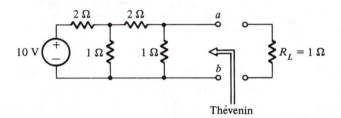
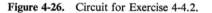

Figure 4-26. Circuit for Exercise 4-4.2.

$$\textit{Ans.} \quad i_L = \tfrac{10}{19} \text{ A}$$

4-4.3. Figure 4-27 shows a ladder circuit. Determine the Thévenin equivalent circuit at terminals a–b.

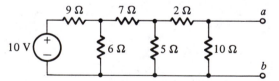

Figure 4-27. Circuit for Exercise 4-4.3.

$$\textit{Ans.} \quad v = 0.833, \; R = 3.505$$

4-5   SUPERPOSITION

A principle that will find frequent application in the chapters to follow is known simply as the *principle of superposition*. Stated simply, if there are several sources in a circuit and a given voltage or current is to be calculated, that voltage or current may be found *by considering each source separately (with other sources "turned off") to find a component of the required voltage or current and then summing all such components.* The only requirement for this to hold is that all resistors in the circuit obey Ohm's law. Such a resistor is called *linear*. When we later generalize this principle, we require that all other elements in the circuit be linear. But we will wait until a later chapter to clarify this generalized meaning of linear.

Recall that a source is turned off when a voltage source produces no voltage and so becomes a short circuit, and when a current source produces no current and so becomes an open circuit. Remember, only the *independent* sources are turned off.

To illustrate the use of the principle of superposition, consider the two-loop circuit given in Fig. 4-28. Let us first determine the current in R_1 using the

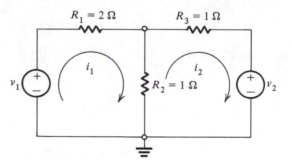

Figure 4-28. Circuit used to illustrate superposition.

loop analysis. Using KVL, we obtain the two loop equations

$$v_1 = (R_1 + R_2)i_1 - R_2i_2 \tag{4-62}$$

$$-v_2 = +(R_2 + R_3)i_2 - R_2i_1 \tag{4-63}$$

Substituting numerical values, we have

$$3i_1 - i_2 = v_1 \tag{4-64}$$

$$-i_1 + 2i_2 = -v_2 \tag{4-65}$$

Our objective is to find the current in the 2-Ω resistor, so we determine i_1, a loop current that is identical with the branch current required:

$$i_1 = \frac{2v_1 - v_2}{5} \tag{4-66}$$

The principle of superposition states that we can determine the current in R_1 in an alternative manner. We first let $v_2 = 0$ and determine the current in R_1 with *only the source v_1 present*. Let us call this current i_{11} as shown in Fig. 4-29(a). Next, we let $v_1 = 0$ and determine the current in R_1 with *only*

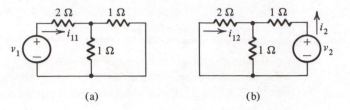

Figure 4-29.   Identification of currents from the two sources.

the source v_2 *present.* This current i_{12} is shown in Fig. 4-29(b). We then sum the two currents to obtain the total current through R_1:

$$i_1 = i_{11} + i_{12} \tag{4-67}$$

We see in Fig. 4-29(a) that the current i_{11} is the current supplied by the source to the circuit,

$$i_{11} = \frac{v_1}{2 + 0.5} \tag{4-68}$$

The current supplied by v_2 in Fig. 4-29(b) is i_2 and given by

$$i_2 = \frac{v_2}{1 + \frac{2}{3}} \tag{4-69}$$

The current in the 2-Ω resistor in this case is obtained by the current-divider relation

$$-i_{12} = \frac{1}{2 + 1} i_2 \tag{4-70}$$

or

$$i_{12} = -\frac{1}{3} \cdot \frac{v_2}{1 + \frac{2}{3}} V_2 \tag{4-71}$$

$$= -\tfrac{1}{5} v_2 \tag{4-72}$$

Substituting in Eq. 4-67, we obtain

$$i_1 = \frac{2v_1 - v_2}{5} \tag{4-73}$$

as in Eq. 4-66.

In our illustration of the superposition principle, the amount of labor is not obviously reduced by its application. However, the concept is a very important one and much of circuit analysis as well as engineering analysis depends on it. Much of our development in later chapters will depend upon the use of superposition.

EXERCISES

4-5.1. Apply the principle of superposition to the circuit of Ex. 4-1.1 by considering i_1 and i_2 operating separately, and determine v_1 and v_2.

Ans. $v_1 = \tfrac{2}{3}$ V, $v_2 = \tfrac{5}{6}$ V

4-5.2. Apply the principle of superposition to the circuit shown in Fig. 4-14 by letting v_1 and v_2 operate separately. Determine the current i if $v_1 = 10$ V, $v_2 = 2$ V, R_1 and $R_2 = 10\ \Omega$, and $k = 2$.

Ans. $i = 0.1$ A

4-5.3. For the circuit shown in Fig. 4-30, make use of the principle of superposition to determine i_1, i_2, v_1, and v_2.

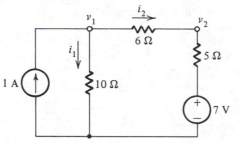

Figure 4-30. Circuit for Exercise 4-5.3.

Ans. $i_1 = \frac{4}{21}$, $i_2 = \frac{17}{21}$, $v_1 = \frac{40}{21}$, $v_2 = \frac{-62}{21}$

4-6 POWER TRANSFER

Circuits are used to *process signals*. By signals we mean voltages or currents. A signal may be supplied directly by a source or it may be a result of some previous *processing*. By processing we mean that an *excitation* or an incoming *input signal* is changed by the circuit in some precise manner, resulting in a *response* or *output signal* at some place in the circuit. The output signal is said to be *delivered to a load* which usually consists of a single resistor, open circuit, and so on. The configuration is shown in Fig. 4-31. The input signals v_1 and i_1

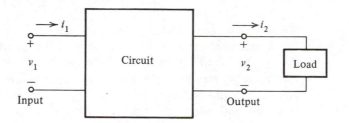

Figure 4-31. The circuit is to transfer power to the load.

are assumed to be supplied by a Thévenin equivalent source network. Examples of signal processing are shown in Fig. 4-32. In Fig. 4-32(a) we have a voltage divider delivering a certain fraction of the input voltage to the load. Similarly, in Fig. 4-32(b) the current divider is used to deliver a certain fraction of the input current to the load.

Frequently, our concern is the power delivered to the load. The load is

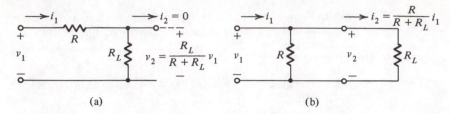

(a) (b)

Figure 4-32. Power transfer in communication circuits.

usually a device that responds to the amount of power delivered. Some common examples are loudspeakers, antennas, earphones, and so forth. We are then interested in *maximizing the amount of power delivered to the load*. Consider the circuit of Fig. 4-33, in which a Thévenin voltage source v_S with an associated

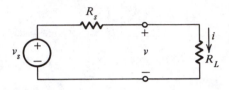

Figure 4-33. Power transfer in a voltage-divider circuit.

source resistance R_S is connected to a load resistance R_L. The current in R_L is

$$i = \frac{v_S}{R_S + R_L} \tag{4-74}$$

and the voltage across it is

$$v = \frac{R_L}{R_S + R_L} v_S \tag{4-75}$$

We find that the power delivered to the load is

$$P = vi = \frac{R_L}{(R_S + R_L)^2} v_S^2 \tag{4-76}$$

If the source resistance is fixed and cannot be altered, the power delivered to the load changes with the value of R_L. For example, when R_L is very small, the power delivered is also small. When R_L is very large compared to the source resistance, R_S may be neglected in the denominator. Thus, $P = v_S^2/R_L$, which is also very small since R_L is very large. Power delivered to the load is thus very small for both very small and very large values of R_L. At some intermediate value of R_L, the power delivered will reach a maximum value.

To obtain the value of R_L for which maximum power is obtained, we differentiate the expression in Eq. 4-76 with respect to R_L and set it to zero,

$$\frac{d}{dR_L} P = v_S^2 \left[\frac{R_S - R_L}{(R_S + R_L)^3} \right] = 0 \tag{4-77}$$

Equation 4-77 is satisfied when $R_L = R_S$ and yields the *maximum power* from Eq. 4-76,

$$P_{\text{max}} = \frac{v_S^2}{4R_S} \tag{4-78}$$

When $R_L = R_S$, the power dissipated in the source resistance is $i^2 R_S$, which by using Eq. 4-74 also becomes $v_S^2/4R_S$. The total power supplied to the whole circuit is thus $v_S^2/2R_S$.

The value of P_{max} is also referred to as the *available power*, since that is the maximum power that the load can receive from a source with a fixed source resistance.

If, on the other end, the load resistance were fixed and the source resistance could be varied, then from Eq. 4-76, the maximum power is obtained when $R_S = 0$ and is

$$P_{max} = \frac{v_S^2}{R_L} \qquad (4\text{-}79)$$

It is thus advantageous to make the source resistance as small as possible.

EXERCISES

4-6.1. For the ladder circuit shown in Fig. 4-34, find the value of R_L that will determine that value of power P_{max}.

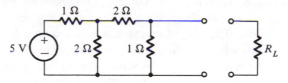

Figure 4-34. Circuit for Exercise 4-6.1.

Ans. $R_L = \frac{8}{11}$, $P_{max} = 0.284$ W

4-6.2. Repeat Exercise 4-6.1 for the circuit shown in Fig. 4-35.

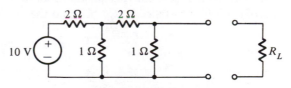

Figure 4-35. Circuit for Exercise 4-6.2.

Ans. $R_L = \frac{8}{11}$, $P_{max} = 0.284$ W

4-6.3. The ladder circuit shown in Fig. 4-36 is to be terminated in load resistor R_L. Determine the value of R_L that will result in maximum power transfer to R_L and determine the power that results when this value of R_L is selected.

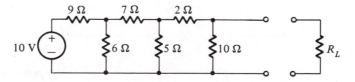

Figure 4-36. Circuit for Exercise 4-6.3.

Ans. $R_L = 3.505$, $P_{max} = 0.0495$

We discuss the feedback concept and its practical use in Chapter 5. For now we merely show how the presence of feedback affects the equivalent resistance of a circuit.

An example of feedback is shown in Fig. 4-37. Whenever a controlled

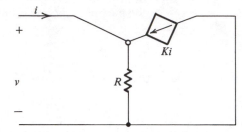

Figure 4-37. Circuit with feedback for which the equivalent resistance is to be found.

source influences the controlling signal, we have *feedback*. The term comes from a controlling signal that is input to a device represented by the controlled source. The output at the controlled source terminals is *fed back* to alter the input, which, in turn, controls the output. In the figure the current-controlled current source is connected to the incoming controlling signal at a node. For a given node voltage the value of the current i is directly influenced by the controlled source.

The Thévenin equivalent of the circuit will be just a resistance, since there are no independent sources to generate any open-circuit voltage at the terminals. The Thévenin equivalent resistance is obtained from observing that the branch current in the resistance is $(i + Ki)$ and the voltage across it is v. Then

$$i + Ki = \frac{v}{R} \tag{4-80}$$

and

$$R_{eq} = R_{in} = \frac{v}{i} = (1 + K)R \tag{4-81}$$

Thus, the effect of feedback in this case is to alter the resistance by a factor of $(1 + K)$.

Another example of feedback is shown in Fig. 4-38. In this case the Thévenin

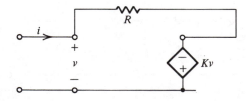

Figure 4-38. Second circuit with feedback.

resistance is obtained by using KVL:

$$v + Kv = Ri \qquad (4\text{-}82)$$

and

$$R_{eq} = \frac{v}{i} = \frac{R}{1 + K} \qquad (4\text{-}83)$$

We thus see that feedback may be used to alter the resistance and control the effective value to practical advantage. Note, however, that since K can be positive or negative, it sometimes leads to negative values of R_{eq}. Physically, of course, resistance cannot be negative. But the active element represented by the controlled source model may give a behavior that is equivalent to a negative resistance for those applications where the model is valid.

A final comment concerns the energy conservation principle. Energy supplied by the independent sources is dissipated in resistances as heat. For circuits consisting only of sources and resistors, the conservation principle holds. The energy supplied must equal that dissipated. In resistive circuits driven by batteries, energy is just the time interval times the power. Thus, power is also conserved in such circuits. Conservation of power (and energy) is not valid, however, when controlled sources are present in the circuit.

Conservation of energy or power requires that the power supplied by the sources be equal to that dissipated in the resistances. This is true in the case of the circuit shown in Fig. 4-33, as can be easily checked. It is also true of all circuits without controlled sources. The R_L in such cases may be assumed to represent the equivalent resistance and the rest follows as before. Such circuits are called *passive* circuits whereas circuits containing controlled sources are called *active*.

Consider now the circuit of Fig. 4-39 containing a controlled source.

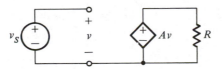

Figure 4-39. Circuit without feedback.

The power dissipated in the resistance is

$$P = \frac{(Av)^2}{R} = \frac{(Av_s)^2}{R} = A^2 \cdot \frac{v_s^2}{R} \qquad (4\text{-}84)$$

However, the power supplied by the independent source is v_S times the current. But the current through the source is zero, since it is open at the terminals. The power supplied by the source is thus zero, but that dissipated in the resistance is given by Eq. 4-84, which is greater than zero. Where does the power and the corresponding energy come from?

The answer lies in the fact that a circuit *model*, like any model, shows only certain relevant and idealized information and suppresses information presumed not to be relevant. Such is the case when independent sources are defined without any source resistances. It is also the case when resistances are assumed to be linear for all signals and their values. Such is also the case when devices are modeled by controlled sources which *show the dependence on the controlling signals but usually suppress independent sources that actually supply power to these devices.* A complete circuit showing these independent sources will, of course, satisfy conservation of power (and energy) but at the price of considerably complicating the circuit to be analyzed. For these reasons, the controlled sources suppress this information for the most part. One example of such a model will be encountered in Chapter 5.

EXERCISES

4-7.1. The circuit shown in Fig. 4-40 contains an active element which is a voltage-controlled voltage source, v_2. Find the input resistance for this circuit.

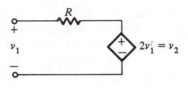

Figure 4-40.   Circuit for Exercise 4-7.1.

Ans. $R_{in} = -R$

4-7.2. The circuit of Fig. 4-41 contains a current-controlled current source, $i_2 = 2i_1$. For this circuit, determine the input resistance.

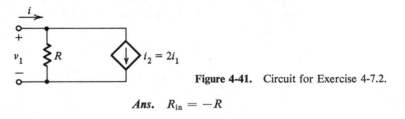

Figure 4-41.   Circuit for Exercise 4-7.2.

Ans. $R_{in} = -R$

PROBLEMS

4-1. The circuit given in the figure shows resistor sizes in ohms and current source values in amperes. For the element sizes given, find the branch currents i_1, i_2, and i_3 and the node-to-datum voltages v_1 and v_2.

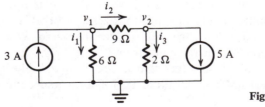

Figure P4-1

4-2. Determine the node-to-datum voltages v_1 and v_2 for the circuit given in the figure.

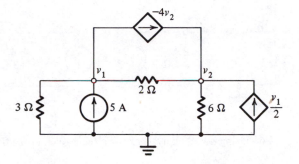

Figure P4-2

4-3. Write the node equations for the circuit given in the figure and solve for the voltages v_1 and v_3.

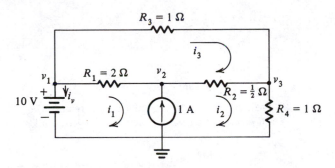

Figure P4-3

4-4. Find the currents i_1, i_2, and i_3 and the voltage v_1 making use of nodal analysis for the circuit given in the figure.

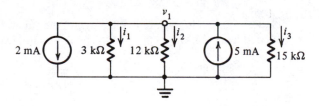

Figure P4-4

4-5. Find the currents i_1, i_2, and i_3 and the two voltages v_1 and v_2 by nodal analysis of the two circuits given in the figure.

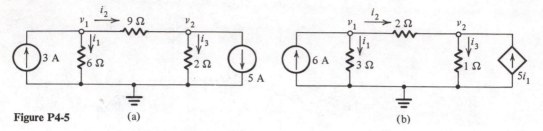

Figure P4-5   (a)   (b)

4-6. Using nodal analysis of the circuit given in the figure, find the voltages v_1 and v_2.

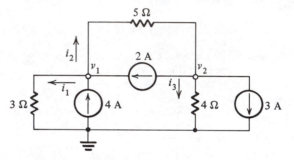

Figure P4-6

4-7. Making use of loop analysis, find i_1 and i_2 for the circuit shown in the figure. Knowing these values, determine v_1, v_2, and v_3.

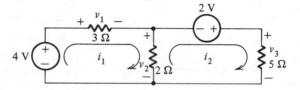

Figure P4-7

4-8. For the circuit given in the figure, find the values of the loop currents i_1 and i_2, and from this knowledge find the node-to-datum voltage v_1.

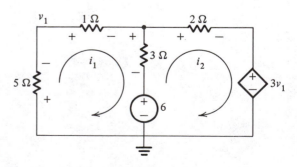

Figure P4-8

4-9. Making use of loop analysis, find the currents i_1 and i_2 and from this information, determine the voltages v_4, v_5, and v_6.

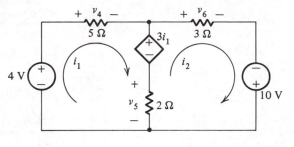

Figure P4-9

4-10. Find expressions for i_1 and i_2 in terms of v, i, R_1, R_2, and R_3 for the circuit
shown in the figure.

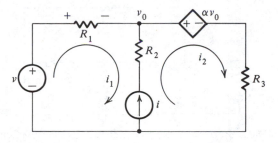

Figure P4-10

4-11. Making use of loop equations, analyze the circuit given for Prob. 4-3 and determine the values for i_1 and i_3.

4-12. Given the circuit shown in the figure, use loop analysis to determine the two
loop currents i_1 and i_3.

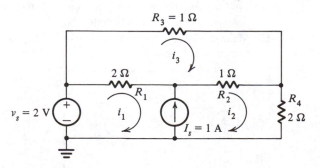

Figure P4-12

4-13. Given the circuit shown in the figure, use nodal analysis to determine values for the voltages v_2 and v_3.

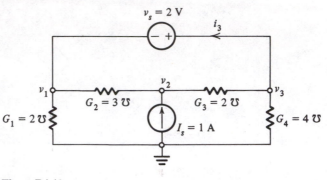

Figure P4-13

4-14. For the resistive circuit shown in the figure, determine the input resistance $R_{in} = v/i$.

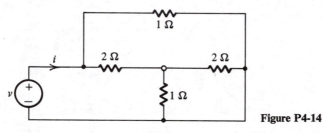

Figure P4-14

4-15. For the simple ladder circuit given in the figure, determine the value of the input resistance R_{in} when the output port is open-circuited.

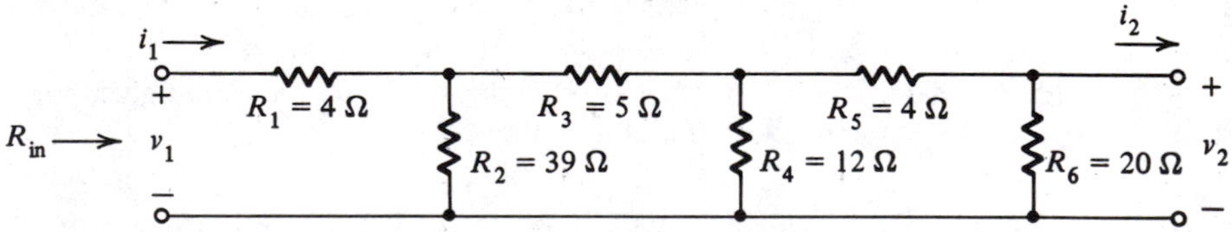

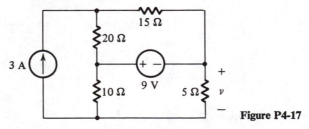

Figure P4-15

4-16. Find the voltage transfer function v_2/v_1 for the ladder circuit of Prob. 4-15 when the output port is open-circuited.

4-17. Make use of superposition to determine the value of the voltage v in the circuit shown in the figure.

Figure P4-17

4-18. Consider the circuit shown in the figure. (a) Find R_L that maximizes the power delivered to R_L and find the power in the load under this condition. (b) How much power is dissipated by the 1-kΩ resistor when R_L is selected to maximize the power delivered to R_L?

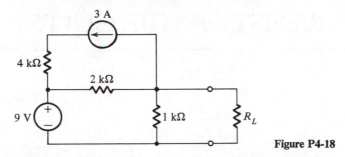

Figure P4-18

4-19. Consider the circuit shown in the figure. (a) Find the value of R such that $v_2 = 2$ V. (b) Find the power in the controlled source and specify where the power is being absorbed or delivered. (c) Find the Thévenin equivalent circuit at terminals a–b.

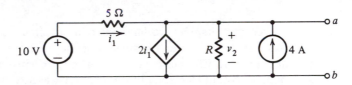

Figure P4-19

5

OP AMP–RESISTOR CIRCUITS

In preceding chapters we have introduced circuit analysis concepts using the resistor as the only circuit element, together with the voltage source and the current source. In Chapter 6 we will add the capacitor and the inductor to the list of passive circuit elements. In this chapter we introduce a circuit element that is very important in circuit design, the *operational amplifier*. This element will be used extensively in the studies of future chapters. Here we consider its use with resistors only to perform the algebraic functions of multiplication, addition, and subtraction.

The operational amplifier is different from the resistor in that it has more terminals. The resistor has two, whereas op-amp circuits have three or more. We begin by a discussion of two-port circuits.

5-1 TWO-PORT CIRCUITS

In preceding chapters we studied circuits in which two terminals were marked 1–1′ and another two 2–2′. It was intended that terminals 1 and 1′ and terminals 2 and 2′ be associated together as a *terminal pair* at which an input signal is connected or an output measured. An equivalent word for terminal pair is *port*. Thus, such circuits are *two-port circuits*. A general form of the two-port circuit is shown in Fig. 5-1 with four variables identified—two currents, two voltages. Most often, terminals 1′ and 2′ will be connected and grounded, in which case the circuit is known as a grounded two-port. Terminals 1–1′ are identified as the *input* port, 2–2′ as the *output* port.

In terms of the four variables identified in Fig. 5-1, we define six ratios of

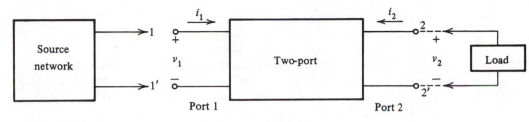

Figure 5-1. General representation of a two-port coupling a source to a load.

variables. The ratios of voltage to current at each of the two ports, v_1/i_1 and v_2/i_2, the *input* and *output* resistances, are known as *driving-point* functions. Ratios of voltage to voltage or current to current, v_2/v_1 and i_2/i_1, are known as *gain* functions, and the mixed ratios, v_2/i_1 and i_2/v_1, are known as *transfer* functions. We will sometimes be interested in the reciprocals of each of these six quantities. The ratios v_2/v_1 and i_2/i_1 are known also as *transfer voltage and current ratios*, respectively.

5-2 THE OP AMP

The *operational amplifier* (hereafter *op amp*) is a two-port circuit developed in the 1940s primarily for use in analog computers, made cheaply and readily available in the late 1960s through integrated-circuit technology. The actual construction of a modern op amp is indeed very complex, involving many transistors, diodes, resistors, and capacitors. For many analysis and design functions, the op amp may be represented by simple models.

The op-amp symbol is shown in Fig. 5-2(a). The B^+ and B^- terminals are where batteries or *power supplies* with positive and negative voltage values (with respect to ground) are connected. Typical values of B^+ and B^- are $+15$ V and -15 V, respectively. The op amp is able to function only when B^+ and B^- power supplies are connected. There are two input terminals, in addition to the ground,

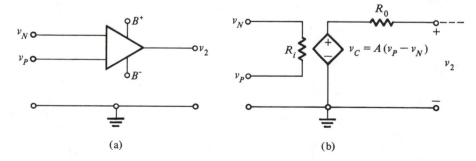

Figure 5-2. (a) Some of the connection terminals to the op amp, and (b) a simple model of the op amp.

identified by their voltages with respect to ground, v_N and v_P. When the op amp is used, one of these terminals is usually grounded either directly or through a resistance. The output voltage with respect to ground is v_2.

Our interest is in the input–output relations, and the model showing these is given in Fig. 5-2(b). Observe that B^+ and B^- are not relevant for the input–output relations and are therefore suppressed. The input voltages control a controlled source which has a voltage

$$v_C = A(v_P - v_N) \tag{5-1}$$

where A is a positive constant called the *op-amp gain*. The resistance between the input terminals is R_i and that at the output terminal is R_o. The output voltage v_2 will depend on the output current, determined by the load, and will be v_C when the output current is zero. The op amp is a *unilateral* device: The input voltages determine the output voltage, but a voltage applied at the output would not influence the input. This unilateral nature is implied by the use of a triangular shape for the op amp, indicating that its functioning goes from input to output. Typical values for A, R_i, and R_o are shown in Table 5-1. The gain is very high for reasons we will explain later. The input resistance is also high, and the output resistance low. These unusual values suggest an even simpler model, identified in Table 5-1 as *ideal*.

Table 5-1

Quantity	Typical Values	Ideal Model
A	10^5	∞
R_i	10^4	∞
R_o	10	0

Since the B^+ and B^- terminals play no role in the input–output model, they are usually suppressed in the symbol, as in Fig. 5-3. The op amp always has a

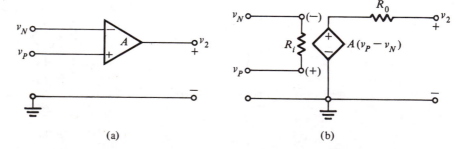

(a)   (b)

Figure 5-3. The $B+$ and $B-$ terminals are assumed connected but not shown in this representation.

negative and a positive sign, with terminals marked − and +. These are *not* polarity marks but *names* or *identifications* for the terminals as supplied by the manufacturer. The voltages at these terminals are shown as v_N and v_P. Figure 5-4 shows the ideal model. In the ideal model, *the output voltage v_2 is identical to v_C since $R_o = 0$, and the currents at the input terminals are always zero since $R_i = \infty$.*

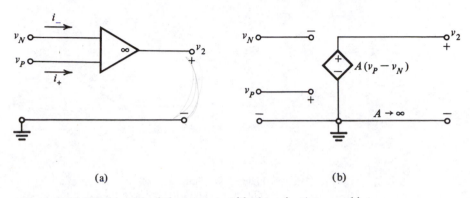

(a)   (b)

Figure 5-4.   Ideal model of the op amp with the gain A approaching infinity.

We should reflect on a device with a gain as large as 10^5. If the device actually behaved as the model, rather than saturating, then 1 V at the input would result in 0.1 million volts at the output! Such a connection, shown in Fig. 5-5 is never used, for the op amp saturates and its behavior is significantly altered. The high gain is used to ensure accuracy, as will be discussed later. As soon as the op amp is "harnessed" by two resistors, the overall gain is controlled by the values of the resistors, as we show next.

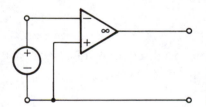

Figure 5-5.   If the + terminal shown in Fig. 5-4(a) is connected to ground, we have the result shown in this figure.

EXERCISES

5-2.1. To the circuit of Fig. 5-3(b), we connect a resistor R_1 in series with the terminal marked v_N, and connect v_P to ground. The circuit then has an input voltage applied to R_1 having a constant value of 10^{-5} V. Using the values given in the second column of Table 5-1, determine the output voltage v_2 with the output terminals open if (a) $R_1 = 10$ kΩ; (b) $R_1 = 10$ MΩ.

Ans.   (a) $-\frac{1}{2}$ V; (b) −1 mV

Having assumed the ideal model described in Section 5-4, we shall deal with op amps having the characteristics summarized in Fig. 5-6. The input currents to the op amp, i_N and i_P, are both zero. The voltage between the terminals of the

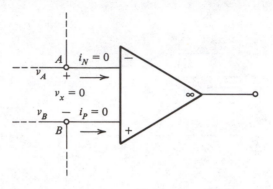

Figure 5-6. Summary of the voltages and currents at the input terminals to the op amp.

op amp is also zero, $v_x = 0$. That implies that the two node voltages, v_A and v_B, are always equal to each other. It is important to remember that these voltages are related by the equation

$$v_A = v_B \tag{5-2}$$

it is also important to remember that, in general, two node equations formulated from KCL will be required, one for node A, and one for node B.

The circuit shown in Fig. 5-7 may be analyzed using KCL. At node x, we sum the currents to obtain

$$\frac{v_2 - v_x}{R_2} + \frac{v_1 - v_x}{R_1} = i \tag{5-3}$$

But in terms of the model described by Fig. 5-6, $v_x = 0$ (being at ground voltage) and $i = 0$. Then Eq. 5-3 becomes

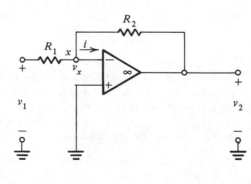

Figure 5-7. Inverting op-amp circuit described by Eq. 5-4.

$$\frac{v_2}{v_1} = -\frac{R_2}{R_1} \tag{5-4}$$

The negative sign in this equation is responsible for the name given to this circuit. It is called the *inverting circuit*, since the polarity of v_2 is always the opposite of v_1.

The circuit shown in Fig. 5-8 is known as a noninverting circuit, for reasons that will soon be apparent. The same analysis with KCL applied at node x gives

$$\frac{v_x}{R_1} + \frac{v_x - v_2}{R_2} = i \tag{5-5}$$

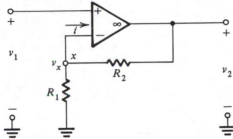

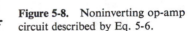

Figure 5-8. Noninverting op-amp circuit described by Eq. 5-6.

As connected, $v_x = v_1$, and $i = 0$ by assumption. Then this equation may be arranged to give

$$\frac{v_2}{v_1} = 1 + \frac{R_2}{R_1} \tag{5-6}$$

From this we see that v_2 is always greater than v_1 and that the circuit does not invert v_1 to give v_2.

The two gain equations, Eqs. 5-4 and 5-6, are obviously important in the design of op-amp circuits. In some applications, the input and output resistances of the circuits are of importance. These may be expressed in terms of R_i and R_o, which are defined in Fig. 5-3. Derivations of the equations are given in courses in electronic circuits* and are summarized in Table 5-2. From this table we see that the two circuits differ considerably in input resistance, R_{in}. For the inverting circuit, the input to the op amp is said to be a "virtual short," so that $R_{in} = R_1$. For the noninverting circuit, R_{in} is a function of the product of A and R_i, and since both quantities are very large, R_{in} may reasonably be assumed to be infinite. We discuss these features further in the next section after the voltage follower is introduced.

*W. G. Oldham and S. E. Schwarz, *An Introduction to Electronics* (New York: Holt, Rinehart and Winston, 1972), Chap. 12.

Table 5-2

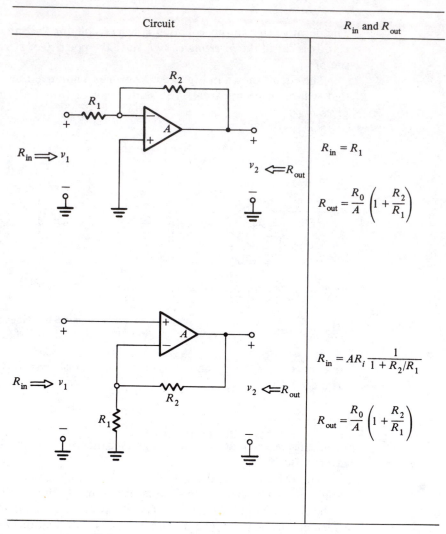

Circuit	R_{in} and R_{out}
	$R_{in} = R_1$
	$R_{out} = \dfrac{R_0}{A}\left(1 + \dfrac{R_2}{R_1}\right)$
	$R_{in} = AR_i \dfrac{1}{1 + R_2/R_1}$
	$R_{out} = \dfrac{R_0}{A}\left(1 + \dfrac{R_2}{R_1}\right)$

EXAMPLE 5-1

For each of the two circuits introduced in this section, design an op amp–resistor circuit to give a gain of 10.

Solution

From the design equations, we see that the design is complete when values for R_1 and R_2 have been selected. In practical circuit design, values should be chosen that are not too low nor too high (the latter for reasons suggested by Exercise 5-2.1). For many applications, a resistance level of $10^4\ \Omega$ is satisfactory. If we select this as the value of R_1, the design is completed by selecting for the inverting circuit from Eq. 5-4,

$$R_1 = 10 \text{ k}\Omega \quad \text{and} \quad R_2 = 100 \text{ k}\Omega \qquad (5\text{-}7)$$

For the noninverting circuit, we make use of Eq. 5-6, and obtain

$$R_1 = 10 \text{ k}\Omega \quad \text{and} \quad R_2 = 90 \text{ k}\Omega \qquad (5\text{-}8)$$

EXERCISES

5-3.1. An amplifier is required that is noninverting with an overall gain of 2. Design the circuit, and make use of the information in Table 5-2 to determine the values of R_{in} and R_{out}, using op-amp parameters given in the second column of Table 5-1.

> *Ans.* $R_1 = R_2 = 10 \text{ k}\Omega$, $R_{in} = 5 \times 10^8 \ \Omega$,
> $R_{out} = 2 \times 10^{-4} \ \Omega$

5-3.2. Repeat Exercise 5-3.1 for an inverting circuit with a gain of 2.

> *Ans.* $R_1 = 10 \text{ k}\Omega$, $R_2 = 20 \text{ k}\Omega$, $R_{in} = 10 \text{ k}\Omega$,
> $R_{out} = 3 \times 10^{-4} \ \Omega$

5-4 VOLTAGE FOLLOWER (OR ISOLATOR)

The *voltage follower* is a circuit in which the output is identical to the input, that is, $v_2 = v_1$. This circuit is shown in Fig. 5-9. From the circuit we recall that

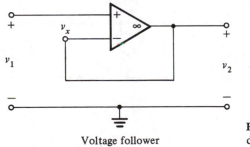

Voltage follower

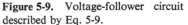

Figure 5-9. Voltage-follower circuit described by Eq. 5-9.

since the voltage across the input to the op amp is zero, then $v_x = v_1$. Further, by connection, $v_x = v_2$, so that

$$\frac{v_2}{v_1} = 1 \qquad (5\text{-}9)$$

It is shown in electronic circuits* that the voltage follower has input and output resistance given by

$$R_{in} = AR_i \quad \text{and} \quad R_{out} = \frac{R_o}{A} \qquad (5\text{-}10)$$

*Oldham and Schwarz, ibid.

so that the input resistance approaches an infinite value, while the output resistance approaches zero. Thus, the controlled source representation of the voltage follower is that shown in Fig. 5-10.

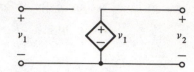

Figure 5-10. Controlled-source representation of the voltage follower.

The application of the voltage follower is to provide isolation or to serve as a "buffer" of the input from the output. It serves to isolate the output of the circuit from the input. Thus, we may "load" the output of the voltage follower without loading the source voltage v_1. Such loading will be explained in terms of an example.

Interference by the load in the signal values is known as *loading*. An example of this for the voltage divider is shown in Fig. 5-11. In (a), the voltage divider is designed to deliver a voltage of

$$v_1 = \frac{R_1}{R_1 + R_2}v_0 \qquad (5\text{-}11)$$

But if the load R_L is connected directly at the output, Eq. 5-11 is no longer valid. Resistor R_L now draws a current and v_1 is now across the parallel combination $R_2 \| R_L$, so that

$$v_1 = \frac{R_2 \| R_L}{R_1 + R_2 \| R_L}v_0 \qquad (5\text{-}12)$$

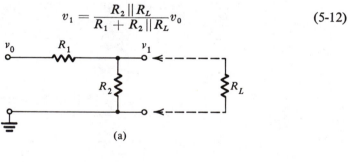

(a)

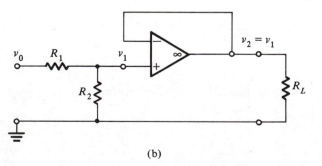

(b)

Figure 5-11. The voltage follower permits isolation of parts of the circuit.

which is different from Eq. 5-11. In Fig. 5-11(b) we show the same voltage divider, but here the load R_L is isolated and does not interfere with v_1. Care should always be exercised that loading does not invalidate our design.

For these reasons, op-amp circuits serve to isolate parts of the overall circuit and, therefore, serve as dividing lines for the analysis. Each side of an op-amp circuit is handled separately and we move from one side of the op-amp circuit to the other by the appropriate gain formula.

The simplest example of such a circuit consists of a cascade of op-amp circuits, shown in Fig. 5-12. The dividing line shows a noninverting amplifier

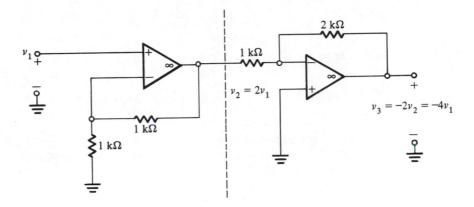

Figure 5-12. These circuits are said to be connected in cascade.

followed by an inverting one. We see that $v_2 = 2v_1$ and $v_3 = -2v_2$, so that $v_3 = -4v_1$.

It is a common practice to show input signals in op-amp circuits without identifying the source network. That information is unnecessary due to the isolating property of op amps. We may assume an ideal voltage source at the input without affecting our analysis.

5-5 ANALOG ADDITION AND SUBTRACTION

Op-amp circuits to add and subtract voltages were highly developed for use in analog computers in the early 1940s. The circuit of Fig. 5-13 has two inputs, v_1 and v_3, and one output, v_2. At node a, KCL may be used to equate currents *in* to those *out*, giving

$$\frac{v_1 - v_a}{R_1} + \frac{v_3 - v_a}{R_3} = \frac{v_a - v_2}{R_2} \tag{5-13}$$

But $v_a = 0$, so this equation simplifies directly to the form

$$v_2 = -\left(\frac{R_2}{R_1}v_1 + \frac{R_2}{R_3}v_3\right) \tag{5-14}$$

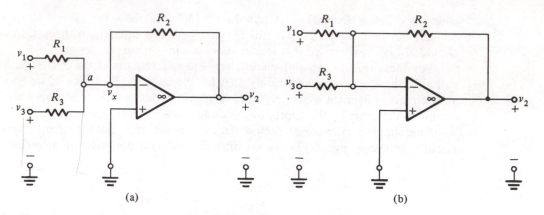

(a)   (b)

Figure 5-13. Two representations of the same circuit described by
Eq. 5-14.

Let $R_1 = R_3$ and $R_2/R_1 = K$ and this equation becomes

$$v_2 = -K(v_1 + v_3) \tag{5-15}$$

so that the circuit both multiplies and adds. If $K = 1$, the circuit operates to
add v_1 and v_3, also inverting in the process. The controlled source equivalent of
the circuit of Fig. 5-13 is shown in Fig. 5-14. As shown in the circuit, the input
resistances are R_1 and R_3.

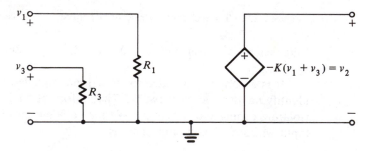

Figure 5-14. Controlled source representation corresponding to
Eq. 5-15.

Equation 5-14 could have been derived by using superposition for the two
independent signals v_1 and v_3. If we turn v_3 off (i.e., $v_3 = 0$), then one end of
R_3 is shorted to ground. The other end of R_3 is also shorted to ground by the
virtual short. So R_3 is out of the circuit and we have an inverting amplifier with

$$v_{21} = -\frac{R_2}{R_1}v_1 \tag{5-16}$$

Similarly, if we set $v_1 = 0$, then R_1 is out of the circuit and

$$v_{23} = -\frac{R_2}{R_3}v_3 \tag{5-17}$$

The output voltage by superposition is

$$v_2 = v_{21} + v_{23} \tag{5-18}$$

$$= -\frac{R_2}{R_1}v_1 - \frac{R_2}{R_3}v_3 \tag{5-19}$$

as before.

The result we have found for two inputs may be extended to any number of inputs. If there are three, for example, then Eq. 5-13 with $v_a = 0$ becomes

$$\frac{v_1}{R_1} + \frac{v_3}{R_3} + \frac{v_4}{R_4} = -\frac{v_2}{R_2} \tag{5-20}$$

As an example, if it is required to design a circuit to realize

$$v_2 = -3v_1 - 2v_3 \tag{5-21}$$

this may be accomplished by making $R_2/R_1 = 3$ and $R_2/R_3 = 2$. If we let $R_2 = 60 \text{ k}\Omega$, the design requires that $R_1 = 20 \ \Omega$ and $R_3 = 30 \text{ k}\Omega$.

The *differential amplifier* circuit shown in Fig. 5-15 is the basic circuit for forming the difference of voltages. The circuit may be analyzed by superposition.

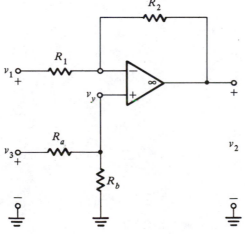

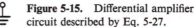
Figure 5-15. Differential amplifier circuit described by Eq. 5-27.

We first set $v_3 = 0$. No current can flow in R_a and R_b, and $v_y = 0$. This is an inverting amplifier, so that

$$v_{21} = -\frac{R_2}{R_1}v_1 \tag{5-22}$$

Next, we set $v_1 = 0$ and we have a noninverting amplifier for the signal v_y at the + terminal of the op amp. Thus,

$$v_{2y} = \left(1 + \frac{R_2}{R_1}\right)v_y \tag{5-23}$$

But v_y is just the fraction of v_3 determined by the voltage divider, so that

$$v_y = \frac{R_b}{R_a + R_b} v_3 \tag{5-24}$$

Then, by superposition,

$$v_2 = v_{21} + v_{2y} \tag{5-25}$$

$$= \left(\frac{\frac{R_2}{R_1} + 1}{\frac{R_a}{R_b} + 1} \right) v_3 - \frac{R_2}{R_1} v_1 \tag{5-26}$$

$$= \frac{R_2}{R_1} \left[\left(\frac{\frac{R_1}{R_2} + 1}{\frac{R_a}{R_b} + 1} \right) v_3 - v_1 \right] \tag{5-27}$$

From this result we see that if $R_a = R_1$ and $R_2 = R_b$, then

$$v_2 = \frac{R_2}{R_1}(v_3 - v_1) \tag{5-28}$$

and this simplified circuit is shown in Fig. 5-16. Further simplification results

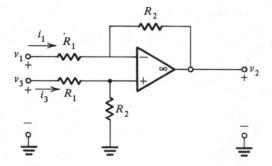

Figure 5-16. Simplified differential amplifer circuit described by Eq. 5-28.

when $R_2 = R_1$ and

$$v_2 = v_3 - v_1 \tag{5-29}$$

This information is summarized in Table 5-3.

These equations permit routine design of difference amplifiers. For example, if we require that

$$v_2 = 2v_3 - 3v_1 \tag{5-30}$$

then Eq. 5-27 may be used to determine that $R_2/R_1 = 3$ and $R_a/R_b = 1$, so that a suitable design would result with $R_2 = 30 \text{ k}\Omega$ and $R_1 = R_a = R_b = 10 \text{ k}\Omega$. For many applications, we will wish to realize Eq. 5-29, and in this case R_1, R_2, R_b, and R_a will have identical values.

Finally, with respect to the circuit of Fig. 5-16, we see that the input resistance at terminal pair 1–0 is the ratio of v_1 and i_1 with $v_3 = 0$ and that is R_1.

Table 5-3

Circuit	Controlled Source Representation	Equation for v_2
		$v_2 = v_1$
		$v_2 = -\dfrac{R_2}{R_1} v_1$
		$v_2 = \left(1 + \dfrac{R_2}{R_1}\right) v_1$
		$v_2 = -\left(\dfrac{R_2}{R_3} v_1 + \dfrac{R_2}{R_1} v_3\right)$
		$v_2 = \dfrac{1 + R_2/R_1}{1 + R_a/R_b} v_3 - \dfrac{R_2}{R_1} v_1$
		$v_2 = \dfrac{R_2}{R_1} (v_3 - v_1)$

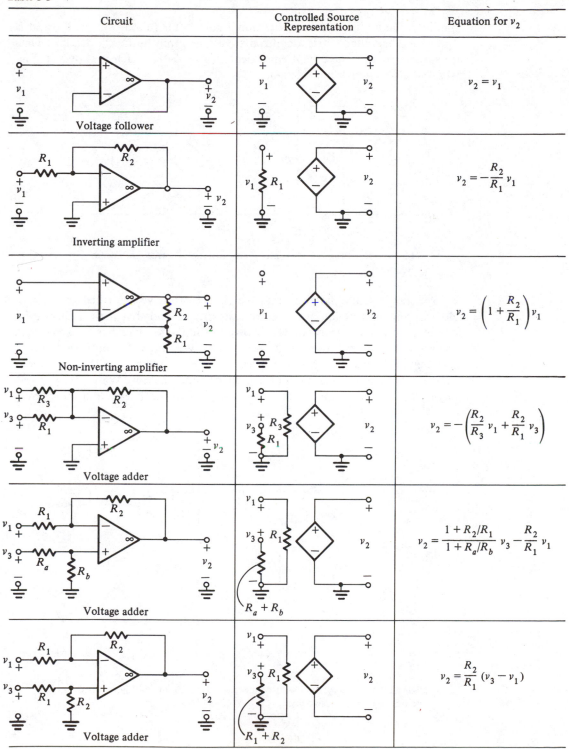

137

Similarly, the input resistance at terminal pair 3–0 is found with $v_1 = 0$ and is $R_1 + R_2$. The voltage v_2 is a controlled voltage given by Eq. 5-28. These facts are summarized by the circuit equivalent to Fig. 5-16, which is shown in Fig. 5-17 and also entered in Table 5-3.

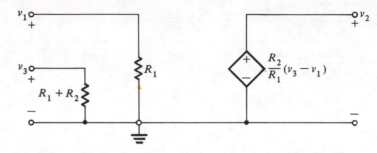

Figure 5-17. Controlled source representation corresponding to Eq. 5-28.

We should make the point that the circuits we have described for voltage addition and subtraction are by no means unique and that many others exist (or can be invented). An example is shown in Fig. 5-18, which has the advantage that the input resistance is infinite (or at least very high) for both the voltage sources, but the disadvantage that three op amps are required in the realization. The circuit is analyzed using the principle of superposition. We let $v_3 = 0$ and note that $v_y = 0$ since the op-amp current $i_4 = 0$. Note that the output voltage,

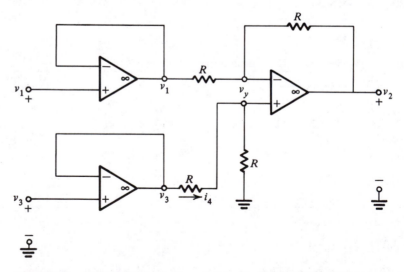

Figure 5-18. Voltage followers connected to the circuit of Fig. 5-16 to provide high input resistance.

v_{21}, due to v_1 is $-v_1$. Similarly, we let $v_1 = 0$, note that $v_y = v_3/2$ because of voltage-divider action, and determine the output voltage, v_{2y}, due to v_3, which is $(v_3/2)(1 + 1) = v_3$. Then the output due to both v_1 and v_3 is

$$v_2 = v_{21} + v_{2y} = v_3 - v_1 \qquad (5\text{-}31)$$

Thus, the circuit is a difference amplifier and performs like that of Fig. 5-16 with $R_2 = R_1 = R$.

EXAMPLE 5-2

Design op-amp circuits to give outputs (a) $v_1 = x + y$, (b) $v_2 = x - y$, where x and y are the voltage inputs. Power dissipation in any resistor should not exceed 1 W with $x = 10$ V and $y = 5$ V.

Solution

(a) The summing amplifier of Fig. 5-13 includes inversion of sign. We therefore add in tandem an inverting amplifier, as in Fig. 5-19. Then

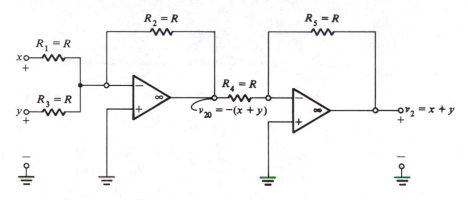

Figure 5-19.   Circuit designed in Example 5-1.

$v_2 = x + y$. Power dissipated in each resistor with $x = 10$ and $y = 5$ is

$$P_1 = \frac{x^2}{R_1} = \frac{100}{R_1} \qquad (5\text{-}32)$$

$$P_3 = \frac{y^2}{R_3} = \frac{25}{R_3} \qquad (5\text{-}33)$$

$$P_2 = \frac{v_{20}^2}{R_2} = \frac{225}{R_2} \qquad (5\text{-}34)$$

and

$$P_4 = P_5 = \frac{225}{R} \qquad (5\text{-}35)$$

Since all resistors equal R, the highest power dissipated is $225/R$. If $R = 1$ kΩ, the power in each resistor is less than 1 W.

(b) The difference amplifier of Fig. 5-16 with $R_1 = R_2 = R$ will give $v_2 = x - y$, with $v_3 = x$ and $v_1 = y$. We show the circuit in Fig. 5-20, where $R_1 = R_2 = R_3 = R_4 = R$. To determine power in each resistor, we must find the voltage across each resistor. In (a) this was easy because $+$ and $-$ terminals were grounded. In the case of Fig. 5-20, these terminals are at voltage v_a. Since $R_3 = R_4 = R$, we have $v_a = x/2$.

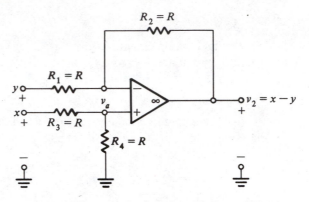

Figure 5-20.   Second circuit designed in Example 5-1.

Then power in each resistor with $x = 10$ and $y = 5$ is

$$P_4 = \frac{v_a^2}{R_4} = \frac{25}{R_4} \tag{5-36}$$

$$P_3 = \frac{(x - v_a)^2}{R_3} = \frac{25}{R_3} \tag{5-37}$$

$$P_1 = \frac{(y - v_a)^2}{R_1} = 0 \tag{5-38}$$

and

$$P_3 = \frac{(v_2 - v_a)^2}{R_2} = 0 \tag{5-39}$$

We see that $R = 1$ kΩ will give power less than 1 W.

EXERCISES

5-5.1. Design an op-amp circuit with the output $6x - 2y$, where x and y are input voltages.

5-5.2. Design an op-amp circuit with the output

$$2x - y - z$$

where x, y, and z are distinct input voltages. Using the assumption that x, y, and z will never exceed 10 V, design the circuit so that no resistor dissipates more than 0.1 W.

PROBLEMS

5-1. For the circuit shown in the figure, it is required that $R_{in} = 1\,\Omega$, and also that

$$V_3 = k_1 V_1 + k_2 V_2$$

Show that the required resistor values are

$$R_a = \frac{1}{k_1}, \qquad R_c = \frac{1}{k_2}, \qquad R_b = \frac{1}{1-(k_1+k_2)}$$

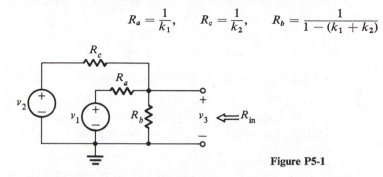

Figure P5-1

5-2. The circuit shown in the figure is to be constructed with an R-net in which the resistors are matched and all have the value of 10 kΩ.
(a) Find V_3 as a function of V_1 and V_2.
(b) Find an expression for V_4 in terms of V_1 and V_2.
(c) If $V_1 = 10$ V and $V_2 = 10$ V, how much power is dissipated by R_1? By R_6?

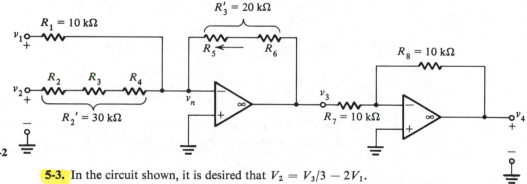

Figure P5-2

5-3. In the circuit shown, it is desired that $V_2 = V_3/3 - 2V_1$.

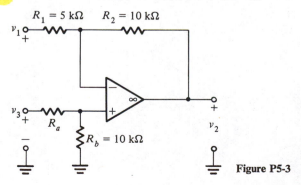

Figure P5-3

141

(a) Find the value of R_a that gives the desired relationship.
(b) Suppose that $V_1 = -10$ V and $V_2 = +10$ V. Find the magnitude of the currents through all resistors and the power dissipated by all resistors.
(c) Suppose that $V_1 = 10$ V and $V_2 = 10$ V. Find the currents and powers as in part (b).

5-4. Design a circuit making use of one operational amplifier such that the output voltage is

$$V_0 = 3V_1 + 0.2V_2$$

where V_1 and V_2 are distinct input voltages.

5-5. Design a circuit using a single op amp with output voltage

$$V_3 = V_1 - 4V_2$$

where V_1 and V_2 are input voltages. Choose resistor values so that no more than 0.25 W is dissipated by any resistor, provided that V_1 and V_2 have magnitudes less than 10 V.

5-6. Design an op-amp circuit with an output voltage equal to $6V_1 - 2V_2$, where V_1 and V_2 are input voltages.

5-7. Design an op-amp circuit with output voltage

$$V_0 = 2V_1 + V_2 - V_3$$

where V_1, V_2, and V_3 are input voltages. Use the assumption that these three voltages will never exceed 10 V in magnitude, and design the circuit so that no resistor dissipates more than 0.1 W.

5-8. For the circuit given in the figure, (a) show that with $G_i = 1/R_i$,

$$\frac{V_2}{V_1} = T = \frac{\dfrac{G_3}{G_2 G_4}(G_1 + G_2)}{\dfrac{G_3 + G_5}{G_4} - \dfrac{G_1}{G_2}}$$

(b) Let all G's be the same except for G_4. Sketch T as a function of $R_4 = 1/G_4$.
(c) Explain the behavior of the circuit when $R_4 = R/2$.

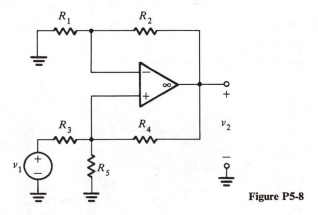

Figure P5-8

5-9. For the circuit given in the figure, find the ratio V_2/V_1 as a function of the four resistor values R_1, R_2, R_3, and R_4.

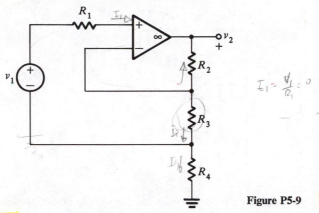

$$I_1 = \frac{v_1}{R_1} = 0$$

Figure P5-9

5-10. For the op-amp circuit shown in the figure, find the input resistance R_{in} or the resistance that is equivalent to the three resistors and the one op amp.

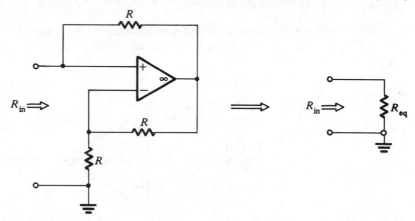

Figure P5-10

5-11. In the circuit given in the figure, $R = 10 \text{ k}\Omega$. Find the value of R_1 such that $V_2 = -100V_1$.

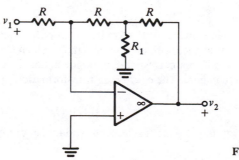

Figure P5-11

5-12. For the circuit given in the figure, determine $T = V_2/V_1$ in terms of the resistor values of the circuit.

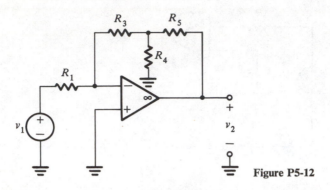

Figure P5-12

5-13. The amplifier circuit shown in the figure is to be designed such that any value of gain between $T = -10$ and $T = +10$, corresponding to $k = 0$ and $k = 1$, respectively, can be achieved by adjusting one potentiometer.
  (a) Derive an expression for V_2/V_1 in terms of r_1, r_2, and k, with $R_1 = 1$ and $R_2 = 11$.
  (b) Obtain values for r_1 and r_2 to accomplish the stated objective.
  (c) For what value of k is the gain zero?

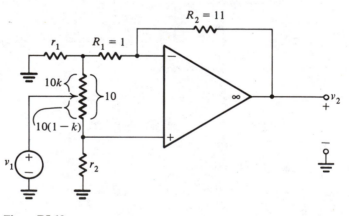

Figure P5-13

5-14. Piezoelectric tilt elements of the kind used to position laser beams and optical scanners can be aligned with an op-amp circuit which converts input voltages in x–y coordinates into a corresponding nonorthogonal (a, b, c) three-axis system as shown in the figure. The coordinate transformation must satisfy the relationships

$$x = c - a, \qquad y = b - \tfrac{1}{2}(a + c), \qquad a + b + c = 0$$

Design an op-amp circuit that accepts voltages x and y as inputs and yields voltages a, b, and c as outputs, using the transformation

$$a = c + x, \qquad b = -(a + c), \qquad c = -\left(\frac{x}{2} + \frac{y}{3}\right)$$

The circuit must be designed using resistors of only a single value of resistance, say 10 kΩ.

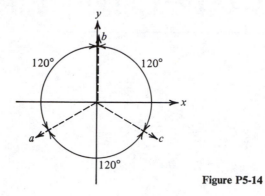

Figure P5-14

6

RESISTOR–CAPACITOR CIRCUITS

In this chapter and the next, we introduce two new elements—the *capacitor* and the *inductor*. Kirchhoff's laws and the loop and node analysis methods are, of course, applicable to all circuits, including those containing capacitors and inductors. The only difference in the analysis comes when we use the *current–voltage relationships* or the *branch constraints* for the circuit elements to express the Kirchhoff equations in terms of loop currents or node voltages. We have seen that for resistors, these relationships are represented by Ohm's law. As we shall see below, the current–voltage relationships for capacitors and inductors are quite different from those for the resistors. We emphasize this difference by first examining the physical behavior of these elements.

So far we have discussed circuits in which the response is instantaneous. The voltage and the current signals in a resistive circuit are proportional to each other. Therefore, as soon as an excitation is applied, the response is immediately present. As soon as the excitation is terminated, so is the response. Other circuit elements do not behave in this simple manner. For example, *in a capacitor the voltage is proportional to the charge, not the current*. The capacitor has the *ability to retain or store charge* and thereby to provide a sort of memory. Thus, it is possible for a capacitor to have voltage across it even when there is no current flowing. It takes a finite amount of time to move charge onto a capacitor, so a buildup of voltage on the capacitor will lag behind the applied voltage.

These elements differ from resistors in that they do not dissipate energy as resistors do and are called *reactive* elements. The presence of reactive elements in a circuit makes circuit behavior much more interesting. Let us turn our attention to the study of these elements.

6-1 THE CAPACITOR

The ideal *capacitor* or *condenser* consists of two parallel conducting plates separated by an ideal insulator as shown in Fig. 6-1. As mentioned above, the voltage

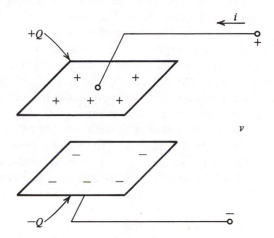

Figure 6-1. Parallel plates constitute a simple capacitor.

across it is proportional to the charge q on the plate

$$q = Cv \tag{6-1}$$

where the constant of proportionality C is called the capacitance. The unit of capacitance is the farad (F) and depends entirely on the geometry of the conducting plates and the physical properties of the insulator. The circuit symbol for the capacitor is shown in Fig. 6-2.

The current in the capacitor is given by the time rate of change of charge

$$i = \frac{dq}{dt} = C\frac{dv}{dt} \tag{6-2}$$

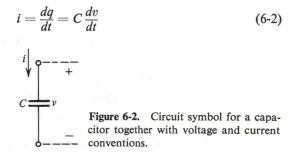

Figure 6-2. Circuit symbol for a capacitor together with voltage and current conventions.

The current–voltage relationship for a capacitor by Eq. 6-2 is

$$i = C \frac{dv}{dt} \tag{6-3}$$

where the polarities are shown in Fig. 6-2. Observe that *the current in the capacitor is present only when the voltage on it changes with time.* Otherwise, the derivative will be zero, and no current will flow. Thus, *for a dc or constant voltage the current is zero and the capacitor behaves like an open circuit.*

Let us examine the mechanism of current flow when there is an insulator separating the two plates. The insulator, of course, prevents any actual current flow through it. The mechanism can be explained in terms of the change in the net charge on the plates. The current entering at the positive terminal in Fig. 6-1 deposits positive net charge on the plate. Let us say that this rate is Q coulombs per second. The current flowing out of the other plate toward the negative terminal takes positive net charge away from this plate. Therefore, the net charge on this plate is negative at the rate of Q coulombs per second. Thus, there is a net transfer of charge of Q coulombs per second from the negative plate to the positive plate. The change in the net charge on the plates represents the change in the voltage on these plates since the voltage is proportional to the charge. Any flow of current is thus associated with a change in net charge and therefore a change in voltage on the capacitor. If there is no change in voltage, the current must indeed be zero.

The buildup of voltage on the capacitor due to current flow can be represented mathematically by integrating Eq. 6-3,

$$v(t) = \frac{1}{C} \int_{-\infty}^{t} i(\tau) \, d\tau \tag{6-4}$$

Since the current i is the derivative of q, we could also write the voltage on the capacitor as

$$v(t) = \frac{1}{C} \int_{-\infty}^{t} \frac{dq}{dt} \, d\tau = \frac{1}{C} q(t) - q(-\infty) = \frac{1}{C} q(t) \tag{6-5}$$

where we have made the perfectly natural assumption that the charge at minus infinity is zero. Equation 6-5 is, of course, a restatement of Eq. 6-1. Equations 6-3 and 6-4 represent the current–voltage relations for a capacitor.

EXAMPLE 6-1

A capacitor is excited by a current source $i_s(t)$ which has the form shown in Fig. 6-3. What is the voltage on the capacitor as a function of time?

Solution

The voltage on the capacitor by Eq. 6-4 is

$$v(t) = \frac{1}{C} \int_{-\infty}^{t} i_s(\tau) \, d\tau \tag{6-6}$$

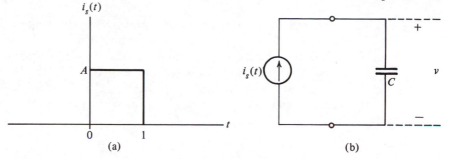

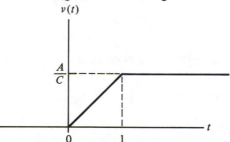

Figure 6-3. A current source producing a pulse of current is connected to a capacitor.

Since $i_s(\tau) = 0$ for all negative time until $t = 0$, the voltage is also zero for all negative time. The current $i_s = A$ between $t = 0$ and $t = 1$ and the voltage for $0 \le t \le 1$ is given by

$$v(t) = \int_{-\infty}^{t} \frac{1}{C} i_s \, d\tau = \frac{1}{C} \int_{-\infty}^{0} 0 \, d\tau + \frac{1}{C} \int_{0}^{t} A \, d\tau = \frac{A}{C} t \qquad (6\text{-}7)$$

which becomes A/C at $t = 1$. For $t > 1$ the current is zero again and the voltage remains unchanged at A/C since

$$v(t) = \frac{1}{C} \int_{-\infty}^{t} i_s \, d\tau = \frac{1}{C} \int_{-\infty}^{0} 0 \, d\tau + \frac{1}{C} \int_{0}^{1} A \, d\tau + \frac{1}{C} \int_{1}^{t} 0 \, d\tau = \frac{A}{C} \qquad (6\text{-}8)$$

The voltage is shown in Fig. 6-4.

Figure 6-4. Capacitor voltage resulting from the arrangement of Fig. 6-3.

Example 6-1 shows that the voltage on the capacitor changes only when the current is flowing through the capacitor. Initially, there is no current flowing, so the voltage remains zero. When the current starts flowing, the voltage starts building up or *charging* and finally reaches the value A/C, where it remains unchanged when the current goes to zero. The current source, of course, represents an open circuit when the current is zero. We thus have an open-circuited capacitor exhibiting *stored* charge and a corresponding voltage. Further, this voltage will remain on the capacitor until the charge is changed again by a new current flowing through it. We can explicitly show this by rewriting Eq. 6-4 as

$$v(t) = \frac{1}{C} \int_{-\infty}^{t_0} i(\tau) \, d\tau + \frac{1}{C} \int_{t_0}^{t} i(\tau) \, d\tau = v_0 + \frac{1}{C} \int_{t_0}^{t} i(\tau) \, d\tau \qquad (6\text{-}9)$$

In this equation the first integral represents the charging of the capacitor until time t_0, at which time the voltage on it is v_0. The second integral represents the increment in the voltage after time t_0. This is a useful equation since frequently we know what the initial voltage on the capacitor is but do not know the past history that was responsible for charging the capacitor to that voltage. For example, let us consider what happens when the capacitor of Example 6-1 is charged again starting at $t = 2$.

EXAMPLE 6-2

Consider the capacitor of Example 6-1. It has been charged to an initial voltage of A/C. What is the voltage if at $t = 2$ a current starts charging it again as shown in Fig. 6-5(a)?

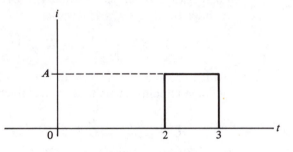

Figure 6-5(a).   Current waveform for Example 6-2.

Solution

The voltage is given by Eq. 6-9, and for $t \leq 3$ is

$$v(t) = v_0 + \frac{1}{C} \int_2^t A\, d\tau = \frac{A}{C} + \frac{A}{C}(t - 2) \tag{6-10}$$

For $t > 3$,

$$v(t) = v_0 + \frac{1}{C} \int_2^3 A\, d\tau + \int_3^t 0\, d\tau \tag{6-11}$$

$$= \frac{A}{C} + \frac{A}{C} = 2\frac{A}{C} \tag{6-12}$$

The voltage will remain at this value as shown in Fig. 6-5(b) until further charging or discharging occurs.

The same result could be obtained by Eq. 6-4, but we must then use integration from minus infinity to t. The use of Eq. 6-9 simplifies matters if we know what the initial voltage is at the start of the present current flow.

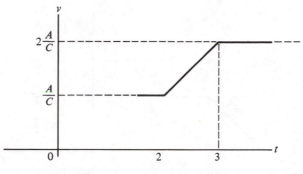

Figure 6-5(b).   Resulting voltage for Example 6-2.

The capacitor is *discharged* by a current flowing in a direction opposite to that in Fig. 6-2. Consider the circuit of Fig. 6-6 with the capacitor initially charged to v_0 volts. The current i is flowing in the direction opposite to our conventions, so that $-i$ is flowing in the conventional direction. Equation 6-9 then becomes

$$v(t) = v_0 - \frac{1}{C} \int_{t_0}^{t} i(\tau)\, d\tau \tag{6-13}$$

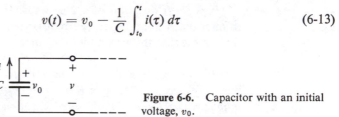

Figure 6-6.   Capacitor with an initial voltage, v_0.

If the value of the current i is positive, the integral is also positive and the voltage v reduces. The situation therefore represents a *discharge* phenomenon.

Let us conclude this section with the observation that *the voltage on a capacitor cannot change instantaneously.* Equation 6-9 shows that the change in an initial value v_0 depends on the integral of the current. The integral is zero if both lower and upper limits are the same. The integral has a value only if the difference between the upper and lower limits has a *nonzero* value. *It thus takes a finite amount of time to change the voltage on a capacitor.*

Let us see what happens if we try to change the capacitor voltage instantly. We will examine this by first changing the voltage slowly and then comparing the results when the change is speeded up. Consider a voltage similar to that of Fig. 6-4 but reaching the value V in τ seconds as shown in Fig. 6-7. Then the current is found from the equation

$$i = C \frac{dv}{dt} \tag{6-14}$$

since the voltage reaches the value V in τ seconds. The derivative between 0 and τ is V/τ and zero elsewhere. So the current has a value of $C(V/\tau)$ between 0

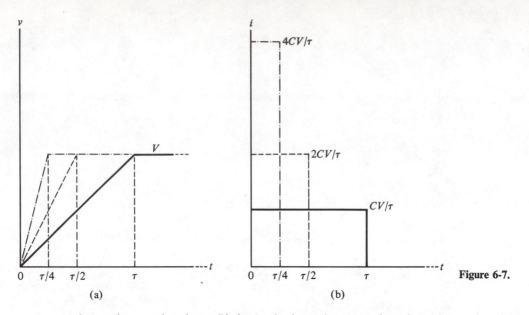

(a)   (b)   **Figure 6-7.**

and τ and zero elsewhere. If, instead, the voltage reaches the value V in $\tau/2$ seconds, then, of course, the slope is doubled. The current then has a value of $2C(V/\tau)$ for a duration of $\tau/2$ seconds. If we change the voltage from 0 to V in one-fourth the time, the current is four times the original for one-fourth the time. Note that the area under the current curve is always CV.

If we change the voltage from 0 to V in τ/K seconds, the current will reach a value of $(K/\tau)CV$ for the duration. In the limit, if we change the voltage instantly, we are letting $\tau/K \longrightarrow 0$ or $K \longrightarrow \infty$. The current will then reach *an infinite value for an infinitesimal duration*. This is an infinite spike called an *impulse*. Impulses are useful *abstractions*, but for now we will only be concerned with physical situations in which impulses are not permissible. Capacitor voltage cannot change instantly because that would require infinite current.

EXERCISES

6-1.1. Find the voltage as a function of time, both as a mathematical expression and as a graph, across a 50-μF capacitor if the voltage across the capacitor at $t = 0$ is 100 mV. The waveform for $i(t)$ is that shown in Fig. 6-8.

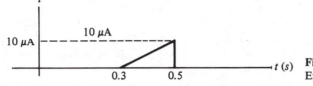

Figure 6-8. Current waveform for Exercise 6-1.1.

Ans.

$$v(t) = \begin{cases} 100 \text{ mV} & 0 \leq t < 0.3 \\ 0.1 + \dfrac{(t - 0.3)^2}{2} \quad \text{V} & 0.3 \leq t < 0.5 \\ 120 \text{ mV} & 0.5 \leq t \end{cases}$$

6-1.2. Repeat Exercise 6-1.1 if $i(t) = (5 \times 10^{-5})e^{-6t}, t \geq 0$.

$$\textit{Ans.} \quad v(t) = 0.1 + \tfrac{1}{6}(1 - e^{-6t})$$

6-1.3. Repeat Exercise 6-1.1 if $i(t) = (10^{-4}) \sin (10 \pi t), t \geq 0$.

$$\textit{Ans.} \quad v(t) = \frac{1}{10\pi}[(\pi + 2) - 2 \cos 10\pi t]$$

6-2 ENERGY AND POWER

We have seen in Section 6-1 that a capacitor is able to store or retain charge and the corresponding voltage. This form of storage represents *storage of energy.* In this section we examine the quantitative values of power and energy in a capacitor. The power delivered to a circuit element is the product of the voltage and the current with standard polarities

$$p(t) = v(t)i(t) \quad \text{W} \tag{6-15}$$

Substituting the value of i from Eq. 6-3, we obtain

$$p(t) = Cv(t)\frac{dv(t)}{dt} = \frac{d}{dt}\left(\frac{1}{2}Cv^2\right) \tag{6-16}$$

Power is the time rate of change of energy or time derivative of energy. Therefore, the quantity in parenthesis in Eq. 6-16 equals the energy stored in the capacitor. We may now write that the energy w stored in the capacitor at any instant of time is given by

$$w = \tfrac{1}{2}Cv^2 \quad \text{W-s} \tag{6-17}$$

The positive value of power (under the polarity conventions) represents power delivered to the capacitor by a charging current and results in stored energy in the capacitor; the negative value represents power delivered by the capacitor to the rest of the circuit during a discharge phenomenon. Note that the energy stored in a capacitor at any instant of time depends only on the value of the voltage at that instant of time. It does not depend on the variation of the voltage prior to that instant of time.

EXAMPLE 6-3

Find the energy stored in a 1-μF capacitor if there is a voltage of 10 V at a given instant of time.

Solution

By Eq. 6-17,

$$w = \tfrac{1}{2}Cv^2 = \tfrac{1}{2} \times 10^{-6} \times 10^2 = 5 \times 10^{-5} \text{ W-s} \tag{6-18}$$

EXERCISES

6-2.1. For Exercise 6-1.1, find the energy stored in the capacitor at $t = 0.4$ s.

$$\textit{Ans.} \quad 0.2756 \ \mu\text{W-s}$$

6-2.2. Repeat Exercise 6-2.1 for the conditions of Exercise 6-1.2.

Ans. 1.582 μW-s

6-2.3. Repeat Exercise 6-2.1 for the conditions of Exercise 6-1.3.

Ans. 0.25 μW-s

6-3 LINEARITY AND SUPERPOSITION

Resistors obeying Ohm's law are called linear because the voltage–current relationships for them are given by straight lines. Superposition is valid for circuits containing linear resistors and sources. We now generalize the concept of linearity to situations where the voltage–current relationship is not a straight-line. Since our primary interest is in superposition, we shall call elements *linear* if the superposition principle is valid for them, and vice versa.

Is it possible to apply superposition to capacitors? The equations governing the capacitor are Eqs. 6-3 and 6-4. If the response to a sum of two signals is equal to the sum of the individual responses, superposition can be applied. First, let us examine the current, i, in the capacitor due to a voltage, $v = v_1 + v_2$, which is the sum of two different voltages, v_1 and v_2. By substituting in Eq. 6-3, we have

$$i = C\frac{d}{dt}(v_1 + v_2) = C\frac{dv_1}{dt} + C\frac{dv_2}{dt} = i_1 + i_2 \qquad (6\text{-}19)$$

Here i_1 is the current in the capacitor if only v_1 is present, and i_2 is the current if only v_2 is present. We see that the current in the capacitor is the sum of the individual currents i_1 and i_2, each obtained in response to one of the sources (v_1 or v_2) with the other set at zero. So superposition holds true.

Let us now use Eq. 6-4 to determine the voltage due to a sum of two currents i_1 and i_2:

$$v = \frac{1}{C}\int_{-\infty}^{t}(i_1 + i_2)\,d\tau = \frac{1}{C}\int_{-\infty}^{t}i_1\,d\tau + \frac{1}{C_2}\int_{-\infty}^{t}i_2\,d\tau = v_1 + v_2 \qquad (6\text{-}20)$$

Again, the voltage is the sum of the individual voltages due to each current. We see that superposition can be applied to capacitors.

If, however, we use Eq. 6-9, we run into some difficulty. The actual voltage due to both the currents is

$$v = \frac{1}{C}\int_{t_0}^{t}(i_1 + i_2)\,d\tau + v_0 \qquad (6\text{-}21)$$

Now, let us sum the individual voltages due to currents i_1 alone and i_2 alone, using Eq. 6-9:

$$v_1 + v_2 = \frac{1}{C}\int_{t_0}^{t}i_1\,d\tau + v_0 + \frac{1}{C}\int_{t_0}^{t}i_2\,d\tau + v_0 \qquad (6\text{-}22)$$

$$\neq v \qquad unless\ v_0 = 0$$

Equation 6-22 shows that superposition cannot be used unless the initial voltage $v_0 = 0$. Incidentally, this was true for Eq. 6-20 because the lower limit of minus infinity guarantees that the initial voltage is zero. We conclude that *superposition is possible for capacitors only if the initial voltage is equal to zero.* Otherwise, superposition cannot be applied. Capacitors governed by the fundamental equations, Eqs. 6-3 and 6-4, are *linear* elements. Note that this is a consequence of the straight-line relationship of Eq. 6-1. Superposition, however, cannot be applied to circuits containing charged capacitors.

EXAMPLE 6-4

A current of 1 mA charges a 0.01-F uncharged capacitor for 1 min, and a current of 2 mA charges it for $\frac{1}{2}$ min; what is the voltage on it?

Solution

Since the capacitor is uncharged, we may use superposition

$$v_1 = \frac{1}{0.01} \int_0^{60} 10^{-3} \, d\tau = 60 \times 10^{-3} \times 100 = 6 \text{ V} \tag{6-23}$$

$$v_2 = \frac{1}{0.01} \int_0^{30} 2 \times 10^{-3} \, d\tau = 2 \times 10^{-3} \times 30 \times 100 = 6 \text{ V} \tag{6-24}$$

$$v = v_1 + v_2 = 12 \text{ V} \tag{6-25}$$

6-4 SERIES AND PARALLEL COMBINATIONS

Consider parallel combinations of two capacitors C_1 and C_2, as shown in Fig. 6-9. The voltage across the two capacitors is the same since they are in parallel. Kirchhoff's laws are, of course, applicable to all circuit elements. The only difference is that instead of Ohm's law, we now use the current–voltage relationships for a capacitor given by Eqs. 6-3 and 6-4 or 6-9. Let us apply KCL to the parallel combination to obtain

$$i = i_1 + i_2 = C_1 \frac{dv}{dt} + C_2 \frac{dv}{dt} \tag{6-26}$$

$$= (C_1 + C_2) \frac{dv}{dt} \tag{6-27}$$

$$= C \frac{dv}{dt} \tag{6-28}$$

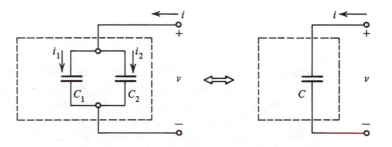

Figure 6-9. Parallel capacitors described by Eq. 6-29.

where C is the sum of C_1 and C_2. Equation 6-28 shows the current–voltage relationship of an equivalent capacitor whose capacitance is given by

$$C = C_1 + C_2 \tag{6-29}$$

Thus, *the equivalent capacitance of a parallel combination is given by the sum of the individual capacitances.* Any initial voltage on the parallel combination will naturally be present on the equivalent capacitor.

Next, consider a series combination of two capacitors, as shown in Fig. 6-10. Here the voltage across the series combination is v and the current through

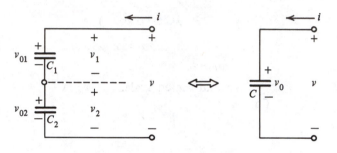

Figure 6-10. Series capacitors described by Eq. 6-33.

the series combination is i. Assume that C_1 has the initial voltage v_{01} and C_2 has the initial voltage v_{02}. By KVL the total voltage drop v across the combination is the sum of voltage drops v_1 and v_2.

$$v = v_1 + v_2 = \frac{1}{C_1} \int_{t_0}^{t} i\, d\tau + v_{01} + \frac{1}{C_2} \int_{t_0}^{t} i\, d\tau + v_{02} \tag{6-30}$$

$$= \left(\frac{1}{C_1} + \frac{1}{C_2} \right) \int_{t_0}^{t} i\, d\tau + (v_{01} + v_{02}) \tag{6-31}$$

$$= \frac{1}{C} \int_{t_0}^{t} i\, d\tau + v_0 \tag{6-32}$$

where we have set

$$\frac{1}{C} = \frac{1}{C_1} + \frac{1}{C_2} \tag{6-33}$$

and

$$v_0 = v_{01} + v_{02} \tag{6-34}$$

Equation 6-32 represents the capacitor equation for an equivalent capacitor with capacitance C and initial voltage v_0 given by Eqs. 6-33 and 6-34. If the initial voltages are zero, the initial voltage on the equivalent capacitor will also be zero. Thus, *for a series combination, the reciprocal of the equivalent capacitance is the sum of the reciprocals of the individual capacitances.*

EXAMPLE 6-5

Find the equivalent capacitance for the circuit of Fig. 6-11(a) assuming that all capacitors are uncharged.

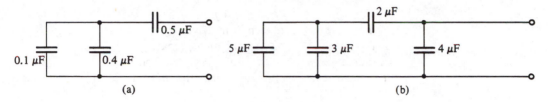

(a)   (b)

Figure 6-11. Capacitive circuits: (a) for Example 6-5 and (b) for Exercise 6-4.1.

Solution

We find the equivalent capacitance of the parallel combination using Eq. 6-29:

$$C_p = 0.1 \times 10^{-6} + 0.4 \times 10^{-6} \tag{6-35}$$

$$= 0.5 \times 10^{-6} \text{ F} \tag{6-36}$$

This equivalent C_p is in series with 0.5 μF, so we use Eq. 6-33:

$$\frac{1}{C_{eq}} = \frac{1}{0.5 \times 10^{-6}} + \frac{1}{0.5 \times 10^{-6}} = \frac{1}{0.25 \times 10^{-6}} \tag{6-37}$$

or

$$C_{eq} = 0.25 \ \mu\text{F} \tag{6-38}$$

EXERCISES

6-4.1. Find the equivalent capacitance of the circuit given in Fig. 6-11(b). Assume that all capacitors are initially uncharged.

Ans.   5.6 μF

6-5 RC CIRCUITS

Consider a capacitor that has been charged to a voltage v_0 and is then allowed to discharge into a resistor as in Fig. 6-12. We, of course, know that as long as there is any charge on C, there is a corresponding voltage on it; consequently, a current flows in the resistor. The current in turn discharges the capacitor and

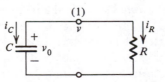

Figure 6-12. *RC* circuit in which the capacitor is charged to voltage v_0.

reduces the voltage on it (see Eq. 6-13) until it reaches zero. At this stage the capacitor is completely discharged and no current can flow. We wish to examine this *discharge* in detail as a function of time.

For the circuit of Fig. 6-12 we can write KCL in terms of the two currents at node 1:

$$i_C + i_R = 0 \qquad (6\text{-}39)$$

Substitute the values of i_C from Eq. 6-3 and i_R from Ohm's law, so that

$$C\frac{dv}{dt} + \frac{v}{R} = 0 \qquad (6\text{-}40)$$

Equation (6-40) in v and its derivative is known as a *differential equation*. When the right-hand side is zero with the unknown variable and its derivatives on the left as in Eq. 6-40, it is called a *homogeneous differential equation*. The equation can be rearranged as

$$\frac{dv}{v} = -\frac{1}{RC}dt \qquad (6\text{-}41)$$

We integrate both sides to obtain

$$\int \frac{1}{v}dv = \int -\frac{1}{RC}dt \qquad (6\text{-}42)$$

or

$$\ell n\, v = -\frac{1}{RC}t + \text{constant} \qquad (6\text{-}43)$$

which becomes

$$v(t) = e^{-(1/RC)t+\text{constant}} = Ae^{-t/RC} \qquad (6\text{-}44)$$

where A is an arbitrary constant as a result of the indefinite integration. *The value of A can be determined if we know the value of v at some instant of time.*

Let us say that the circuit is connected at $t = 0$. At that time the value of v is the same as the initial voltage on the capacitor v_0:

$$v(0) = v_0 \qquad (6\text{-}45)$$

Substituting Eq. 6-45 into Eq. 6-44, we have

$$v(0) = v_0 = A \qquad (6\text{-}46)$$

so that Eq. 6-44 becomes

$$v(t) = v_0 e^{-t/RC} \qquad (6\text{-}47)$$

Equation 6-47 gives the manner in which the capacitor discharges and it is plotted in Fig. 6-13. At the start when $t = 0$, the voltage is v_0, which decays exponentially as t becomes larger and larger and vanishes as $t \to \infty$. The behavior of $v(t)$ depends entirely on the RC parameters of the circuit and is independent of any external excitation. The exponential behavior of the discharge mechanism is completely determined by the values of R and C. The same exponential will occur regardless of the initial value v_0. Only the coefficient will have

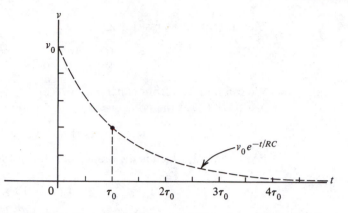

Figure 6-13. Plot of Eq. 6-47 for multiple values of $\tau_0 = RC$.

different values for different situations. For these reasons it is called the *natural response* and, as we have seen, it is the solution to the homogeneous equation.

The product $RC = \tau_0$ is called the *time constant* of the circuit and indicates how fast the exponential decays or how fast the circuit settles down to its *quiescent state*. For example, when $t = \tau_0$, $v = v_0 e^{-1} = (1/e)v_0$. When $t = 2\tau_0$, $v = (1/e^2)v_0$ and when $t = K\tau_0$, we find that $v = (1/e^K)v_0$. Thus, the time constant τ_0 indicates the time it takes to reduce the value of the exponential by a factor of $1/e = 0.37$. In four time constants the exponential reduces to only 2% of its original value. It is therefore safe to say that *the exponential vanishes in about four time constants.*

EXAMPLE 6-6

Find the time constant τ_0 and the capacitor voltage and current for the circuit of Fig. 6-14 which is connected at $t = 0$ by a switch as shown.

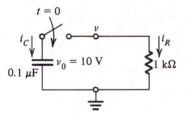

Figure 6-14. Circuit for Example 6-6.

Solution

The time constant is

$$\tau_0 = RC = 10^3 \times 10^{-7} = 10^{-4} = 100 \ \mu s \tag{6-48}$$

From Eq. 6-47, we have

$$v(t) = v_0 e^{-t/RC} = 10e^{-t \times 10^4} \ V \tag{6-49}$$

The current in the resistor is

$$i_R(t) = \frac{v(t)}{R} = \frac{10e^{-t \times 10^4}}{1000} = 0.01e^{-t \times 10^4} \qquad (6\text{-}50)$$

We can find $i_C(t)$ from either Eq. 6-39 or Eq. 6-3. From Eq. 6-39, which is just the KCL at the node, we have

$$i_C(t) = -i_R(t) = -0.01e^{-t \times 10^4} \text{ A} \qquad (6\text{-}51)$$

From Eq. 6-3, which is the branch constraint for the capacitor, we have

$$i_C(t) = C\frac{dv}{dt} = -10^{-7} \times 10 \times 10^4 e^{-t \times 10^4} \qquad (6\text{-}52)$$

$$= -0.01e^{-t \times 10^4} \qquad (6\text{-}53)$$

as before.

Charged capacitors are encountered frequently and care must be exercised in handling them. As Eq. 6-50 shows, the maximum discharge current is v_0/R at $t = 0$. If R is very small, a very large current results. This is the case of many electrical shocks when capacitors are touched or shorted by human beings unaware of the state of the charge. Before touching any capacitors, it is wise to discharge it through a resistor for at least four time constants.

We can use the analysis above in a variety of circuits as long as they can be replaced by equivalent circuits containing only single capacitor–single resistor combinations. Some examples will illustrate the concept.

EXAMPLE 6-7

Find the time constant τ_0 for the circuit of Fig. 6-15.

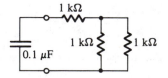

Figure 6-15.   Circuit for Example 6-7.

Solution

The resistive part of the circuit is equivalent to a resistor

$$R_{eq} = 10^3 + \frac{10^3 \times 10^3}{2 \times 10^3} \; 1500 \; \Omega \qquad (6\text{-}54)$$

Then the time constant is given by

$$\tau_0 = R_{eq}C = 1500 \times 10^{-7} = 150 \; \mu s \qquad (6\text{-}55)$$

EXAMPLE 6-8

Find v_2 in the circuit of Fig. 6-16.

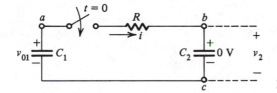

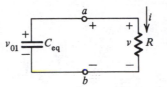

Figure 6-16. Circuit for Example 6-8.

Solution

We first replace the circuit by an equivalent circuit containing a single equivalent capacitor and a single resistor, as shown in Fig. 6-17. The node

Figure 6-17. Simplification of the circuit of Example 6-8.

c is suppressed, so we cannot determine v_2 directly from this circuit. But we can determine the current i flowing in the resistor. This current also flows into C_2, and we can thus determine v_2. The equivalent capacitance is

$$C_{eq} = \frac{C_1 C_2}{C_1 + C_2} \tag{6-56}$$

and the initial charge on it is v_{01}, since the initial charge on C_2 is zero.

The voltage v is obtained from the equivalent circuit, Eq. 6-47, as

$$v = v_{01} e^{-t/RC_{eq}} \tag{6-57}$$

and the current is

$$i = \frac{v}{R} = \frac{v_{01}}{R} e^{-t/RC_{eq}} \tag{6-58}$$

Returning to the original circuit of Fig. 6-16, we obtain

$$v_2 = v_{20} + \frac{1}{C_2} \int_0^t i(\tau)\, d\tau \tag{6-59}$$

$$= 0 + \frac{1}{C_2} \int_0^t \frac{v_{01}}{R} e^{-\tau/RC_{eq}}\, d\tau \tag{6-60}$$

$$= \frac{v_{01}}{C_2} \cdot C_{eq}(1 - e^{-t/RC_{eq}}) \tag{6-61}$$

The circuit discharges C_1 and charges C_2 until they both have the same quiescent or *steady-state* voltages. We see that v_2 is initially zero and charges until in steady state it has the value $v_{01}C_{eq}/C_2$. The same voltage value is on C_1 when the steady state is reached.

EXERCISES

6-5.1. The circuit given in Fig. 6-18 is connected together by the switch at time $t = 0$.
  (a) Find the time constant of the overall circuit.
  (b) Find i, v_1, and v_2 as function of time for $t > 0$.
  (c) Plot $i(t)$ for $0 \leq t \leq 500$ ms.

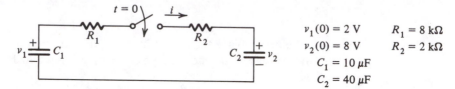

$$v_1(0) = 2 \text{ V} \qquad R_1 = 8 \text{ k}\Omega$$
$$v_2(0) = 8 \text{ V} \qquad R_2 = 2 \text{ k}\Omega$$
$$C_1 = 10 \ \mu\text{F}$$
$$C_2 = 40 \ \mu\text{F}$$

Figure 6-18. Circuit for Exercise 6-5.1.

Ans.
(a) 8×10^{-2} s; (b) $i(t) = -6 \times 10^{-4}e^{-t/8 \times 10^{-2}}$, $v_1(t)$
$= 6.8 - 4.8e^{-t/8 \times 10^{-2}}$, $v_2 = 6.8 + 1.2e^{-t/8 \times 10^{-2}}$

6-5.2. Find the current for a circuit connected at $t = 0$ by the switch as shown in Fig. 6-19. An ideal capacitor will hold its stored charge indefinitely if left uncon-

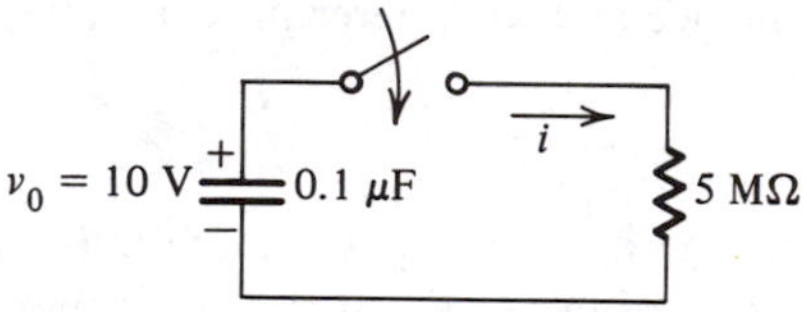

$v_0 = 10$ V $0.1 \ \mu$F i 5 MΩ

Figure 6-19. Circuit for Exercise 6-5.2.

nected (in an open circuit). However, in actual capacitors some leakage current passes through the insulator separating the two plates of the capacitor. Such a capacitor in which the leakage current is not negligible can be modeled as a parallel combination of an ideal capacitor and a resistor. Suppose that the 0.1-μF leakage-free capacitor is replaced by a leaky capacitor that can be modeled as an ideal 0.1-μF capacitor in parallel with a 10-MΩ resistor. Change the circuit given above to reflect the use of the leaky capacitor and find $i(t)$ for this second case.

Ans. $i(t) = 2 \times 10^{-6}e^{-3t}$; the capacitor discharges faster

6-6 DRIVEN RC CIRCUITS

Let us now assume that the capacitor, initially charged to v_0 volts, is now connected to a battery source of V volts through a resistor R, as in Fig. 6-20.
  If $V = v_0$, the voltage across R is zero, no current will flow, and the capa-

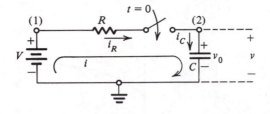

Figure 6-20. A battery is connected to an *RC* series circuit.

citor will maintain its voltage v at $v_0 = V$. In other words, the *quiescent or steady state* (for the voltage v on the capacitor) is the battery voltage V.

If $V = 0$, we have the situation of Fig. 6-12, and the voltage v is a decaying exponential as in Eq. 6-47. If V were of some other value, the capacitor will discharge or charge in a decaying exponential with a time constant equal to RC until its voltage reaches V, the battery voltage, at which time no further current can flow and the steady state is reached. In other words, the voltage v consists of a *steady-state part or component* of value V, and a possible *component with a decaying exponential with RC as the time constant*. Then we can write intuitively that

$$v(t) = V + Ae^{-t/RC} \qquad (6\text{-}62)$$

where we have left A, the coefficient of the decaying exponential, unspecified.

The value of A is determined from the initial value of v (i.e., v_0). This is easily accomplished by setting $t = 0$ in Eq. 6-62 and setting $v(0) = v_0$; thus,

$$v(0) = v_0 = V + A \qquad (6\text{-}63)$$

or

$$A = v_0 - V \qquad (6\text{-}64)$$

Substituting the value of A into Eq. 6-62,

$$v(t) = V + (v_0 - V)e^{-t/RC} \qquad (6\text{-}65)$$

We will soon derive the same response mathematically, but first let us examine the result.

As discussed above, we see in Eq. 6-65 that the steady-state value of v as $t \longrightarrow \infty$ is V, the exponential is not present if $v_0 = V$, and the initial value of v at $t = 0$ is v_0 as required. The result above is correct for all values of v_0 and V. The *steady-state response* or component is also called *forced* or *driven* response, since it is present only in response to the *driving force* or *excitation*. The decaying exponential represents the natural response and it vanishes at $t \longrightarrow \infty$. It is also called the *transient response*, since it is transitory and ultimately vanishes when the circuit settles down to its driven or steady-state response. It is generally safe to assume that the transient component has vanished after about four time constants and that only the steady-state response is present thereafter.

The capacitor voltage given by Eq. 6-65 is plotted in Fig. 6-21. If $(v_0 - V)$ is positive, the capacitor *discharges* from v_0 to V and then remains constant, as

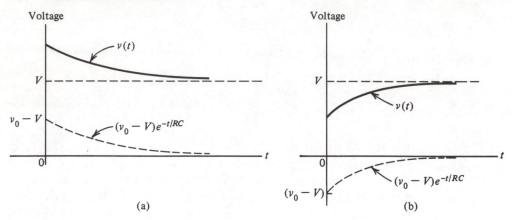

(a)   (b)

Figure 6-21.   Capacitor voltage described by Eq. 6-65.

shown in Fig. 6-21(a). If $(v_0 - V)$ is negative, the capacitor is actually *charged* starting from an initial value of v_0 until it reaches the battery voltage V and then remains constant as in Fig. 6-21(b). The current in the resistor will be positive in the direction shown if the capacitor is actually charging and negative if the capacitor is discharging. The current in the capacitor is obtained from Eq. 6-65 and the KCL at node (2) as follows:

$$i_C(t) = i_R(t) = \frac{V - v}{R} = \left(\frac{V - v_0}{R}\right)e^{-t/RC} \tag{6-66}$$

The plot of the current is shown in Fig. 6-22 assuming that $(V - v_0)$ is positive.

The steady-state value of the current is zero, as is to be expected when the capacitor reaches the battery voltage. The *steady-state condition under constant excitation is called dc.* For the dc case, the capacitor voltage is constant and the

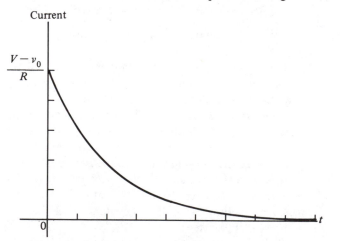

Figure 6-22.   Capacitor current described by Eq. 6-66.

current through it is zero, as seen above and in Eq. 6-3. The capacitor therefore behaves like an *open circuit* under dc excitation. The initial value of the current can also be derived by inspection of the circuit. Initially, the voltage on the capacitor is v_0 and it cannot change instantly. Therefore, initially the voltage across the resistor is $(V - v_0)$ and the current is $(V - v_0)/R$. The initial current is thus found by freezing the capacitor voltage at v_0 (i.e., replacing the capacitor by a voltage source of v_0 volts). *The capacitor behaves initially like a corresponding voltage source. An uncharged capacitor behaves initially like a short circuit.*

Alternatively, we can derive the response for the circuit of Fig. 6-20 by writing KCL at node 2,

$$i_C - i_R = 0 \tag{6-67}$$

or

$$C\frac{dv}{dt} + \frac{v - V}{R} = 0 \tag{6-68}$$

or

$$C\frac{dv}{dt} + \frac{v}{R} = \frac{V}{R} \tag{6-69}$$

This is a *nonhomogeneous* differential equation which must be solved for the unknown $v(t)$. The known constant on the right is the forcing function. The solution consists of two parts: the solution to the homogeneous equation obtained by setting the right-hand side of Eq. 6-69 to zero, and the particular solution obtained due to the forcing function. We now derive these two solutions.

The homogeneous equation is obtained simply as

$$C\frac{dv}{dt} + \frac{v}{R} = 0 \tag{6-70}$$

which may be solved as before. An easier way is to assume a solution of the exponential type

$$V_{\text{tr}} = Ae^{st} \tag{6-71}$$

where s is to be determined and A is arbitrary. Then, by substitution into Eq. 6-70,

$$\left(Cs + \frac{1}{R}\right)Ae^{st} = 0 \tag{6-72}$$

Since A is arbitrary and the exponential cannot be zero for all time t, we have

$$Cs + \frac{1}{R} = 0 \tag{6-73}$$

or

$$s = -\frac{1}{RC} \tag{6-74}$$

The assumed solution in Eq. 6-71 becomes

$$v_{tr} = Ae^{-t/RC} \tag{6-75}$$

as is to be expected.

The particular solution is obtained in a similar fashion by assuming a form of solution and substituting it in Eq. 6-69 to determine its value. Since the forcing function is constant, we assume a constant for a forced solution,

$$v_{ss} = K \tag{6-76}$$

We now substitute this in Eq. 6-69 to obtain

$$0 + \frac{K}{R} = \frac{V}{R} \tag{6-77}$$

or

$$v_{ss} = K = V \tag{6-78}$$

The total solution is the sum of the two solutions in Eqs. 6-71 and 6-78,

$$v(t) = V + Ae^{-t/RC} \tag{6-79}$$

We now use the initial value $v(0) = v_0$ to obtain that $A = v_0 - V$, and the complete solution is the same as before,

$$v(t) = V + (v_0 - V)e^{-t/RC} \tag{6-80}$$

On the other hand, if we had written the KVL for the circuit of Fig. 6-20, then we have

$$V = Ri + v_0 + \frac{1}{C}\int_0^t i \, d\tau \tag{6-81}$$

where i is the loop current assumed in the clockwise direction. This equation is easily converted to a differential equation by differentiating both sides to obtain

$$R\frac{di}{dt} + \frac{1}{C}i = 0 \tag{6-82}$$

There is no forcing function, so the forced response is zero. The equation is easily solved for the transient part by assuming that $i = Ae^{st}$, substituting, and solving for s to obtain the complete solution

$$i(t) = Ae^{-t/RC} \tag{6-83}$$

The initial value $i(0)$ is obtained either by inspection of the circuit or from Eq. 6-81 as

$$i(0) = \frac{V - v_0}{R} \tag{6-84}$$

and the complete response becomes

$$i(t) = \frac{V - v_0}{R}e^{-t/RC} \tag{6-85}$$

the same as in Eq. 6-66. We can now determine $v(t)$ by Eq. 6-3.

Let us summarize the steps involved in analyzing any circuit with a single equivalent capacitor and a single equivalent resistor:

1. Find the transient response either from the homogeneous equation or *from knowledge of the time constant*. This is of the form $Ae^{-t/RC}$.
2. Find the steady-state response either from the differential equation or *from the circuit by open-circuiting all capacitors* if the excitation is constant. This is a constant, say K.
3. The complete response is the sum $K + Ae^{-t/RC}$, with A yet to be determined.
4. Determine the initial value. This is either given in the circuit, in the Kirchhoff equation, or is obtained *by replacing all charged capacitors with corresponding voltage sources*. Use the initial value to determine A in the complete response above.

EXAMPLE 6-9

Design a charging circuit to charge a 100-μF capacitor to the battery voltage of 10 V. The maximum current should not exceed 2 mA. What is the approximate charging time for a full charge?

Solution

We assume a charging circuit of the form shown in Fig. 6-20. By Eq. 6-85, the maximum current is at $t = 0$ and has the value

$$i_{max} = \frac{V}{R} \tag{6-86}$$

where we have naturally assumed an uncharged capacitor with $v_0 = 0$. Then

$$R = \frac{V}{i_{max}} = \frac{10}{2 \times 10^{-3}} = 5 \times 10^3 \ \Omega \tag{6-87}$$

The time constant is

$$\tau_0 = RC = 5 \times 10^3 \times 100 \times 10^{-6} = 0.5 \ \text{s} \tag{6-88}$$

Let us assume about four time constants to reach steady state. Then the charging time is about $4 \times 0.5 = 2$ s.

EXAMPLE 6-10

Find i_1 and v_2 in the circuit of Fig. 6-23 for $t \geq 0$.

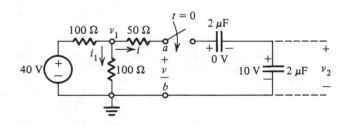

Figure 6-23.   Circuit for Example 6-10.

Solution

The easiest way to approach this is to replace the resistive circuit to the left of *a–b* by its Thévenin equivalent, and to the right of *a–b* by the equivalent capacitance as in Fig. 6-24. The steady-state value of v at *a–b* is obtained

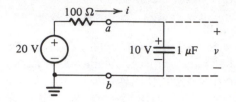

Figure 6-24. Thévenin equivalent of the circuit of Fig. 6-23.

by open-circuiting the capacitor and is 20 V. The initial value of v is 10 V. The time constant is

$$R_{eq}C_{eq} = 100 \times 10^{-6} = 10^{-4} \tag{6-89}$$

The voltage response at *a–b* is the sum of the steady-state and the transient response:

$$v(t) = 20 + Ae^{-10^4 t} \tag{6-90}$$

At $t = 0$, $v(0) = 10$, so that

$$10 = 20 + A; \quad A = -10 \tag{6-91}$$

and the response is

$$v(t) = 20 - 10e^{-10^4 t} \tag{6-92}$$

The current response at *a–b* is

$$i(t) = \frac{20 - v(t)}{100} \tag{6-93}$$

$$= 0.1e^{-10^4 t} \tag{6-94}$$

Now we return to the original circuit for which we know the current and the voltage at *a–b*. The voltage v_1 is, by KVL,

$$v_1 = 50i + v \tag{6-95}$$

$$= 5e^{-10^4 t} + 20 - 10e^{-10^4 t} \tag{6-96}$$

$$= 20 - 5e^{-10^4 t} \tag{6-97}$$

and $i_1 = v_1/100$, or

$$i_1 = 0.20 - 0.05e^{-10^4 t} \tag{6-98}$$

The current i flows through the capacitors so that the voltage across either of them is easily determined by the capacitor voltage–current relation of Eq. 6-9. Then

$$v_2 = 10 + \frac{10^6}{2} \int_0^t 0.1e^{-10^4 \tau} \, d\tau \tag{6-99}$$

$$= 10 + 5(1 - e^{-10^4 t}) \tag{6-100}$$

We could have obtained the steady-state value of v_1 directly by open-circuiting the capacitors. This is 20 V by voltage-divider action. But to obtain the transient response, we need the single C–single R equivalent for determining the time constant. Thus, it is always more convenient to work with the simple equivalent circuit first and then work backwards to the original circuit.

EXERCISES

6-6.1. In the circuit of Fig. 6-25, the switch is open until time $t = 100$ and is closed for all time thereafter. Find the voltage $v(t)$ for all time greater than 100 if $v(100) = -3$, and sketch that voltage as a function of time.

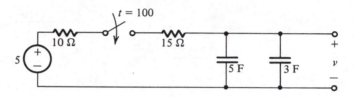

Figure 6-25. Circuit for Exercise 6-6.1.

Ans. $v(t) = 5 - 8e^{-(t-100)/200}$

6-6.2. Use node analysis to obtain the pair of differential equations involving node voltages v_1 and v_2 in the circuit of Fig. 6-26.

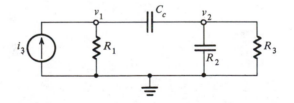

Figure 6-26. Circuit for Exercise 6-6.2.

6-7 SIMPLE CAPACITOR–OP AMP CIRCUITS

Two of the simplest op-amp circuits containing capacitors are shown in Figs. 6-27 and 6-28. We will analyze them using the virtual-short phenomenon, which states that the $+$ and $-$ terminals of the op amp are effectively short-circuited and open-circuited at the same time. For our circuits, then, the $+$ and $-$ terminals of the op amp are effectively grounded and the current $i_- = i_+ = 0$.

In Fig. 6-27 we write KCL at node a to obtain

$$i_1 = -i_2 \tag{6-101}$$

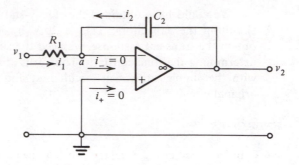

Figure 6-27. Integrator op-amp circuit.

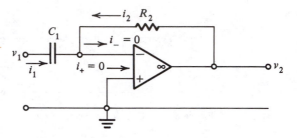

Figure 6-28. Differentiator op-amp circuit.

and since node a is effectively grounded, we have

$$\frac{v_1}{R_1} = -C_2 \frac{dv_2}{dt} \tag{6-102}$$

or

$$v_2(t) = -\frac{1}{R_1 C_2} \int_{-\infty}^{t} v_1 \, d\tau = v_2(0) - \frac{1}{R_1 C_2} \int_{0}^{t} v_1 \, d\tau \tag{6-103}$$

Usually, the integrator is initially reset by discharging C_2, in which case $v_2(0) = 0$, and

$$v_2(t) = -\frac{1}{R_1 C_2} \int_{0}^{t} v_1 \, d\tau \tag{6-104}$$

The output is thus proportional to the integral of the input. Hence, this circuit is known as an (op-amp) *integrator*.

Similarly, for the circuit in Fig. 6-28, we write the KCL as

$$i_1 = -i_2 \tag{6-105}$$

or

$$C_1 \frac{dv_1}{dt} = -\frac{v_2}{R_2} \tag{6-106}$$

and

$$v_2 = -R_2 C_1 \frac{dv_1}{dt} \tag{6-107}$$

Slope = S_0

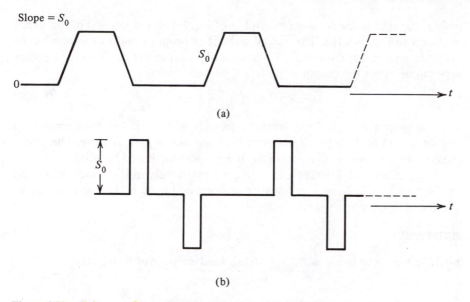

(a)

(b)

Figure 6-29. Pulse waveform with slopes of value $\pm S_0$, and the derivative of this waveform.

Here the output is proportional to the derivative of the input, and the circuit is called an op-amp *differentiator*.

A set of pulses is shown in Fig. 6-29(a). Suppose that this signal is input to a differentiator as v_1. Each pulse rises and falls with slope $\pm S_0$. The differentiator action, except for a constant multiplier, is to take the derivative of the input. This is shown in Fig. 6-29(b). The output is then the signal in Fig. 6-29(b) multiplied by $-R_2 C_1$. The output results in much sharper positive and negative pulses than those at the input. The circuit thus serves as a *pulse sharpener*.

A different set of positive pulses, shown in Fig. 6-30(a), is input to an inte-

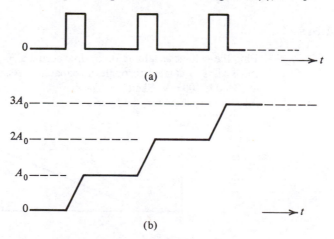

(a)

(b)

Figure 6-30. Integral of a train of three pulses.

grator. Except for the constant multiplier of Eq. 6-104, the integral of the signal is shown in Fig. 6-30(b). The output voltage is proportional to the number of pulses in a given time interval t. If the area of each pulse is A_0, then for k pulses arriving in time t the output is

$$v_2(t) = -\frac{1}{R_1 C_2} k A_0 \qquad (6\text{-}108)$$

The output voltage at any instant tells us how many pulses have arrived up to that instant at the input. The circuit thus serves as a *pulse counter*. The pulses must all be positive or all negative for the proportionality to be valid.

Integrators and differentiators are used extensively in instrumentation and in simulation. They are also used in analog computers as building blocks to simulate behavior of dynamic systems.

EXERCISES

6-7.1. In the circuit in Fig. 6-31, $R = 10$ kΩ. Find the value of R_1 such that

$$V_2 = -100 V_1$$

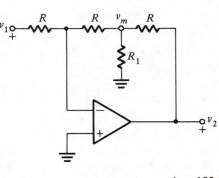

Figure 6-31.   Circuit for Exercise 6-7.1.

Ans. 102 Ω

PROBLEMS

6-1. Find the voltage as a function of time across a 50-μF capacitor for the following three cases of current through the capacitor. Assume that the capacitor voltage at $t = 0$ is 100 mV. Sketch $v_C(t)$.

(a) $i(t)$ is as shown in the figure.

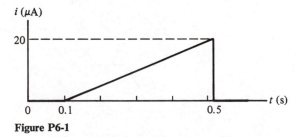

Figure P6-1

(b) $i(t) = 5 \times 10^{-5}e^{-6t}$.

(c) $i(t) = 10^{-4} \sin 10\pi t$.

6-2. If the current waveform shown in the figure is applied to a 2-μF capacitor, find the capacitor voltage $v_C(t)$ and prepare a sketch showing this waveform. Assume that the initial capacitor voltage is zero.

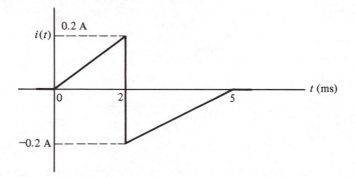

Figure P6-2

6-3. If the current waveform shown in the figure is applied to a 2-μF capacitor, sketch the voltage across the capacitor, $v_C(t)$. Assume that the capacitor voltage is zero at $t = 0$.

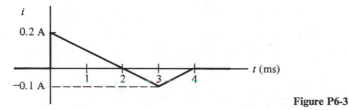

Figure P6-3

6-4. Find the equivalent capacitance C_{eq} for the circuit shown in the figure. Let all capacitors be uncharged.

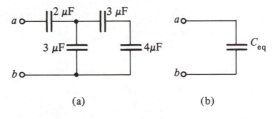

(a)   (b)

Figure P6-4

6-5. Suppose that for the circuit shown in the figure, $v_1(0) = 10$ V and $v_2(0) = 5$ V. At what time will the voltage $v(t)$ have a value of 0.75 V?

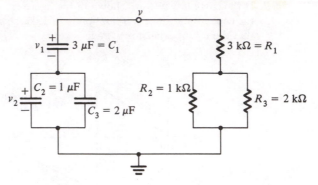

Figure P6-5

6-6. For each of the three waveforms considered in Prob. 6-1, determine the energy stored in the capacitor at $t = 0.4$ s.

6-7. For Prob. 6-2, determine the energy stored in the capacitor at the two times, $t = 2$ s and $t = 5$ s.

6-8. For Prob. 6-3, determine the energy stored in the capacitor at the two times, $t = 2$ s and $t = 5$ s.

6-9. Consider the circuit given in the figure. At time $t = 0$, the voltage across the capacitor is zero and the switch is moved to position a. The switch is left in position a until $t = 10$ s and then moved to position b and locked into position. Find the voltage across the capacitor at $t = 20$ s.

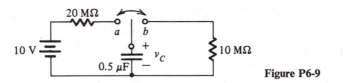

Figure P6-9

6-10. The switch in the circuit of the figure is opened at $t = 0$. Find the voltage $v_2(t)$ for all time greater than zero.

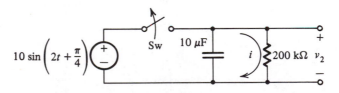

Figure P6-10

6-11. In the circuit shown in the figure, the capacitors are initially uncharged. From $t = -1$ until $t = 0$, the current from the source has a sinusoidal waveform as shown in (b) of the figure. At $t = 0$, the switch is closed. Determine an expression for the capacitor voltage $v_C(t)$ and sketch this waveform for all time.

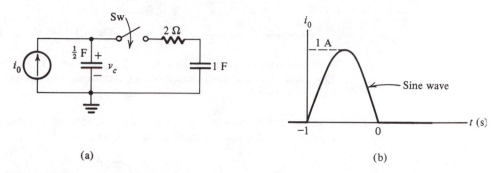

(a)   (b)

Figure P6-11

6-12. In the circuit shown in the figure, the switch has been open for a long time and at $t = 0$ it is closed, connecting the battery to the remainder of the circuit. (a) Write the differential equation for $v_2(t)$. (b) Determine the initial value and the final value of the voltage $v_2(t)$.

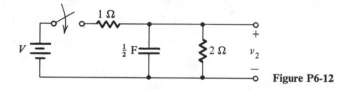

Figure P6-12

6-13. The circuit shown in the figure is connected by the closing of the switch at time $t = 0$. (a) What is the overall time constant of the circuit? (b) Determine $i(t)$, $v_1(t)$, and $v_2(t)$ as functions of time for $t > 0$. (c) Sketch the three waveforms as determine from part (b).

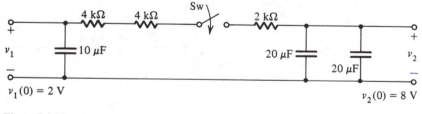

Figure P6-13

6-14. An ideal capacitor is assumed to have no loss and so will hold its charge indefinitely if left unconnected. In actual capacitors, some leakage current passes through the insulator separating the two plates of the capacitor. Such a capacitor in which the leakage current is not negligible can be modeled as a parallel combination of an ideal capacitor and a resistor. Suppose that the 0.1-μF capacitor shown in the circuit can be modeled as an ideal 0.1-μF capacitor in parallel with

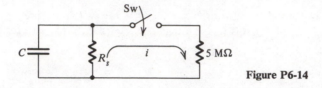

Figure P6-14

a 10-MΩ resistor. (a) Find the current $i(t)$ using the leaky capacitor model. (b) Does the capacitor discharge more or less rapidly as predicted by the leaky capacitor model?

6-15. For the circuit given in the figure, use nodal analysis to determine differential equations for the two voltages v_1 and v_2. Do not solve the differential equations.

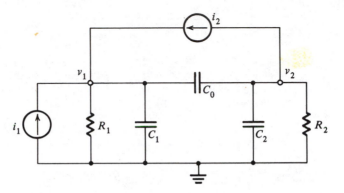

Figure P6-15

6-16. Consider the circuit shown in the figure, in which a battery of 10 V is connected to the circuit by the closing of a switch at $t = 0$. Assuming that the initial voltage of the capacitor is zero, determine expressions for $i_C(t)$ and $v_C(t)$ as identified on the figure, and sketch the waveforms for these two functions, showing any significant features.

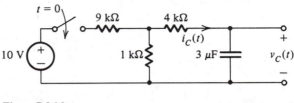

Figure P6-16

6-17. In the circuit shown in the figure, a steady state is reached with the switch open. At $t = 0$, the switch is closed. For the element values given in the figure, determine (a) $v_a(0-)$ and $v_b(0-)$ and (b) $v_a(0+)$ and $v_v(0+)$, where $-$ and $+$ refer to just before and just after the closing of the switch.

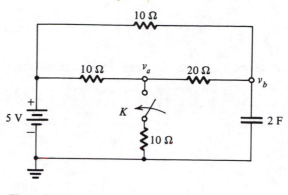

Figure P6-17

6-18. The figure shows an op amp–capacitor circuit. Determine an expression for v_2 as a function of the input voltages v_1 and v_3.

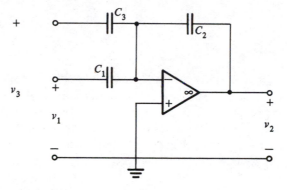

Figure P6-18

6-19. In the circuit of Prob. 6-18, suppose that C_3 is replaced by a resistor R_3. In this case, determine v_2 as a function of v_1 and v_3.

7

RESISTOR–INDUCTOR CIRCUITS

The *inductor* is an element in which the current takes time to build up. *For an inductor, the current is proportional to the magnetic field, not the voltage across it.* Since it takes time and energy to change magnetic field, it takes a certain amount of time for current to build up in an inductor. *Inductors retain or store magnetic energy,* whereas capacitors retain or store electrical energy in the form of charge.

7-1 INDUCTORS

Physically, inductors consist of coils of wire wound around a medium or core as shown in Fig. 7-1 (a). When a current flows through any wire, a magnetic field circling the wire is set up. The *magnetic flux*, Φ, is then proportional to the current through the wire. By winding the wire in a coil as shown, the flux due to each turn is reinforced and, ideally, the total flux is now proportional to both the current and the number of turns in the coil:

$$\Phi = kNi \tag{7-1}$$

where N is the number of turns. This flux, in turn, links all the N turns of the coil so that the total *flux linkage*, λ, is

$$\lambda = N\Phi = kN^2i \tag{7-2}$$

In the ideal case, then, the flux linkage is proportional to the current and may be expressed as

$$\lambda = Li \tag{7-3}$$

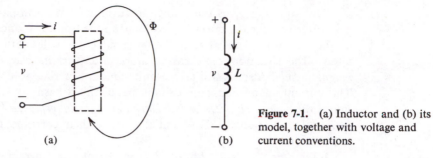

Figure 7-1. (a) Inductor and (b) its model, together with voltage and current conventions.

(a)　(b)

where the constant of proportionality, L, is called the *inductance*. The unit of inductance is in *henrys* (H) and the value of the inductance is proportional to N^2, as seen from Eq. 7-2.

The current–voltage relationship for the inductor is derived from *Faraday's law of induction*. The law states that the *induced voltage in a wire or coil is equal to the time rate of change (i.e., the time derivative) of the flux linkage*. Thus, the induced voltage appearing at the coil terminals is

$$v = \frac{d\lambda}{dt} = L \frac{di}{dt} \tag{7-4}$$

The symbol for the inductor and the polarities defined by the current–voltage relationship of Eq. 7-4 are shown in Fig. 7-1 (b). A small but sudden change in the magnetic flux or in the current gives rise to a very large derivative and, therefore, a very large voltage on the inductor. Thus, the induced voltage tends to oppose change in the current or the magnetic field and it will take a certain amount of time to build up the current in an inductor. This tendency to retain the existing current or the field represents a *storage of magnetic energy*. Equation 7-4 also states that the voltage is present on an inductor only if the current is changing with time. Otherwise, the derivative is zero and so is the voltage. Thus, under constant, or dc, excitation when the current is constant, the voltage on the inductor is zero and it behaves like a *short circuit*.

There is a direct analogy between the capacitor and the inductor. If we interchange L and C, and i and v, identical behavior is obtained. For example, the buildup of current in an inductor is given by an expression analogous to that of the voltage buildup in a capacitor. We integrate Eq. 7-4 to obtain

$$i = \frac{1}{L} \int_{-\infty}^{t} v(\tau) \, d\tau \tag{7-5}$$

This may also be rewritten in terms of an initial current at $t = 0$ as follows:

$$i = \frac{1}{L} \int_{-\infty}^{0} v \, d\tau + \frac{1}{L} \int_{0}^{t} v \, d\tau \tag{7-6}$$

$$= i_0 + \frac{1}{L} \int_{0}^{t} v \, d\tau \tag{7-7}$$

This equation gives the change in current from an initial value of i_0 to a new value i under the effect of a voltage applied on the inductor. If the voltage on the inductor is kept at zero, the same current i_0 will continue to flow indefinitely. The situation represents *a storage of current and the corresponding magnetic field*. Also, it will take a finite amount of time to change the current. The current in an inductor cannot be changed instantly. When discussing initial values in a circuit, *the inductors may thus be replaced by current sources of initial current values*. Zero initial current then corresponds initially to an open circuit.

Equations 7-4 and 7-5 or 7-7 represent the current–voltage relations for the inductor. These equations are identical to the ones for the capacitor, except that now L replaces C and the i and v are interchanged. Therefore, the same mathematical behavior is to be expected for the equations. In particular, superposition is valid for the equations as before *except* when the initial current $i_0 \neq 0$. In our generalized sense the inductor is a linear element where it is understood that the initial current is zero.

Similarly, power and energy delivered to an inductor are obtained as

$$p(t) = v(t)i(t) \quad \text{W} \tag{7-8}$$

which by use of Eq. 7-4 becomes

$$p(t) = Li\frac{di}{dt} = \frac{d}{dt}\left(\frac{1}{2}Li^2\right) \tag{7-9}$$

The energy stored in the inductor is given by integrating the power,

$$w(t) = \tfrac{1}{2}Li^2 \quad \text{W-s} \tag{7-10}$$

In Eq. 7-10, with the polarities of Fig. 7-1, a positive value of w implies energy stored in the inductor, and a negative value will mean the energy supplied by the inductor.

Equivalent inductances of series and parallel combinations of inductances are obtained in a manner similar to that used for capacitances. Consider the series case first as shown in Fig. 7-2 (a). Then, since the same current is passing through L_1 and L_2, we use the KVL and Eq. 7-4 to write

$$v = v_1 + v_2 = L_1\frac{di}{dt} + L_2\frac{di}{dt} \tag{7-11}$$

$$= (L_1 + L_2)\frac{di}{dt} \tag{7-12}$$

$$= L\frac{di}{dt} \tag{7-13}$$

Thus, the equivalent inductance is given by

$$L = L_1 + L_2 \tag{7-14}$$

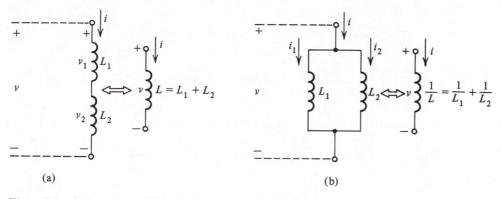

Figure 7-2. Series and parallel inductors with their equivalents.

for the series combination. The parallel combination of two inductances, assuming zero initial currents, is analyzed in a similar manner by using Eq. 7-5 and the KCL:

$$i = i_1 + i_2 = \frac{1}{L_1} \int_0^t v \, d\tau + \frac{1}{L_2} \int_0^t v \, d\tau \tag{7-15}$$

$$= \frac{1}{L} \int_0^t v \, d\tau \tag{7-16}$$

and the equivalent inductance of the parallel combination [see Fig. 7-2 (b)] is given by

$$\frac{1}{L} = \frac{1}{L_1} + \frac{1}{L_2} \tag{7-17}$$

If the parallel combination has any initial currents present, these must, of course, be taken into account. The initial current for the equivalent inductance will be the algebraic sum of the initial currents in L_1 and L_2. For the series case, any common initial current will also be present in the equivalent inductance.

EXAMPLE 7-1

Find the equivalent inductance of the circuit shown in Fig. 7-3 if the initial currents are all zero.

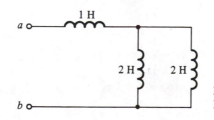

Figure 7-3. Inductive circuit for Example 7-1.

Solution

The parallel inductances result in an equivalent inductance of 1 H. This is in series with the 1 H inductance, so the equivalent inductance of the circuit is 2 H.

EXAMPLE 7-2

An inductance of 0.1 H has a current of 1 A flowing through it. This current is reduced to zero in 10^{-4} s as shown in Fig. 7-4. What is the voltage induced in the inductance?

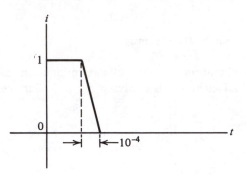

Figure 7-4. Current waveform for Example 7-2.

Solution

The voltage is given by Eq. 7-4. The slope di/dt during the 10^{-4} s of transition is -10^4 and zero elsewhere. The voltage $v = L(di/dt)$ is thus -1000 V during the 10^{-4} s of current turn-off time, and zero elsewhere, as shown in Fig. 7-5. This principle is used in automobile ignition systems to generate the very high voltages needed to fire or spark the spark plugs.

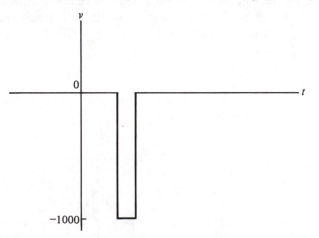

Figure 7-5. Solution for Example 7-3.

EXAMPLE 7-3

Find the current in an inductor of 0.1 H if the voltage impressed on it is 1 V for 1 s and zero the rest of the time.

Solution

The current is given by Eq. 7-5:

$$i = \frac{1}{0.1} \int_0^1 1 \, d\tau + \frac{1}{0.1} \int_1^t 0 \, d\tau = 10 \text{ A} \qquad (7\text{-}18)$$

The current in Example 7-3 would continue to flow indefinitely. The inductor is short-circuited since the voltage source is set to zero. Physical inductors differ considerably from the ideal ones, more so than is the case for the capacitors. Real inductors have a resistance due to the resistance of the wire used in the coil. A more realistic model of an inductor would be the ideal inductance in series with a resistance, as shown in Fig. 7-6.

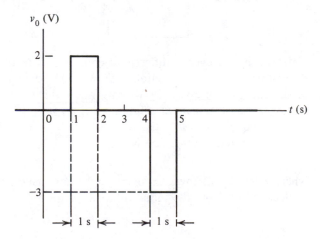

Figure 7-6. Improved model of an inductor accounting for loss.

Any initial current in a shorted real inductor will decay to zero since the flow of current through the resistance will dissipate the stored magnetic energy. The rate at which this decay occurs is an important parameter which depends on the inductance and the resistance in the circuit. We study this in Section 7-2.

EXERCISES

7-1.1. Figure 7-7 shows the voltage applied to an inductor with $L = \frac{1}{2}$ H. On the same coordinates, sketch the current in the inductor, $i_L(t)$.

Figure 7-7. Waveform for Exercise 7-1.1.

7-1.2. A current source, $i_0(t)$, is connected to a $\frac{1}{3}$-H inductor. If the waveform of $i_0(t)$ is that of Fig. 7-8, sketch the voltage, $v_L(t)$, using the same coordinate system as was used for $i_0(t)$.

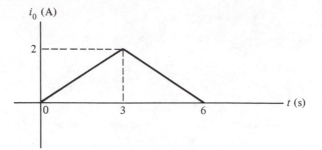

Figure 7-8. Waveform for Exercise 7-1.2.

7-2 LR CIRCUITS

Consider the circuit in Fig. 7-9 with an inductor in series with a resistor and an initial current in the inductor at $t = 0$ of i_0. By use of KVL, we write

$$v_R + v_L = 0 \tag{7-19}$$

or, by using Ohm's law and Eq. 7-4,

$$iR + L\frac{di}{dt} = 0 \tag{7-20}$$

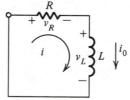

Figure 7-9. Series RL circuit.

This is a homogeneous differential equation as in the case of the discharging capacitor discussed in Section 6-5. We can solve the equation as before by separating the variables. Or we can follow the approach of Section 6-6, where we assume an exponential solution with an unknown exponent and substitute it in the equation to find the value of the exponent. We follow this approach and assume that the solution is

$$i = Ae^{st} \tag{7-21}$$

where s is the unknown exponent and the constant A is to be determined as in Section 6-5 from the initial value. Substituting Eq. 7-21 into Eq. 7-20 yields

$$(sL + R)Ae^{st} = 0 \tag{7-22}$$

or

$$sL + R = 0 \tag{7-23}$$

from which we see that $s = -R/L$ and the solution in Eq. 7-21 becomes

$$i(t) = Ae^{-(R/L)t} \tag{7-24}$$

We can complete the solution by applying the initial condition $i(0) = i_0$. Then

$$i(0) = i_0 = A \tag{7-25}$$

and

$$i(t) = i_0 e^{-(R/L)t} \tag{7-26}$$

The time constant of the circuit is L/R and the behavior of exponential decay is analogous to that of the CR circuit. The current vanishes with increasing time and may be safely assumed to have decayed in about four time constants. The current response of Eq. 7-26 is shown in Fig. 7-10.

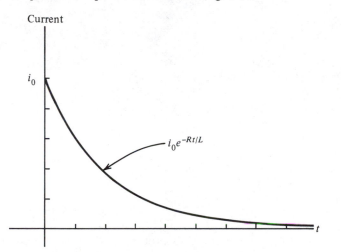

Figure 7-10. Current response of the circuit of Fig. 7-9.

The presence of a driving source can be handled as before by adding the steady-state response to the transient response of Eq. 7-24. In Fig. 7-11, the series RL circuit is excited by a constant source. The initial value of the current is i_0.

We can use either of the two alternatives discussed in Section 6-6 to determine the steady-state response. We can write the loop differential equation, assume an unknown constant solution, substitute in the equation, and determine

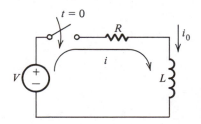

Figure 7-11. Series RL circuit driven by a voltage source.

the value of the constant (steady-state) solution. Or we can assume steady-state conditions for the circuit and derive the steady-state response from physical reasoning. We will use the latter approach.

In our circuit, the excitation is a constant and so under steady-state conditions (dc), the current is a constant. The inductor therefore behaves, in steady-state, like a short circuit, since the current is constant. The entire voltage appears across the resistor, so the steady-state value of the current is

$$i_{ss} = \frac{V}{R} \tag{7-27}$$

The transient response is as given in Eq. 7-24. The complete response is the sum of the steady-state and the transient components:

$$i(t) = \frac{V}{R} + Ae^{-(R/L)t} \tag{7-28}$$

We determine the constant A by substituting the initial value of the current at $t = 0$ in Eq. 7-28. Thus,

$$i(0) = i_0 = \frac{V}{R} + A \tag{7-29}$$

or

$$A = i_0 - \frac{V}{R} \tag{7-30}$$

and the complete response becomes

$$i(t) = \frac{V}{R} + \left(i_0 - \frac{V}{R}\right)e^{-(R/L)t} \tag{7-31}$$

This response is shown in Fig. 7-12 with the assumption that $i_0 = 0$. The current in our circuit then starts at zero and slowly builds up toward the steady-state

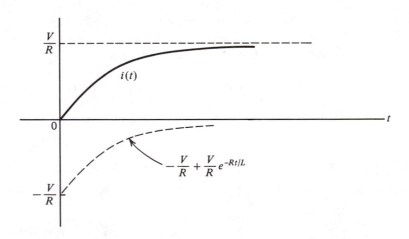

Figure 7-12.   Response of the circuit of Fig. 7-11 as given by Eq. 7-31.

value. The analysis can, of course, be applied to any circuit equivalent to the single L–single R circuit.

EXAMPLE 7-4

Find the current i in the circuit of Fig. 7-13.

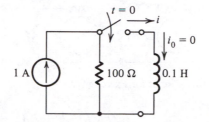

Figure 7-13. Circuit of Example 7-4.

Solution

We use source transformation to obtain the circuit in Fig. 7-14. The initial current $i(0) = 0$. The time constant is $L/R = 0.1/100 = 10^{-3}$. The transient response is $Ae^{-10^{3}t}$. The steady-state response, obtained by shorting the inductor, is $100/100 = 1$. The total response, for all $t \geq 0$, is

$$i(t) = 1 + Ae^{-10^{3}t} \tag{7-32}$$

Since $i(0) = 0$, we have $A = -1$. The response becomes

$$i(t) = 1 - e^{-10^{3}t} \quad (t \geq 0) \tag{7-33}$$

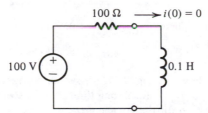

Figure 7-14. Equivalent of the circuit of Fig. 7-13.

Inductors are not commonly used in op-amp circuits. Among the many reasons that may be cited are that inductors tend to be bulky, and they change values with temperature and time. But the principal reason is that the integrated-circuit technology used in modern electronics does not lend itself to using inductors, and that the desired circuit functions can always be performed without them.

On the other hand, inductors play a key role in power circuits. Inductive circuits also play a dominant role whenever high power, or high voltage or current, is involved, since electronic devices are not suited for this purpose. In particular, *transformers*, which are used just about everywhere in our modern technological society, are magnetically coupled inductors or coils. We introduce these elements in Section 7-3.

EXERCISES

7-2.1. Use loop analysis to obtain two differential equations for i_1 and i_2 for the circuit given in Fig. 7-15.

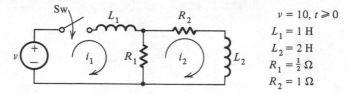

$$v = 10, t \geqslant 0$$
$$L_1 = 1 \text{ H}$$
$$L_2 = 2 \text{ H}$$
$$R_1 = \tfrac{1}{2}\ \Omega$$
$$R_2 = 1\ \Omega$$

Figure 7-15.   Circuit for Exercise 7-2.1.

$$\textit{Ans. } L_1 \frac{di_1}{dt} + R_1(i_1 - i_2) = v,$$

$$-R_1 i_1 + L_2 \frac{di_2}{dt} + (R_1 + R_2)i_2 = 0$$

7-2.2. In the circuit given in Fig. 7-16, the switch is closed at $t = 0$. Determine the voltage across the inductor as a function of time.

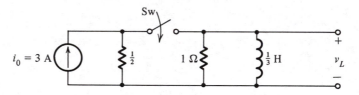

Figure 7-16.   Circuit for Exercise 7-2.2.

$$\textit{Ans. } v_L = e^{-t}(t \geq 0)$$

7-2.3. In the circuit shown in Fig. 7-17, the switch changes from a to b at $t = 0$. The switch was in position a for a long time prior to switching to position b. Determine an expression for the voltage $v_2(t)$.

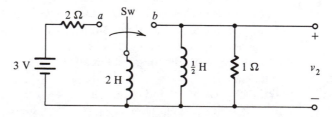

Figure 7-17.   Circuit for Exercise 7-2.3.

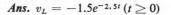

$$\textit{Ans. } v_L = -1.5e^{-2.5t} \ (t \geq 0)$$

The fact that a (changing) magnetic flux linking any coil generates an induced voltage in the coil is the basis of many significant developments in electrical engineering. For example, electrical generators, transformers, and numerous instruments are consequences of this basic phenomenon. A major attraction of the phenomenon is the unique feature, known as *magnetic coupling*, which results when there is magnetic flux linking or *coupling* the coil *but* the part producing the magnetic flux is *not electrically connected* to the part in which the voltage is induced.

Consider two coils located physically close to one another but *not* connected, as shown in Fig. 7-18. The coils are wound on a magnetic core in the figure,

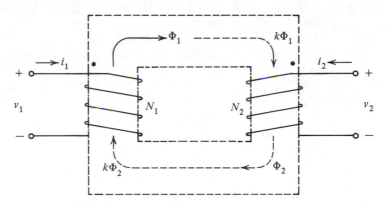

Figure 7-18. Magnetically coupled circuit.

but this is not essential to our discussion. The number of turns in the two coils are N_1 and N_2, as shown. Let us assume initially that only the current i_1 flows in the coil with N_1 turns and that $i_2 = 0$. The current i_1 generates a flux Φ_1. A fraction, k, of this flux also links the second coil of N_2 turns. The flux $k\Phi_1$ linking the N_2 turns coil induces a voltage in this coil and represents magnetic coupling of the two coils. The core material is usually selected to make the fraction k, also called the *coefficient of coupling*, as close to unity as possible for efficient coupling.

We have seen in Section 7-1 that the flux Φ_1 is proportional to $N_1 i_1$, and the flux linkage for the first coil is $N_1 \Phi_1$. The voltage v_1 induced in the first coil is then

$$v_1 \propto N_1^2 \frac{di_1}{dt}\bigg|_{i_2=0} \tag{7-34}$$

or

$$v_1 = L_1 \frac{di_1}{dt}\bigg|_{i_2=0} \tag{7-35}$$

and the constant of proportionality, L_1, is the inductance of the coil. Since there are two coils here, we call this the self-inductance. The self-inductance L_1 is thus proportional to N_1^2.

We have also $k\Phi_1$ linking N_2 turns of the second coil. The flux linkage in this case is $kN_2\Phi_1$. The voltage v_2 induced in this coil is

$$v_2 \propto kN_1N_2 \frac{di_1}{dt}\bigg|_{i_2=0} \tag{7-36}$$

or

$$v_2 = M \frac{di_1}{dt}\bigg|_{i_2=0} \tag{7-37}$$

The constant of proportionality, M, in this case is called the *mutual inductance*. It is proportional to kN_1N_2. The polarity of the voltage induced in the second coil depends on the way the coils are wound and it is usually indicated by *dots*, as in the figure. The dots signify that the induced voltages in the two coils (due to a single current or flux) have the same polarities at the dotted ends of the coils. Thus, due to i_1 flowing as shown by the conventions of Section 7-1, the induced voltage v_1 must be positive at the dotted end of coil 1; then v_2 is also positive at the dotted end in coil 2.

The same reasoning applies if i_2 were flowing in coil 2 and $i_1 = 0$ in coil 1. The flux $\Phi_2(\alpha N_2)$ is produced by i_2 and links N_2 turns of coil 2. The induced voltage v_2 is

$$v_2 \propto N_2^2 \frac{di_2}{dt}\bigg|_{i_1=0} \tag{7-38}$$

or

$$v_2 = L_2 \frac{di_2}{dt}\bigg|_{i_1=0} \tag{7-39}$$

where L_2 is the *self-inductance* of coil 2 and is proportional to N_2^2. Next, we assume that the same fraction k of Φ_2 will link the coil N_1. Then the induced voltage v_1 is

$$v_2 \propto kN_1N_2 \frac{di_2}{dt}\bigg|_{i_1=0} \tag{7-40}$$

or

$$v_1 = M \frac{di_2}{dt}\bigg|_{i_1=0} \tag{7-41}$$

The mutual inductance M is the same in this case and is proportional to kN_1N_2. The polarities of v_1 and v_2 follow the dot convention, as before. The voltage polarity is positive at the dotted end of the inductor L_2 when the current direction for i_2 is as shown. Therefore, the voltage induced in coil 1 must be positive at the dotted end also.

We have seen in Section 7-1 that the inductors are linear and that superposition may be used for circuits containing them, provided that the initial currents are zero. We assume no initial currents in the two coils. Now, if both currents i_1 and i_2 are present, we *may* use superposition to determine the terminal voltages at the two coils. Thus, summing Eqs. 7-35 and 7-41, and Eqs. 7-37 and 7-39, we have

$$v_1 = L_1 \frac{di_1}{dt} + M \frac{di_2}{dt} \tag{7-42}$$

$$v_2 = M \frac{di_1}{dt} + L_2 \frac{di_2}{dt} \tag{7-43}$$

The circuit symbols for the *coupled coils* together with the polarities and the dot markings are shown in Fig. 7-19 (a). Notice that the dots in Fig. 7-19 (b) are placed at the opposite ends in the two coils. Let us examine this case. Due

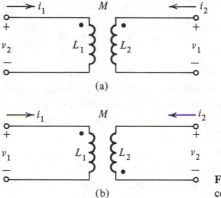

(a)

(b)

Figure 7-19. Dot convention for coupled coils.

to i_1, with $i_2 = 0$, the dotted end in coil 1 is positive, so the induced voltage coil 2 is positive at the dot, which is the reverse of the designated polarity for v_2. Similarly, due to i_2, with $i_1 = 0$, the dotted ends have negative polarities for the induced voltages. The mutually induced voltages in both cases have polarities that are the reverse of the terminal voltages, and the equations are

$$v_1 = L_1 \frac{di_1}{dt} - M \frac{di_2}{dt} \tag{7-44}$$

$$v_2 = -M \frac{di_1}{dt} + L_2 \frac{di_2}{dt} \tag{7-45}$$

The currents i_1 and i_2 are shown symmetrically flowing into the coils at both ends, for convenience in writing equations. These directions are, of course, arbitrary.

Analysis of circuits containing coupled coils is best handled by loop analysis. We illustrate by a simple example in the case of the circuit shown in Fig. 7-20. We have two loops as shown, and KVL gives us

$$v_0 = R_1 i_1 + v_1 \tag{7-46}$$

$$v_2 = -R_2 i_2 \tag{7-47}$$

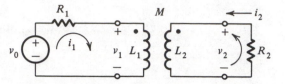

Figure 7-20. Coupled circuit analyzed in Eqs. 7-48 and 7-49.

We have chosen the second loop in the counterclockwise direction to make it easier to express v_1 and v_2 by Eqs. 7-42 and 7-43. Then, substituting for v_1 and v_2 in Eq. 7-46 and 7-47 yields

$$v_0 = R_1 i_1 + L_1 \frac{di_1}{dt} + M \frac{di_2}{dt} \qquad (7\text{-}48)$$

$$0 = M \frac{di_1}{dt} + L \frac{di_2}{dt} + R_2 i_2 \qquad (7\text{-}49)$$

Equations 7-48 and 7-49 are two simultaneous differential equations that must be solved in the two unknown currents i_1 and i_2. The solution of such equations is quite involved and we will not pursue it in this text. Transform analysis techniques, introduced later, will make solutions of such equations relatively straightforward.

Much more important, however, is the version of coupled coils known as the *transformer*. Transformers are designed so that very little current is needed to generate a lot of flux, and almost all the flux links both coils. Circuits containing transformers are quite easy to analyze, as we shall see below. Most transformers are so efficient today that they are modeled by the *ideal transformer* which is the model that we use in this text.

There are two idealizations involved in the ideal transformer. The first one assumes that *all* the flux produced by the current in one coil links with the second coil (i.e., $k = 1$). Many practical transformers achieve this quite well in that $k \simeq 0.99$. If there is no leakage of flux so that $k = 1$, then due to Φ_1 alone, the induced voltages are $N_1 \, d\Phi_1/dt$ and $N_2 \, d\Phi_1/dt$ in coils 1 and 2, respectively. Similarly, due to Φ_2 alone, the induced voltages are $N_1 \, d\Phi_2/dt$ and $N_2 \, d\Phi_2/dt$ in coils 1 and 2, respectively. Then, by superposition, with the dot conventions of Fig. 7-19 (a)

$$v_1 = N_1 \frac{d\Phi_1}{dt} + N_1 \frac{d\Phi_2}{dt} \qquad (7\text{-}50)$$

$$v_2 = N_2 \frac{d\Phi_1}{dt} + N_2 \frac{d\Phi_2}{dt} \qquad (7\text{-}51)$$

Equations 7-50 and 7-51 show that

$$\frac{v_1}{v_2} = \frac{N_1}{N_2} = n \qquad (7\text{-}52)$$

or the voltages in the two coils are *directly proportional to their turns* and *the voltage ratio equals the turns ratio.*

The second idealization is that negligible current is needed to create the flux, and therefore the power into the coupled coils equals the power out. Then

$$v_1 i_1 = -v_2 i_2 \tag{7-53}$$

where the negative sign is needed since in our representation we show the power going in at both ends. By substituting Eq. 7-52 into Eq. 7-53, we obtain

$$\frac{i_1}{i_2} = -\frac{v_2}{v_1} = -\frac{1}{n} \tag{7-54}$$

or the *current ratio in the two coils equals the reciprocal of the turns ratio.* The negative sign merely indicates the reversal of direction for one of the two currents. The idealization that no power is needed to create the flux in a transformer is, of course, approximate but comes close in many practical cases, where 98 to 99 % of the power in equals the power out.

The symbol for the *ideal transformer,* with positive polarity marks representing the dots, is shown in Fig. 7-21. It is the most frequently encountered

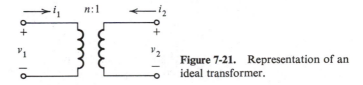

Figure 7-21. Representation of an ideal transformer.

version of coupled coils, both in practice and in circuit analysis. One unfortunate consequence of the idealizations is that while transformer action is a result of *changing magnetic flux,* the equations governing the ideal transformer, Eqs. 7-52 and 7-54, do not incorporate this fact. Otherwise, however, the equations represent a very convenient and accurate transformer behavior.

Equations 7-34 to 7-41 have shown that for coupled coils, in general, $L_1 \propto N_1^2$, $L_2 \propto N_2^2$, and $M \propto k N_1 N_2$. Therefore,

$$\frac{M}{\sqrt{L_1 L_2}} = k \le 1 \tag{7-55}$$

In the case of transformers there is ideal or unity coupling and $k = 1$. Further, the flux linkage is proportional to the current producing it. In the transformer, no current is needed to produce the flux linkage and the corresponding constant of proportionality (i.e., the self-inductances L_1 and L_2 as well as M from Eq. 7-55) are therefore *infinite* in ideal transformers. Analysis of circuits with transformers must *never use coupled-coil equations but rely only on transformer equations.*

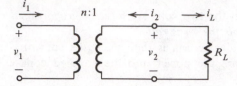

Figure 7-22. Coupled coil terminated in resistor R_L.

Consider the circuit of Fig. 7-22, in which a load R_L is connected at the output end of the transformer. The input end of the transformer is called the *primary* and the output end the *secondary*. In the figure we wish to determine the input signals. For convenience we show $i_L = -i_2$, so that Eq. 7-54 becomes $i_1/i_L = 1/n$. Now, by Ohm's law, $v_2 = R_L i_L$. By transformer equations, then,

$$v_1 = nv_2 = nR_L i_L \qquad (7\text{-}56)$$

$$= n^2 R_L i_1 \qquad (7\text{-}57)$$

or

$$\frac{v_1}{i_1} = n^2 R_L \qquad (7\text{-}58)$$

Thus, the load resistance R_L appears at the primary as an equivalent resistance of $n^2 R_L$. Such use of transformers is frequently made to ensure maximum power transfer to the load.

EXAMPLE 7-5

Design a transformer for maximum power transfer if the generator resistance is 900 Ω and the load is 100 Ω.

Solution

We have seen in Chapter 4 that maximum power is delivered to the load if the load resistance equals the generator resistance. We use a transformer so that the *effective* load resistance at the primary is 900 Ω. By Eq. 7-58, this gives us

$$900 = n^2 R_L = n^2 \times 100 \qquad (7\text{-}59)$$

or

$$n = 3 \qquad (7\text{-}60)$$

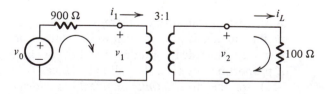

Figure 7-23. Circuit for Example 7-5.

The primary–secondary turns ratio is 3 for maximum power transfer. The circuit is shown in Fig. 7-23.

This is the key to circuit analysis with transformers. We merely use transformer equations to go from one side to the other. Analysis is performed separately on each side of the transformer. Suppose that we are interested in the load voltage and current. We approach this in two different ways.

First, we write KVL for the two loops in terms of transformer terminal voltages and currents,

$$V_0 = 900i_1 + v_1 \qquad (7\text{-}61)$$

$$v_2 = 100i_L \qquad (7\text{-}62)$$

Now, we substitute from the transformer equations that $v_1 = nv_2$ and $i_1 = i_L/n$,

$$v_0 = 900\frac{i_L}{3} + 3v_2 \qquad (7\text{-}63)$$

$$= 300i_L + 300i_L \qquad (7\text{-}64)$$

or

$$i_L = \frac{v_0}{600} \qquad (7\text{-}65)$$

and

$$v_2 = 100i_L = \frac{v_0}{6} \qquad (7\text{-}66)$$

Alternatively, we could replace the circuit at the primary end by the equivalent resistance $n^2R_L = 900 \ \Omega$, as in Fig. 7-24. Then, $v_1 = \frac{1}{2}v_0$ and $i_1 = v_0/1800$.

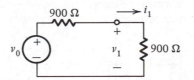

Figure 7-24.   Equivalent to the circuit of Fig. 7-23.

Then, in the original circuit by transformer action, $v_2 = v_1/n = v_0/6$ and $i_L = ni_1 = v_0/600$, as before. The two approaches are identical and either may be followed.

Uses of transformers are too numerous to mention. In electronics they are used to *step up* to higher voltages, to *step down* to lower voltages, as well as in efficient power delivery to the load. The power industry uses them extensively to efficiently deliver power to homes and factories. For example, transformers are used to step up the generator voltage to a more efficient, higher voltage for long-distance transmission, which is then transformed to a lower voltage for load consumption.

EXERCISE

7-3.1. Obtain a pair of differential equations for the loop currents i_1 and i_2 for the circuit containing mutual inductance of Fig. 7-25.

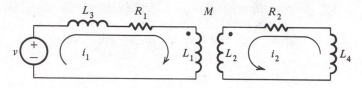

Figure 7-25. Circuit for Exerise 7-3.1.

$$\textit{Ans. } (L_1 + L_3)\frac{di_1}{dt} + M\frac{di_2}{dt} + R_1 i_1 = v,$$

$$M\frac{di_1}{dt} + (L_2 + L_4)\frac{di_2}{dt} + R_2 i_2 = 0$$

In Chapter 8 we turn our attention to a class of circuits that plays an important role in many applications, second-order circuits. These are circuits that contain both an inductor and capacitor, and are characterized by a second-order differential equation. Several terms and concepts of practical importance are also discussed.

PROBLEMS

7-1. The voltage waveform shown in the figure is applied to a 100-mH inductor. Sketch the waveform of the current through the inductor. Assume that the initial current has a value of zero.

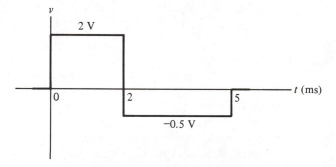

Figure P7-1

7-2. The voltage across a 100-mH inductor has the waveform shown in the figure. Sketch the waveform of the current in the inductor that results. Assume that the initial value of current is zero.

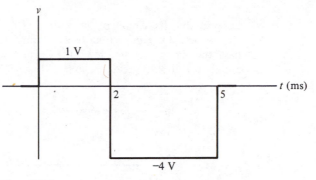

Figure P7-2

7-3. The voltage across a 100-mH inductor has the waveform shown in the figure. As a result, there is a current in the inductor. For this problem determine an expression for this current, and sketch its waveform. Assume that the initial value of the current is zero.

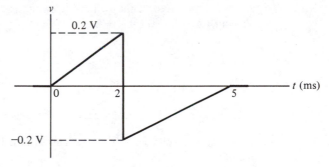

Figure P7-3

7-4. Through means external to the circuit being considered, an inductor is fluxed such that the current is 5 A. At the time $t = 0$, this inductor is connected to a resistor. If the resistor has a value of 1 kΩ and the inductor a value of 50 mH, determine the time at which the current in the inductor will have a value of 1 A.

7-5. The circuit given in the figure is in the steady state when the switch is opened at $t = 0$. Find expressions for $i_L(t)$ and $v_L(t)$ as identified on the figure. Sketch $i_L(t)$ and $v_L(t)$ for all t.

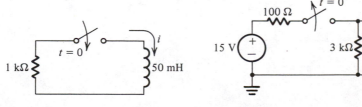

Figure P7-4   **Figure P7-5**

7-6. Suppose that the circuit of the figure has remained connected with the switch closed for an extremely long time. At time $t = 0$ the switch is opened. Determine the current through the 9-kΩ resistor (a) just before the switch is closed, and (b) for all time greater than 0.

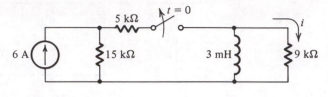

Figure P7-6

7-7. Consider the circuit of Prob. 7-5, this time with the switch initially open and so the inductor unfluxed. At the time $t = 0$ the switch is closed. Under these conditions, find expressions for $i_L(t)$ and $v_L(t)$, and sketch the two functions for all time.

7-8. In the circuit given in the figure, the switch is closed at $t = 0$. Determine and sketch $i_L(t)$ and $v_L(t)$ for $t > 0$.

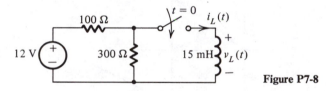

Figure P7-8

7-9. Calculate the voltage v_2 in the circuit shown in the figure. Assume that the transformer is ideal.

7-10. In the circuit of the figure, the switch has been open for all time. At $t = 0$, the switch is closed. Determine the voltage $v_2(t)$ for $t > 0$, evaluating all constants in the equation.

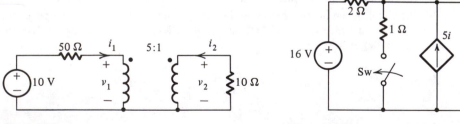

Figure P7-9   **Figure P7-10**

8

RISE TIME, RINGING, AND SUSTAINED OSCILLATIONS

Chapters 6 and 7 dealt with first-order differential equations that described current or voltage in circuits with one capacitor or one inductor. Next, we extend these studies to second-order differential equations, which describe circuits of slightly greater complexity. Will it then be necessary to study third-order differential equations and yet higher-order systems? In fact, many higher-ordered systems can be approximated by second-order systems—a process known as reduced-order modeling—and for this reason a study of both first-order and second-order systems becomes of special importance to the electrical engineer. Suppose that a step of voltage is applied to the input of a system, and some parameter within that system is changed. For each parameter setting, the step-input response is recorded. The variation of the response may be that shown in the various parts of Fig. 8-1. The response given by (a) is sluggish, becoming less so in (b) and (c). In (d) and (e) the response is more rapid and there is a damped oscillation. Finally, as shown in (f), it is possible for sustained oscillations.

It is important that we be able to describe these various response waveforms so that we can communicate with each other. Such descriptions are the topics of this chapter.

8-1 STEP RESPONSE OF FIRST-ORDER DIFFERENTIAL EQUATIONS

In Chapter 7 we considered a RL series circuit suddenly connected to a voltage source, as depicted in Fig. 8-2. The sudden connection is performed by the closing of the switch. The action applies a voltage V_0 instantly to the circuit.

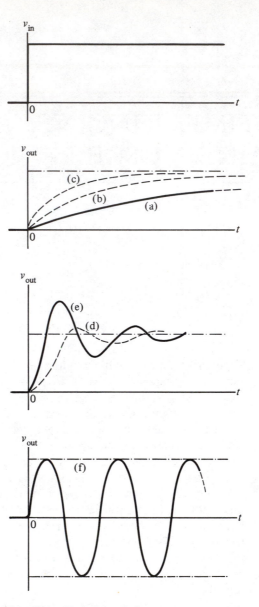

Figure 8-1. Variation of the response to a step input with changes in some parameter.

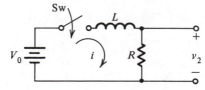

Figure 8-2. Circuit described by Eq. 8-1.

This is called a step voltage, since the voltage suddenly jumps from zero to V_0. The response of the circuit is consequently termed the *step response*. Usually, the step is of unit value, but this merely scales the response appropriately without changing the time function.

Let us assume that the switch is closed at $t = 0$ in Fig. 8-2. Then the differential equation for the circuit is

$$L\frac{di}{dt} + Ri = V_0 \qquad (t \geq 0) \tag{8-1}$$

We first divide by L, and then by methods discussed in Chapter 7 we obtain the solution

$$i(t) = \frac{V_0}{R}(1 - e^{-(R/L)t}) \qquad (t \geq 0) \tag{8-2}$$

The voltage identified as v_2 is obtained by multiplying $i(t)$ of Eq. 8-2 by R:

$$v_2(t) = V_0(1 - e^{-\sigma_1 t}) \qquad (t \geq 0) \tag{8-3}$$

where $\sigma_1 = R/L = 1/T_1$ and T_1 is the *time constant* of the circuit. We seek a method of describing the speed of response of the circuit. This response is shown in Fig. 8-3(a). In theory, v_2 reaches the final value only as time approaches infinity. One way of describing this speed of response is the time constant of the circuit, T_1. We see that a small time constant implies a short time for the response to reach a prescribed value, in this case $v_2(T_1) = 0.63V_0$. In contrast, a long time constant implies a sluggish response.

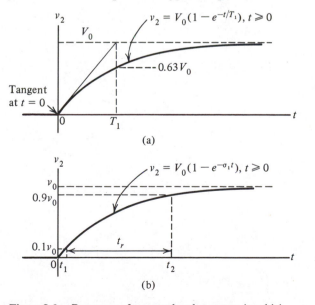

Figure 8-3. Response of a second-order system in which we define in (a) the time constant and in (b) the rise time, t_r.

Another measure of the speed of response is the *rise time*, which applies to responses in general. The rise time is a *defined* quantity. If we let t_1 be the time at which the response has reached 10% of its final value, and let t_2 similarly be the time required for the response to reach 90% of its final value [see Fig. 8-3 (b)], then by definition

$$t_r = t_2 - t_1 \tag{8-4}$$

For t_1 we have

$$0.1V_0 = V_0(1 - e^{-\sigma_1 t_1}) \quad \text{from which } t_1 = \frac{1}{\sigma_1} \ell n \frac{10}{9} \tag{8-5}$$

Similarly, for t_2

$$0.9V_0 = V_0(1 - e^{-\sigma_1 t_2}) \quad \text{from which } t_2 = \frac{1}{\sigma_1} \ell n\, 10 \tag{8-6}$$

Substituting the last two equations into Eq. 8-4 gives

$$t_2 - t_1 = \frac{\ell n\, 10}{\sigma_1} - \frac{\ell n\, 10}{\sigma_1} + \frac{\ell n\, 9}{\sigma_1} \tag{8-7}$$

or

$$t_r = \frac{1}{\sigma_1} \ell n\, 9 = \frac{2.2}{\sigma_1} = 2.2 T_1 \tag{8-8}$$

By this definition, the rise time is 2.2 time constants.

Another defined quantity is the *time delay*, t_D, defined as the time required for the response to reach 50% of its final value, as shown in Fig. 8-4. From Eq. 8-3,

$$v_2(t_D) = 0.5V_0 \tag{8-9}$$

so that

$$0.5V_0 = V_0(1 - e^{-\sigma_1 t_D}) \tag{8-10}$$

and from this

$$t_D = \frac{\ell n\, 2}{\sigma_1} = \frac{0.692}{\sigma_1} = 0.692 T_1 \tag{8-11}$$

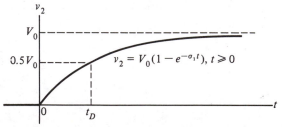

Figure 8-4.   Definition of time delay, t_D.

EXAMPLE 8-1

A response is described by Eq. 8-3 with $\sigma_1 = 1/RC$. For a circuit in which $R = 1\ k\Omega$ and $C = 100\ pF$, find the rise time and time delay.

Solution

The two quantities are

$$t_r = 2.2RC = 2.2 \times 10^3 \times 100 \times 10^{-12} = 220 \text{ ns} \qquad (8\text{-}12)$$

and

$$t_D = 69.2 \text{ ns} \qquad (8\text{-}13)$$

8-2 DAMPED OSCILLATIONS (RINGING)

When some circuits are energized, perhaps by the closing of a switch connecting a source of energy to the circuit, the response is as depicted in Fig. 8-5.

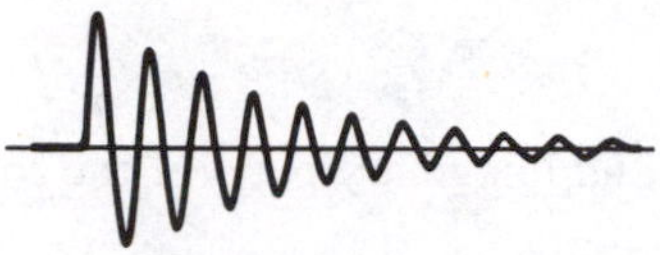

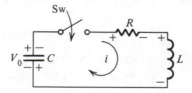

Voltage with time

Figure 8-5. Damped oscillation of voltage as a function of time.

Under this circumstance, the circuit is said to *ring*, by which we mean that a voltage or a current oscillates with time with a decreasing amplitude. This waveform is also described as a damped sinusoid or damped oscillations. Such waveforms occur in practical circuits, sometimes by design and sometimes by accident. How do we describe this damped sinusoid response?

Consider the series *RLC* circuit shown in Fig. 8-6. From Kirchhoff's voltage law, we write

$$\frac{1}{C}\int_0^t i\,dt + Ri + L\frac{di}{dt} = V_0 \qquad (8\text{-}14)$$

Figure 8-6. *RLC* circuit driven by the charge of the capacitor.

In terms of the capacitor charge, $q(t)$, the current is $i = dq/dt$, so that Eq. 8-14 becomes

$$L\frac{d^2q}{dt^2} + R\frac{dq}{dt} + \frac{1}{C}q = \frac{q_0}{C} \qquad (8\text{-}15)$$

If we now divide through by L, we have a standard form of a second-order differential equation;

$$\frac{d^2q}{dt^2} + \frac{R}{L}\frac{dq}{dt} + \frac{1}{LC}q = \frac{q_0}{LC} \qquad (8\text{-}16)$$

We note that solving for $q(t)$ will also give us the capacitor voltage, since $v_C(t) = q(t)/C$.

The characteristic equation for the differential equation 8-16 is

$$s^2 + \frac{R}{L}s + \frac{1}{LC} = 0 \tag{8-17}$$

The roots of this equation will be used as they were in our earlier study of RC and RL circuits with the difference that we are now dealing with a second-order equation rather than one of first order. The solution of Eq. 8-17 is found from the quadratic equation. Because all of the coefficients are positive, we know that the roots may be of three possible forms: (1) real and unequal, (2) real and equal, or (3) conjugate complex. The last possibility corresponds to the case of damped oscillations and so is the only one we consider. Let us identify the roots as

$$s_1, s_2 = -\sigma_1 \pm j\omega_1 \tag{8-18}$$

In terms of the real and imaginary parts of the roots, the characteristic equation is

$$(s + \sigma_1 + j\omega_1)(s + \sigma_1 - j\omega_1) = (s + \sigma_1)^2 + \omega_1^2 \tag{8-19}$$

As in our earlier studies, the roots of the characteristic equation correspond to factors which sum to give the solution of the homogeneous equation. These factors are $K_1 e^{s_1 t}$ and $K_2 e^{s_2 t}$ so that the homogeneous solution is

$$q(t) = K_1 e^{(-\sigma_1 + j\omega_1)t} + K_2 e^{(-\sigma_1 - j\omega_1)t} \tag{8-20}$$

Here K_1 and K_2 are determined by the initial condition, $q(0)$. Suppose that this is such that $K_1 = K_2 = K/2$ so that the last equation simplifies to

$$q(t) = Ke^{-\sigma_1 t}\left(\frac{e^{j\omega_1 t} + e^{-j\omega_1 t}}{2}\right) \tag{8-21}$$

The bracketed term in this equation is simplified by making use of Euler's identity which is

$$e^{+j\omega_1 t} = \cos \omega_1 t \pm j \sin \omega_1 t \tag{8-22}$$

This equation is familiar from studies in algebra and will be explored in greater depth in Section 9-5. Combining the last two equations, we obtain

$$q(t) = Ke^{-\sigma_1 t} \cos \omega_1 t \tag{8-23}$$

If the initial charge on the capacitor is q_0, then $q(0) = K = q_0$. Noting that $v_C(t) = q(t)/C$, we write

$$v_C(t) = V_0 e^{-\sigma_1 t} \cos \omega_1 t \tag{8-24}$$

The form of this voltage is shown in Fig. 8-7. The quantity σ_1 is known as the damping, ω_1 is the frequency of oscillation, and $\pm e^{-\sigma_1 t}$ is called the *envelope* of the waveform. If we expand Eq. 8-19 to

$$s^2 + 2\sigma_1 s + \sigma_1^2 + \omega_1^2 = 0 \tag{8-25}$$

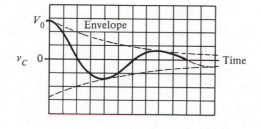

Figure 8-7. Response of the capacitor voltage, V_C, and the envelope of the response.

then we may match coefficients of this equation with Eq. 8-17; thus, we see that

$$2\sigma_1 = \frac{R}{L} \quad \text{and} \quad \sigma_1^2 + \omega_1^2 = \frac{1}{LC} \tag{8-26}$$

Solving for σ_1 and ω_1, we obtain

$$\sigma_1 = \frac{R}{2L} \tag{8-27}$$

and

$$\omega_1^2 = \frac{1}{LC} - \left(\frac{R}{2L}\right)^2 \tag{8-28}$$

To simplify these equations even further, we let the frequency given by Eq. 8-28 when $R = 0$ be ω_0. The quality factor of a coil has traditionally been defined in terms of the Q of the coil. This definition has been generalized, but it may be used in the form

$$Q = \frac{\omega_0 L}{R} \tag{8-29}$$

for the circuit under study. From Eq. 8-29 we solve for $R/2L$ as

$$\frac{R}{2L} = \frac{\omega_0}{2Q} \tag{8-30}$$

Hence,

$$\sigma_1 = \frac{\omega_0}{2Q} \tag{8-31}$$

and

$$\omega_1 = \omega_0 \sqrt{1 - \left(\frac{1}{2Q}\right)^2} \tag{8-32}$$

Hence, we have two forms in which we can express σ_1 and ω_1: In Eqs. 8-27 and 8-28, they are expressed in terms of the circuit parameters, R, L, and C; in Eqs. 8-31 and 8-32, they are expressed in terms of the two parameters ω_0 and Q.

EXAMPLE 8-2

In the circuit given in Fig. 8-6, it is given that $R = 50\ \Omega$, $L = 10\text{mH}$, $C = 1\mu\text{F}$, and $V_0 = 1\text{V}$. We are to determine values for σ_1 and ω_1.

Solution

From these parameters, we determine that $\omega_0 = 10,000$ rad/s and $Q = 2$. From Eqs. 8-31 and 8-32, we find that

$$\sigma_1 = 2500 \quad \text{and} \quad \omega_1 = 9682 \text{ rad/s or } 1541 \text{ Hz} \qquad (8\text{-}33)$$

Now the envelope of the ringing response can be described in terms of a time constant, $T_1 = 1/2500$s. Hence, the envelope of the ringing response reduces to 37% of its initial value in 400 μs.

8-3 STEP RESPONSE FOR A SECOND-ORDER CIRCUIT

The circuit shown in Fig. 8-8 differs from that of Fig. 8-6 in that a voltage source (battery) has been inserted, and the voltage of the capacitor, $v_C(0-)$ = 0. The differential equation that describes the circuit is Eq. 8-14, and hence

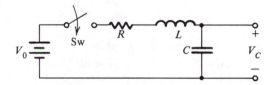

Figure 8-8. Same circuit as in Fig. 8-7 with the addition of a battery.

the solution is that found in Section 8-2 with adjustment for the initial and final value of the voltage of the capacitor. The result is

$$v_C(t) = V_0(1 - e^{-\sigma_1 t} \cos \omega_1 t) \qquad (8\text{-}34)$$

where σ_1 and ω_1 have identical definitions to those given in Section 8-2. The response $v_C(t)$ given by this equation is shown in Fig. 8-9. Clearly, this response differs from that considered for a first-order circuit, shown in Fig. 8-3. The first-order response asymptotically approaches the final value, while that shown in Fig. 8-9 exhibits ringing after first attaining a maximum value that is in excess of the final value.

The definitions given for rise time and time delay in Section 8-1 apply to this response, as well as to the response of a higher-order circuit, and the definitions are as illustrated in Fig. 8-9. Consider first the rise time. The response at time t_2 is

$$v_C(t_2) = 0.9V_0 = V_0(1 - e^{-\sigma_1 t_2} \cos \omega_1 t_2) \qquad (8\text{-}35)$$

or

$$0.1 = e^{-\sigma_1 t_2} \cos \omega_1 t_2 \qquad (8\text{-}36)$$

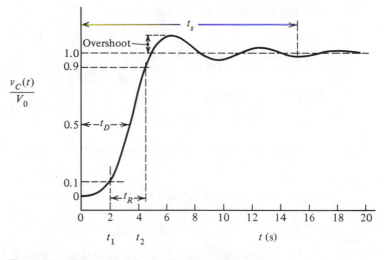

Figure 8-9.   Response of the circuit of Fig. 8-8, in which several quantities are defined, such as overshoot and settling time.

This equation has two unknowns, but from Eq. 8-31, we see that there is a simple relationship between these unknowns,

$$\frac{\sigma_1}{\omega_0} = \frac{1}{2Q} \tag{8-37}$$

and by Eq. 8-32,

$$\omega_0 = \frac{\omega_1}{\sqrt{1 - (1/2Q)^2}} \tag{8-38}$$

Combining these two equations and simplifying, we have

$$\sigma_1 = \frac{1}{\sqrt{(2Q)^2 - 1}}\omega_1 \tag{8-39}$$

To illustrate, suppose that we select the value of $Q = \sqrt{\frac{5}{2}}$ such that $\sigma_1 = \frac{1}{2}\omega_1$; then Eq. 8-36 becomes

$$0.1 = e^{-(\omega_1/2)t_2} \cos \omega_1 t_2 \tag{8-40}$$

Now the solution of a transcendental equation is no simple matter. There are two ways to determine the required value of $\omega_1 t_2$:

1. Plot $\cos \omega_1 t_2$ and $0.1e^{(\omega_1/2)t_2}$ and determine the value of $\omega_1 t_2$ at which the two curves intersect.
2. Use a cut-and-try procedure using a scientific calculator, and so determine $\omega_1 t_2$.

Using either procedure, the solution is found to be

$$\omega_1 t_2 = 1.37 \tag{8-41}$$

In a similar procedure, we find that at $t = t_1$,

$$0.9 = e^{-(\omega_1/2)t_1} \cos \omega_1 t_1 \tag{8-42}$$

for which the solution is found to be

$$\omega_1 t_1 = 0.18 \tag{8-43}$$

Then the rise time is

$$t_r = \frac{1.37}{\omega_1} - \frac{0.18}{\omega_1} = \frac{1.19}{\omega_1} = \frac{0.595}{\sigma_1} \tag{8-44}$$

The time delay, t_D, is found by the same procedure. The equation

$$0.5 = e^{-(\omega_1/2)t_D} \cos \omega_1 t_D \tag{8-45}$$

may be solved to give the value $\omega_1 t_D = 0.77$, so that

$$t_D = \frac{0.77}{\omega_1} = \frac{0.385}{\sigma_1} \tag{8-46}$$

The equations we have found for t_r and t_D in Eqs. 8-44 and 8-46 depend on a specified relationship between σ_1 and ω_1. For our study it was $\sigma_1 = \frac{1}{2}\omega_1$. For any other relationship, the steps in the solution must be repeated. The results we have given are intended to illustrate the relationship of t_r and t_D for a given time response. These characteristic times will ordinarily be measured in the laboratory rather than determined analytically.

8-4 OVERSHOOT AND SETTLING TIME

The characteristic ringing of the output voltage of the circuit of Fig. 8-8 results in *overshoot*, as shown in Fig. 8-9. We define overshoot as the difference between the maximum value of v_C and the final value of this voltage. Thus,

$$\text{overshoot} = v_{C_{\text{max}}} - v_C(\infty) \tag{8-47}$$

More often, we are interested in the percent overshoot, which is

$$\text{percent overshoot} = \frac{\text{overshoot}}{v_C(\infty)} 100\% \tag{8-48}$$

Specifications may be written in terms of this quantity. For example, it may be specified that the percent overshoot should not exceed 5%. In general, excessive overshoot is not desirable.

To illustrate the calculation of overshoot for a given Q, we return to Eq. 8-34 with the condition that $Q = \sqrt{\frac{5}{2}}$ or that $2\sigma_1 = \omega_1$; that is,

$$v_C(t) = V_0(1 - e^{-(\omega_1/2)t} \cos \omega_1 t) \tag{8-49}$$

We may determine the time at which the maximum is attained by differentiating this equation and setting the result to zero. The result is

$$\frac{dv_C(t)}{dt} = V_0 \omega_1 e^{-(\omega_1/2)t}(\sin \omega_1 t + \tfrac{1}{2} \cos \omega_1 t) = 0 \tag{8-50}$$

Solution of this equation gives

$$\omega_1 t_{max} = 2.68 \tag{8-51}$$

or

$$t_{max} = \frac{2.68}{\omega_1} \tag{8-52}$$

Then the overshoot is

$$\text{overshoot} = V_0(1 - e^{-1.34} \cos 2.68) - V_0 \tag{8-53}$$

$$= 0.233 V_0 \tag{8-54}$$

The overshoot as a percentage is given by Eq. 8-48 and is 23.3%.

The *settling time* is identified as t_s on Fig. 8-9. It is defined as the time beyond which the step response does not differ from the final value by more than 5%. It is a measure of the amount of time that it takes for the ringing of the response to essentially disappear, or for the response to settle down to the final value. Although it is difficult to calculate analytically, it is easy to measure in the laboratory, and is frequently given as a specification for a circuit or a system.

8-5 SUSTAINED OSCILLATIONS

We next return to the circuit of Fig. 8-6 without the resistor, which then becomes the circuit shown in Fig. 8-10. The circuit thus consists of an initially charged capacitor in series with an unfluxed inductor. In making $R = 0$, we also assume

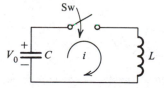

Figure 8-10. Circuit of Fig. 8-7 but without loss (i.e., $R = 0$).

that there is no loss in the inductor, that it represents an ideal element. Although this may not be true with ordinary inductors, it is true of superconducting inductors. With $R = 0$, the characteristic equation, Eq. 8-17, becomes

$$s^2 + \frac{1}{LC} = 0 \tag{8-55}$$

or

$$s_1, s_2 = \pm j\sqrt{\frac{1}{LC}} = \pm j\omega_0 \tag{8-56}$$

From this we confirm the fact that $R = 0$ implies that $\sigma_1 = 0$ in Eq. 8-24, and hence the capacitor voltage is

$$v_C(t) = V_0 \cos \omega_0 t \tag{8-57}$$

Thus, the capacitor voltage is sinusoidal and since $\sigma_1 = 0$, it is not damped.

It is known as a sustained oscillation. According to Eq. 8-57, the oscillations continue for all time in the future.

The variation of voltage or current as $\cos \omega_0 t$ or $\sin \omega_0 t$ is of great importance in our studies, and will be expanded in Chapter 9. Here we observe that the quantity ω_0 is the frequency of oscillation in rad/s. Since there are 2π radians in one cycle of the cosine function, it follows that $\omega_0 = 2\pi f_0$ where f_0 is frequency in cycles/sec or Hertz, abbreviated Hz. Thus

$$f_0 = \frac{\omega_0}{2\pi} \quad \text{Hz} \tag{8-58}$$

Substituting the value of ω_0 from Eq. 8-56, we have

$$f_0 = \frac{1}{2\pi\sqrt{LC}} \tag{8-59}$$

The period of the cosine function is the reciprocal of the frequency f_0 so that

$$T = \frac{1}{f_0} \quad \text{s} \tag{8-60}$$

These quantities are shown in Fig. 8-11.

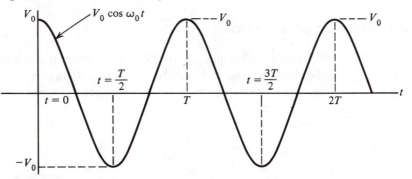

Figure 8-11.   Some characteristics of a sinusoidal waveform.

EXAMPLE 8-3

Suppose that the elements in the circuit are $L = 10$ mH and $C = 1$ μF. Then we may determine ω_0 from Eq. 8-56, f_0 from Eq. 8-59, and T from Eq. 8-60.

Solution

These quantities are

$$\omega_0 = \left(\frac{1}{LC}\right)^{1/2} = (10^{-6} \times 10^{-2})^{-1/2} = 10^4 \text{ rad/s}$$

$$f_0 = \frac{10,000}{2\pi} = 1592 \text{ Hz} \tag{8-61}$$

$$T = \frac{1}{f_0} = 2\pi \times 10^{-4} \text{ s} = 628 \text{ }\mu\text{s}$$

8-6 TRANSIENT VERSUS STEADY-STATE ANALYSIS

It is the presence of capacitors or inductors (or both) in circuits that gives rise to transients; otherwise, in purely resistive circuits the response is instantaneous. Mathematically, the transients result when the circuit equations are differential equations. For the first-and second-order circuits, we were able to determine the response. This is harder to do in more complicated circuits, circuits containing several capacitors and inductors.

The term *transient analysis* is used to identify the complete circuit analysis since the transient component of the complete response is relevant, in addition to the steady-state component. Frequently, only the steady-state component is important because the transient is assumed to vanish relatively fast or the particular circuit application is concerned with just the steady-state response. In such cases, we only need to find the steady-state response and the analysis is termed *steady-state analysis*.

Steady-state behavior of circuits is the key to much of our understanding of circuits. We will, therefore, concentrate on steady-state analysis from now on. We have already seen that the steady-state analysis of circuits with constant excitations (i. e. , dc analysis) is easily carried out. In that case we open-circuit all capacitors and short-circuit all inductors in order to find the desired response. Most of our circuits, however, are not driven by constant sources but by *sinusoidal* sources. For these reasons, in Chapter 9 we turn our attention to *sinusoidal steady-state analysis*.

PROBLEMS

8-1. Suppose that the rise time of the circuit given in the figure is 5.5 μs. Find the response $v_2(t)$ if $i(0) = 2 \times 10^{-3}$ A.

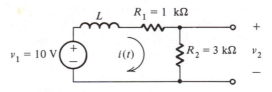

Figure P8-1

8-2. Suppose that the step response of a circuit is of the form

$$v(t) = 1 - e^{-\sigma_1 t} \cos \omega_1 t$$

and the characteristic equation of the circuit is

$$s^2 + \frac{R_1 R_2 C + L}{L R_2 C} s + \frac{R_1 + R_2}{L R_2 C} = 0$$

Find the rise time if $R_1 = 100 \ \Omega$, $R_2 = 1900 \ \Omega$, $C = 19.7 \ \mu$F, and $L = 268$ mH.

8-3. A new definite of rise time has been proposed: the time between $t = 0$ and the time that $v_2 = v_1$ for the first time, as shown by the figure. Let $v_2(t)$ have the form of Eq. 8-49 with the particular choice made that $o_1 = \frac{1}{2}\omega_1$. Determine this rise time.

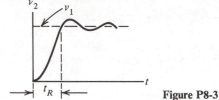

Figure P8-3

8-4. Still another definition of rise time is proposed. It is the time from $t = 0$ until v_2 has reached its maximum value, $t = t_{max}$. Find this rise time if the circuit is adjusted so that $o_1 = \frac{1}{4}\omega_1$.

8-5. Assuming that the response v_2 is given by Eq. 8-49, determine the settling time, t_s, as defined in the text.

8-6. The capacitor voltage for an oscillator is given in Fig. 8-11. Using the same coordinates as Fig. 8-11, sketch the capacitor current, $i_C(t)$.

8-7. In the circuit shown in the figure, the circuit is not energized until the switch is closed at $t = 0$. Find $v_2(t)$ using the numerical values given for element sizes.

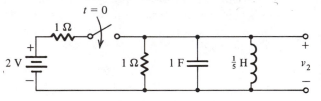

Figure P8-7

8-8. For the circuit given in the figure: (a) Write the differential equation in terms of $i_L(t)$ for $t > 0$. (b) Write the characteristic equation, and find the roots and write the form of the solution for $i_L(t)$. (c) Find the steady-state inductor current and capacitor voltage. (d) Find the time derivative of $i_L(t)$ at $t = 0+$, assuming that the circuit is in the zero state when the switch is closed (all initial conditions are equal to zero).

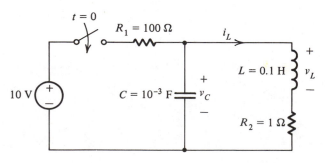

Figure P8-8

8-9. With the switch closed, the relationship between the response $v_0(t)$ and the 10-V input is described by the differential equation

$$LC \frac{d^2 v_0}{dt^2} + \left(\frac{L}{R_2} + R_1 C \right) \frac{dv_0}{dt} + \frac{R_1 + R_2}{R_2} v_0 = 10$$

Assume that the circuit is in the zero state when the switch is closed. If $R_1 = R_2 = 10 \, \Omega$, $L = 0.5$ H, and $C = 0.005$ F, find

(a) The steady-state responses I_{ss} and V_{oss}.

(b) The initial conditions $\frac{di}{dt}(0+)$ and $\frac{dv_0}{dt}(0+)$.

(c) Find the voltage $v_0(t)$ for $t > 0$.

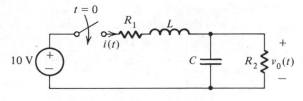

Figure P8-9

9

SINUSOIDS AND PHASORS

We now begin the study of steady-state analysis. Hereafter, it must always be understood that all circuits are in steady-state conditions, that is, the transients have all vanished. In other words, the circuits with their excitations have been in existence for an infinitely long time and we need concern ourselves only with the steady-state response. From here on, the term "response" will refer to the steady-state response.

We also concentrate on excitations that are specific functions of time. A description of the manner in which a signal v changes with time is called the *waveform*. If a voltage of a given waveform is applied to a circuit, the waveform at the various nodes of the circuit will generally differ from the waveform applied. We may regard circuits as "processors" of waveforms, and circuit analysis as the determination of these waveforms. Similarly, design is the devising of a circuit that for a given input v_1 will produce a required output v_2. *Signal processing* and *waveform processing* mean the same thing, of course.

9-1 THE SINUSOID

The most important signal or waveform in electrical engineering is the *sine wave*, or a waveform that has a *sinusoidal variation in value with time*. Suppose that we connect a recording instrument to the nearest wallplug to record the voltage. The instrument will record the waveform shown in Fig. 9-1. How would you describe this waveform? It has a time variation such as $\sin \theta$ or $\cos \theta$, which is studied in trigonometry. The *amplitude* or *magnitude*, V_m, is constant. The waveform repeats or goes through *cycles*. The time that elapses in going through

214

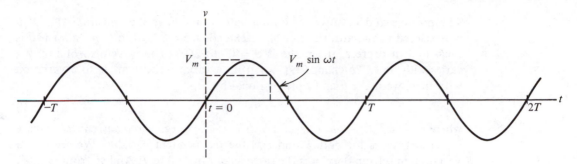

Figure 9-1. Sine function, defined for all positive and negative time. The amplitude is V_m and the period is T.

a cycle is known as the *period*, T. If we measure the period, we can calculate the number of *cycles per second* or the *frequency*, $f = 1/T$. For reasons now difficult to remember, the unit of cycles per second for frequency was replaced by *hertz* (abbreviated Hz) in the 1960s, and the change is now complete. If we are required to write an equation for the waveform of Fig. 9-1, we must do one additional thing. We must tick off a place that we decide is the reference or the place where $t = 0$, the origin.

Next, let us take one cycle of the sine wave comparable to that of Fig. 9-1 and examine it in more detail. This is displayed in Fig. 9-2 and is a plot of the equation

$$v = V_m \sin \theta \qquad (0 \leq \theta \leq 2\pi) \qquad (9\text{-}1)$$

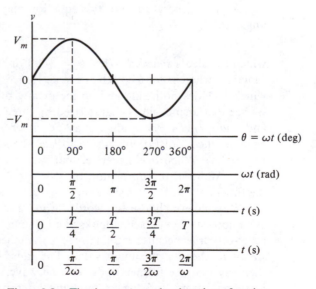

Figure 9-2. The sine wave may be plotted as a function of θ, ωt, or t.

The positive and negative maximum values of v have the *magnitude* V_m. If θ is measured in radians, the sine wave extends from $\theta = 0$ to $\theta = 2\pi$, or if θ is measured in degrees, the range is $\theta = 0°$ to $360°$. The waveform of Fig. 9-1 varies with time. To change Eq. 9-1 from a variation with angle to a variation with time, we make θ and t proportional by

$$\theta = \omega t \tag{9-2}$$

where $\omega = d\theta/dt$ is the rate at which θ is changing in radians/second. Figure 9-2 shows scales for both θ and t as the independent variable. We see that a variation in θ from 0 to 2π is the same as in t from 0 to T. And so from Eq. 9-2,

$$2\pi = \omega T \quad \text{or} \quad T = \frac{2\pi}{\omega} \tag{9-3}$$

Now frequency was found from the waveform of Fig. 9-1 in terms of the period as

$$f = \frac{1}{T} \tag{9-4}$$

Combining these two equations gives us

$$\omega = 2\pi f \tag{9-5}$$

Both f and ω are known as frequency, but ω *has the dimensions of rad/s, while f is measured in Hz*. Thus, there are a number of equivalent ways of writing the time-variable equivalent of Eq. 9-1,

$$v = V_m \sin \omega t = V_m \sin 2\pi ft = V_m \sin \frac{2\pi}{T}t \tag{9-6}$$

A more general form of this equation may be written by including a phase angle, θ,

$$v = V_m \sin (\omega t + \theta) \tag{9-7}$$

which is called a *sinusoid*. When $\theta = \pm 90°$ in this equation, v becomes a cosine function, whereas for $\theta = 0$ it is a sine. Hence, sinusoid is a more general term which includes both sine and cosine or any other θ in Eq. 9-7. From Eq. 9-7 we see that a sinusoid is charaterized by three quantities: *frequency, magnitude* and *phase angle*. Phase angle may be either positive or negative. When θ is positive, then v of Eq. 9-7 is said to *lead* the sine wave of Eq. 9-6 as shown in Fig. 9-3(a). When θ is negative, that waveform *lags* the sine wave as in Fig. 9-3(b). Addition or subtraction of $360°$ (or 2π radians) to θ does not change the function v in Eq. 9-7 since $\sin (x \pm 2\pi) = \sin x$.

Why did we choose the words "lead" and "lag"? If we regard Fig. 9-3(a) as a photo of a horse race, it looks like $V_m \sin \omega t$ is ahead of $V_m \sin (\omega t + \theta)$. The reverse is true because $V_m \sin (\omega t + \theta)$ reaches its maximum value *before* $V_m \sin \omega t$ reaches its maximum value. The same is true for all other values. We then say that the former leads the sine wave. It is helpful to think of lead as implying before and lag as implying after in comparing one sinusoid with another.

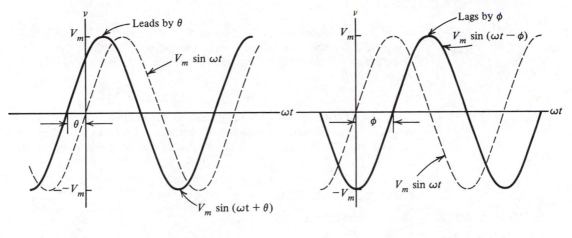

(a)   (b)

Figure 9-3. The reference waveform is showed dashed and another sinusoid is shown which leads in (a) and lags in (b).

EXAMPLE 9-1

In the United States, the frequency of our power system is usually 60 Hz, whereas in many other countries it is 50 Hz. Find the values of ω and T that correspond to these f's.

Solution

From Eq. 9-5, the values of ω are 377 and 314, and from Eq. 9-4, the corresponding periods are 0.016667 and 0.02 s.

EXAMPLE 9-2

We are required to relate the two waveforms

$$v_1 = 100 \sin (377t - 45°) \tag{9-8}$$

and

$$v_2 = 50 \cos (377t + 60°) \tag{9-9}$$

Solution

For comparison, we must change v_2 from a cosine function to a sine function. This is,

$$v_2 = 50 \sin (377t + 90° + 60°) = 50 \sin (377t + 150°) \tag{9-10}$$

Comparing this and v_1, we see that v_2 leads v_1 by 195°. Or we can subtract 360° from the angle of Eq. 9-10 and write it as

$$v_2 = 50 \sin (377t - 210°) \tag{9-11}$$

and from this we can say that v_2 lags v_1 by 165°.

EXERCISES

9-1.1. In making a measurement, it is decided that the reference voltage is $10 \sin t = v_1(t)$. The voltage $v_2(t)$ reaches a peak value which is 1.5 times the reference 2.5 ms before the reference and has the same value 20 ms later. Write the equation for $v_2(t)$.

$$\textit{Ans. } v_2(t) = 15 \sin (314t + 45°) \text{ V}$$

9-1.2. Using the same experimental arrangement as for Ex. 9-1.1, it is found that $v_3(t)$ has half the value of the reference and reaches its peak value 0.25 ms after the reference. The voltage has the same value 2.5 ms later. Write the equation for $v_3(t)$.

$$\textit{Ans. } v_3(t) = 5 \sin (2513t - 36°) \text{ V}$$

9-2 THE ADDITION OF SINUSOIDS OF THE SAME FREQUENCY

Suppose that it is required that we add two voltages which are *sinusoids of same frequency:*

$$v_1 = 5 \sin 5t \tag{9-12}$$

and

$$v_2 = 5\sqrt{2} \sin (5t + 45°) \tag{9-13}$$

to obtain a voltage $v_3 = v_1 + v_2$. We first expand Eq. 9-13 by expanding the sine of a sum of angles, using a relationship from trigonometry,

$$\sin (x + y) = \sin x \cos y + \cos x \sin y \tag{9-14}$$

Then v_2 becomes

$$v_2 = 5 \sin 5t + 5 \cos 5t \tag{9-15}$$

Summing Eqs. 9-12 and 9-15, we obtain v_3 as

$$v_3 = 10 \sin 5t + 5 \cos 5t \tag{9-16}$$

At this point, we define two quantities, C and θ, such that

$$C \cos \theta = 10 \quad \text{and} \quad C \sin \theta = 5 \tag{9-17}$$

Then Eq. 9-16 becomes

$$v_3 = C(\sin 5t \cos \theta + \cos 5t \sin \theta) \tag{9-18}$$

Using Eq. 9-14 once more,

$$v_3 = C \sin (5t + \theta) \tag{9-19}$$

where C and θ are yet to be determined. To do so, we square the two parts of Eq. 9-17 and add them to obtain

$$C^2(\cos^2 \theta + \sin^2 \theta) = 10^2 + 5^2 \tag{9-20}$$

Since $\cos^2 \theta + \sin^2 \theta = 1$, C is $\sqrt{125} = 11.18$. Next, we divide the two parts of Eq. 9-17, so that

$$\frac{C \sin \theta}{C \cos \theta} = \tan \theta = \frac{5}{10} \text{ or } \theta = 26.6° \tag{9-21}$$

Then Eq. 9-19 is

$$v_3 = 11.18 \sin (5t + 26.6°) \tag{9-22}$$

In this particular example, we see that the sum of two sinusoidal functions of the *same frequency* is another sinusoidal function of that frequency. This has been a specific example, but the method may be generalized. If we are given any number of terms of the form

$$v_i = V_{mi} \sin (\omega t + \theta_i) \tag{9-23}$$

then the sum of n of these,

$$v_t = v_1 + v_2 + \ldots + v_n \tag{9-24}$$

always gives v_t with the same frequency and of the general form

$$v_t = V_{mt} \sin (\omega t + \theta_t) \tag{9-25}$$

The computation of V_{mt} and θ_t is tedious, and a better method will be presented shortly, but it may be carried out step by step by adding two terms at a time in the manner of the example, or by adding all terms at once. The result may be expressed in terms of a sine function or a cosine function. Care must be exercised in determining θ as in Eq. 9-21 to be sure that the angle is in the correct quadrant; in other words, we must distinguish between $\frac{5}{10}$ and $-\frac{5}{10}$, since the first implies 26.6° and the second 206.6°.

The result we have just obtained is illustrated in Fig. 9-4, where the two solid-line sinusoids are added to give the dashed-line sinusoid.

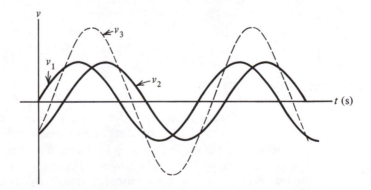

Figure 9-4. Addition of two sinusoids to give a third sinusoid (shown dashed).

EXAMPLE 9-3

Show that

$$A \cos \omega t + B \sin \omega t = \sqrt{A^2 + B^2} \cos (\omega t - \phi) \qquad (9\text{-}26)$$

where $\tan \phi = B/A$.

Solution

We follow a procedure similar to the above, but use instead the trigonometric relationship

$$\cos (x - y) = \cos x \cos y + \sin x \sin y \qquad (9\text{-}27)$$

We let

$$C \cos \phi = A \quad \text{and} \quad C \sin \phi = B \qquad (9\text{-}28)$$

This results directly in the required result of Eq. 9-26. It may also be expressed in the form

$$\sqrt{A^2 + B^2} \sin (\omega t + 90° - \phi) \qquad (9\text{-}29)$$

In general, we make use of trigonometric relationships for $\sin (x \pm y)$ and $\cos (x \pm y)$, and also between the sine and cosine functions to manipulate these equations.

EXERCISES

9-2.1. Express the following summation in the general form $A \sin (\omega t + \theta)$:

$$i(t) = 3 \sin t + 4 \cos t$$

Ans. $i(t) = 5 \sin (t + 0.927)$

9-2.2. Repeat Exercise 9-2.1 for the function

$$i(t) = 4 \sin t + 3 \cos t$$

Ans. $i(t) = 5 \sin (t + 0.6435)$

9-2.3. Repeat Exercise 9-2.1 for the function

$$i(t) = 15 \cos \left(3t + \frac{\pi}{4}\right) + 7 \sin \left(3t + \frac{\pi}{8}\right)$$

Ans. $i(t) = 13.92 \sin (3t + 1.873)$

9-3 SINUSOIDAL STEADY STATE

Let us now examine the *form* of the steady-state response in a circuit driven by a sinusoidal source. In all our discussion we assume that the circuit has been in existence for an indefinitely long time and that the transients have vanished. That is, the circuit has reached *steady state* and we speak of our analysis as the *sinusoidal steady state*. A sinusoid alternates from positive to negative values and the term ac (*alternating-current*) *analysis* is also used in this case. This is in contrast to dc or *direct-current analysis* when the excitation is *constant*. These terms, dc and ac, always imply steady-state conditions.

For our sinusoidal steady-state analysis, we write KVL or KCL (for the steady-state response). In the KVL or KCL equations, some known sinusoid must equal the sum of certain unknown branch voltages or currents. This sum can be a sinusoid only if the component branch voltages or currents are sinusoids of the same frequency. Alternatively, if some terms in an equation are sinusoids of a given frequency, they all must be sinusoids of the same frequency for their algebraic sum to be zero.

If a branch voltage for any circuit element is a sinusoid of a given frequency, then so is the branch current, and vice versa. This is clear for a resistor, since the current and voltage are related by a constant. For the capacitor and the inductor, the branch constraints involve derivatives or integrals. The derivative of a sinusoid is a sinusoid of the same frequency, and same is true for the integral. Thus, any voltage or current, anywhere in a circuit, under steady-state conditions and driven by a sinusoidal source, must be a sinusoid of the same frequency. The same conclusion is valid, by superposition, when several sinusoidal sources of the same frequency are present in the circuit. For op amps or controlled sources, if either the controlling or the controlled signal is a sinusoid, so is the other, and with the same frequency since the two are related by a constant. Our result is thus quite generally true for all circuits. This result is very important: *In the sinusoidal steady state, all voltages and currents are sinusoids of the same frequency as the excitation.*

Sinusoidal steady-state analysis thus becomes quite straightforward. We assume, for our unknowns, sinusoids of the given frequency but unknown magnitudes and phases, substitute them in the equation, and solve for the magnitudes and phases. An example will illustrate the procedure.

Consider the simple circuit of Fig. 9-5. We wish to determine the current

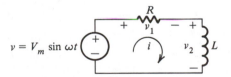

Figure 9-5.   *RL* circuit driven by a sinusoidal voltage source.

i under steady-state conditions. Since the source is sinusoidal, we know that i is a sinusoid of the same frequency, say

$$i = I_m \sin (\omega t + \theta) \tag{9-30}$$

where I_m and θ are unknown. Then, by KVL,

$$v = v_1 + v_2 \tag{9-31}$$

$$= Ri + L \frac{di}{dt} \tag{9-32}$$

or

$$V_m \sin \omega t = RI_m \sin (\omega t + \theta) + \omega L I_m \cos (\omega t + \theta) \tag{9-33}$$

This equation must now be solved for I_m and θ. We first sum the terms on the right and equate the magnitudes and phases on the two sides. For our circuit, let us assume that $V_m = 10$, $\omega = 100$, $R = 10$, and $L = 0.1$. Then the Eq. 9-33 becomes

$$10 \sin (100t + 0°) = 10 I_m \sin (100t + \theta)$$
$$+ 10 I_m \cos (100t + \theta) \tag{9-34}$$
$$= 10\sqrt{2}\, I_m \sin (100t + \theta + 45°) \tag{9-35}$$

Comparing the two sides,

$$I_m = \frac{1}{\sqrt{2}} = 0.707 \tag{9-36}$$

and

$$\theta = -45° \tag{9-37}$$

Then, by Eq. 9-30,

$$i = 0.707 \sin (100t + 45°) \tag{9-38}$$

The procedure is general and may be applied to the sinusoidal steady-state analysis of any circuit. The only trouble is that the trigonometric summation of sinusoids gets very laborious for any but the most trivial cases. Derivatives and integrals of sinusoids are also troublesome. Special techniques, developed to facilitate the sinusoidal steady-state analysis, are discussed next. These techniques are universally used in circuit analysis and represent some of the most fundamental concepts.

EXERCISES

9-3.1. Repeat the analysis of the circuit of Fig. 9-5 to find $i(t)$ if L is changed from the value given to $L = 0.2$ H.

Ans. $i(t) = 0.447 \sin (100t - 63.43°)$

9-3.2. In the circuit of Fig. 9-5, the inductor L is replaced by a capacitor of value $C = 0.001$ F. Find $i(t)$.

Ans. $0.707 \sin (100t + 45°)$

9-4 STEINMETZ'S ANALOG: THE PHASOR

Steinmetz introduced and popularized another way of adding voltages and currents of the same frequency. In his method, the following two quantities are considered to be analogs:

$$v(t) = V_m \sin(\omega t + \theta) \quad \text{and} \quad V = V_m \underline{/\theta} \tag{9-39}$$

In each form, V_m is known as the magnitude and θ as the phase. Since we have learned from Section 9-3 that frequency will be common throughout the circiut, it is not explicitly included in $V_m \underline{/\theta}$ but *it is implied*. The quantity $V_m \underline{/\theta}$ is known as a *phasor*—in this case a voltage phasor. It is shown graphically, as

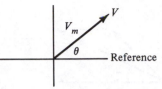

Figure 9-6. Phasor of magnitude V and phase θ.

in Fig. 9-6, as *an arrow of magnitude V_m at an angle θ with respect to a reference, normally taken as $\theta = 0°$.* The time function is always designated by the lower case letter and the phasor analog by the capital letter, as shown in the equation.

The manner in which Steinmetz's analog is used will be illustrated in terms of Fig. 9-7. We start with the problem of summing voltages, each of which is

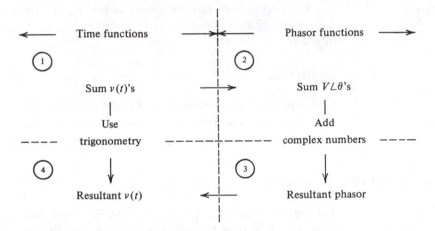

Figure 9-7. Steps in applying the Steinmetz algorithm.

a sinusoid characterized by a magnitude and a phase. This is shown in position 1 of the figure. In previous sections we made use of trigonometric relationships in obtaining the answer. This is implied by going from position 1 to position 4. As an alternative method, we convert all v's to equivalent phasor form, as in Eq. 9-39, going from position 1 to position 2. The summation of phasors is found by using the rules discussed below to give a resultant phasor, shown in position 3. We go from position 3 to position 4 by using the analog of Eq. 9-39 in reverse—from phasor back to time function. More steps but, as we will show, easier computation.

To illustrate, let us return to the problem of adding v_1 and v_2 at the beginning of Section 9-2. From Eq. 9-12, the phasor $V_1 = 5\underline{/0°}$ and from Eq. 9-13, the phasor $V_2 = 5\sqrt{2}\,\underline{/45°}$. The phasor sum is

$$V_3 = 5\underline{/0°} + 5\sqrt{2}\,\underline{/45°} \tag{9-40}$$

These phasors V_1 and V_2 are shown in Fig. 9-8. They are summed by *phasor addition*, perhaps more familiar as *vector addition*, to give V_3, which has a

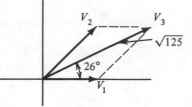

Figure 9-8. Phasor addition of V_1 and V_2 to give V_3.

magnitude of $\sqrt{125}$ and a phase angle of 26.6°. Then using the analog equation in reverse, we obtain

$$v_3 = \sqrt{125} \sin (5t + 26.6°) \qquad (9\text{-}41)$$

which is Eq. 9-22.

We digress from our discussion at this point to make sure that we fully understand phasor addition, which turns out to be the addition of *complex numbers*.

EXERCISES

9-4.1. Use phasors to determine the solution to Exercise 9-2.1.

9-4.2. Use phasors to determine the solution to Exercise 9-2.2.

9-4.3. Use phasors to determine the solution to Exercise 9-2.3.

9-5   COMPLEX NUMBERS

The plane in which $V_m\underline{/\theta}$ was shown in Fig. 9-6 is called the *complex number plane* and V is shown there in *polar form*. To discuss complex numbers, we switch notation at this point to the complex number

$$z = |z|\underline{/\theta} \qquad (9\text{-}42)$$

and consider its representation in Fig. 9-9(a).

As mentioned above, this is called the polar form of representation. The complex number may also be expressed in rectangular form in terms of its projection on the two axes,

$$z = x + jy \qquad (9\text{-}43)$$

where x and y are the projections on the two axes, and $j = \sqrt{-1}$, an imaginary number. Manipulations of complex numbers are facilitated by the use of the imaginary number j. In Eq. 9-43, the complex number has two components, a real number, x, and an imaginary one, jy. The real numbers x and y are then expressed as

$$x = \operatorname{Re} z; \qquad y = \operatorname{Im} z \qquad (9\text{-}44)$$

where *Re* means the *real part of* and *Im* means the *imaginary part of*. From the

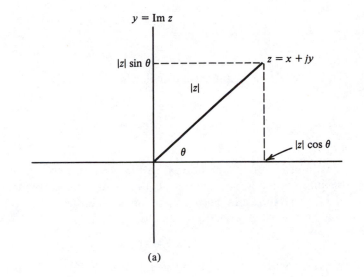

(a)

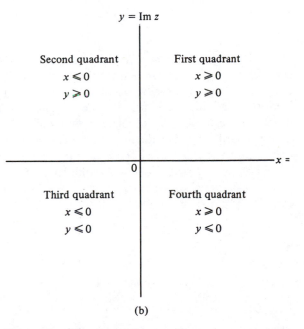

Second quadrant
$x \leqslant 0$
$y \geqslant 0$

First quadrant
$x \geqslant 0$
$y \geqslant 0$

Third quadrant
$x \leqslant 0$
$y \leqslant 0$

Fourth quadrant
$x \geqslant 0$
$y \leqslant 0$

(b)

Figure 9-9. Geometry of the complex number $z = x + jy$.

figure we see that to convert from polar to rectangular form, we use the equations

$$x = |z| \cos \theta \tag{9-45}$$

and

$$y = |z| \sin \theta \tag{9-46}$$

To convert from rectangular to polar form,

$$|z| = \sqrt{x^2 + y^2} \tag{9-47}$$

and

$$\theta = \tan^{-1} \frac{y}{x} \tag{9-48}$$

This information is summarized in Fig. 9-9. Part (b) of the figure summarizes the information about the four quadrants of complex plane.

The relationship between the polar and the rectangular representations may be made explicit through the use of *Euler's identity*,

$$e^{j\theta} = \cos \theta + j \sin \theta \tag{9-49}$$

We can rewrite Eq. 9-43 using Eqs. 9-45, 9-46, and 9-49 to give

$$z = |z| \cos \theta + j |z| \sin \theta \tag{9-50}$$

$$z = |z|(\cos \theta + j \sin \theta) \tag{9-51}$$

$$z = |z| e^{j\theta} \tag{9-52}$$

which is called the *exponential representation* of the complex number. The exponential representation is written in terms of magnitude and phase just like the polar form but is easily converted to the rectangular form by use of Euler's identity. We will, therefore, use the exponential and the rectangular forms whenever we need to manipulate complex numbers. The polar form will be used primarily as a shorthand notation for the exponential form and the corresponding designations will be used interchangeably.

The rules for operations with complex numbers make use of the fact that $j^2 = -1$. Otherwise, the rules are the same as for real numbers. Consider two complex numbers in the two forms

$$z_1 = x_1 + jy_1 \quad \text{and} \quad z_2 = x_2 + jy_2 \tag{9-53}$$

or

$$z_1 = |z_1| \underline{/\theta_1} \quad \text{and} \quad z_2 = z_2 \underline{/\theta_2} \tag{9-54}$$

Addition or subtraction is accomplished as follows:

$$z_1 + z_2 = (x_1 + jy_1) + (x_2 + jy_2) \tag{9-55}$$

$$= (x_1 + x_2) + j(y_1 + y_2) \tag{9-56}$$

For multiplication using rectangular representation, we multiply the terms normally and replace j^2 by -1 to give

$$z_1 z_2 = (x_1 + jy_1)(x_2 + jy_2) \tag{9-57}$$
$$= (x_1 x_2 - y_1 y_2) + j(x_1 y_2 + x_2 y_1) \tag{9-58}$$

For multiplication using polar representation, we use the exponentials

$$|z_1| e^{j\theta_1} \cdot |z_2| e^{j\theta_2} = |z_1||z_2| e^{j(\theta_1 + \theta_2)} \tag{9-59}$$
$$= |z_1||z_2| \underline{/\theta_1 + \theta_2} \tag{9-60}$$

Of the two, it is usually easier to multiply two complex numbers in polar or exponential form. It is important, however, to become familiar with both.

Finally, for division, we first define the conjugate of a complex number

$$z^* = x - jy \qquad \text{if } z = x + jy \tag{9-61}$$

and note from Eq. 9-60 that

$$zz^* = |z|^2 \underline{/0°} = |z|^2 \tag{9-62}$$

To obtain the quotient

$$\frac{z_1}{z_2} = \frac{x_1 + jy_1}{x_2 + jy_2} \tag{9-63}$$

as a complex number in the standard form, we multiply the numerator and the denominator by *the conjugate of the denominator*. Then the denominator is real, and we group the real and the imaginary terms to give

$$\frac{z_1}{z_2} = \frac{(x_1 x_2 + y_1 y_2) + j(x_2 y_1 - x_1 y_2)}{x_2^2 + y_2^2} \tag{9-64}$$

The real and imaginary parts of the numerator are easily divided by the real number in the denominator to give the result in standard form. Division in polar form is accomplished as follows:

$$\frac{z_1}{z_2} = \frac{|z_1| e^{j\theta_1}}{|z_2| e^{j\theta_2}} = \frac{|z_1|}{|z_2|} e^{j(\theta_1 - \theta_2)} \tag{9-65}$$

$$= \frac{|z_1|}{|z_2|} \underline{/\theta_1 - \theta_2} \tag{9-66}$$

which is usually easier to use. Again, both forms are quite important and must be learned thoroughly. These results are summarized in Table 9-1. Phasor manipulations are easily accomplished using complex numbers, and therein lies the importance of phasor representation.

Returning to the example of Eq. 9-40, we see that the addition is accomplished in the rectangular form as follows:

$$V_3 = (5 + j0) + (5 + j5) = 10 + j5 \tag{9-67}$$

or

$$V_3 = (10^2 + 5^2)^{1/2} \underline{/\tan^{-1} \tfrac{5}{10}} = \sqrt{125} \underline{/26.6°} \tag{9-68}$$

as before. This is shown in Fig. 9-8.

Table 9-1 Properties of Complex Numbers

Rectangular Form	*Polar/Exponential Form*
$z = x + jy$	$z = \lvert z \rvert \underline{/\theta}; \quad z = \lvert z \rvert e^{j\theta}$
	$= \lvert z \rvert \cos\theta + j\lvert z \rvert \sin\theta$

1. Relationship to other form:

$x = \lvert z \rvert \cos\theta, \quad y = \lvert z \rvert \sin\theta$ $\qquad \lvert z \rvert = (x^2 + y^2)^{1/2}, \quad \theta = \tan^{-1}\dfrac{y}{x}$

2. Complex conjugate

$z^* = x - jy$ $\qquad\qquad z^* = \lvert z \rvert e^{-j\theta} = \lvert z \rvert \underline{/-\theta}$

$zz^* = (x^2 + y^2) = \lvert z \rvert^2$ $\qquad zz^* = \lvert z \rvert e^{j\theta} \cdot \lvert z \rvert e^{-j\theta} = \lvert z \rvert^2$

3. Addition

$z_1 + z_2 = (x_1 + x_2) + j(y_1 + y_2)$ $\qquad z_1 + z_2 = \lvert z_1 \rvert \cos\theta_1 + \lvert z_2 \rvert \cos\theta_2$

$\qquad\qquad + j(\lvert z_1 \rvert \sin\theta_1 + \lvert z_2 \rvert \sin\theta_2)$

4. Multiplication

$z_1 z_2 = (x_1 x_2 - y_1 y_2) + j(x_1 y_2 + x_2 y_1)$ $\qquad z_1 z_2 = \lvert z_1 \rvert \lvert z_2 \rvert \underline{/\theta_1 + \theta_2}$

5. Division

$\dfrac{z_1}{z_2} = \dfrac{x_1 x_2 + y_1 y_2 + j(x_2 y_1 - x_1 y_2)}{\lvert z_2 \rvert^2}$ $\qquad \dfrac{z_1}{z_2} = \dfrac{\lvert z_1 \rvert}{\lvert z_2 \rvert}\underline{/\theta_1 - \theta_2}$

EXAMPLE 9-4

Multiply $(1 + j2)$ by $(2 - j2)$.

Solution

In rectangular form

$$\text{product} = 2 + 4 + j4 - j2 = 6 + j2 \qquad (9\text{-}69)$$

and in polar form

$$\text{product} = (5)^{1/2}\underline{/63.4^\circ} \times 2(2)^{1/2}\underline{/-45^\circ} = 2(10)^{1/2}\underline{/18.43^\circ} \qquad (9\text{-}70)$$

See Fig. 9-10.

EXAMPLE 9-5

Divide $z_1 = 1 + j2$ by $z_2 = 2 - j2$.

Solution

In rectangular form

$$\frac{z_1}{z_2} = \frac{1 + j2}{2 - j2}\frac{2 + j2}{2 + j2} = \frac{-2 + j6}{8} = \frac{-1}{4} + j\frac{3}{4} \qquad (9\text{-}71)$$

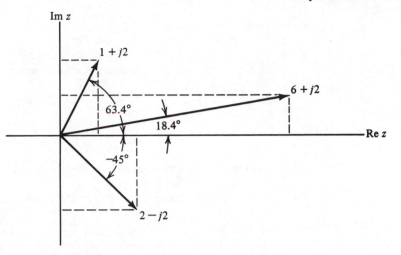

Figure 9-10.   Phasor multiplication of Example 9-4.

and in polar form

$$\frac{z_1}{z_2} = \frac{(5)^{1/2} \underline{/63.4°}}{2(2)^{1/2} \underline{/-45°}} = \frac{(10)^{1/2}}{4} \underline{/108.4°} \tag{9-72}$$

These quantities are shown in Fig. 9-11.

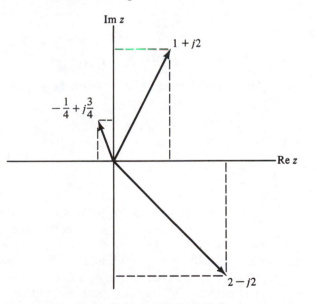

Figure 9-11.   Phasor division of Example 8-5.

EXERCISES

9-5.1. Let $C_1 = 6.5 + j2$, $C_2 = 8 + j13$, $C_3 = 12/\underline{105°}$, and $C_4 = 6/\underline{30°}$. Evaluate the following expressions and state the result in polar form.
(a) $C_1(C_3 + C_4)$
(b) $(C_2 + C_3)/C_4$
(c) $C_1 C_2(C_1 + C_3)$

> **Ans.** (a) $100.22/\underline{98.95°}$; (b) $4.179/\underline{48.74°}$; (c) $1453.52/\underline{151.47°}$

9-5.2. Verify the following identities involving the conjugate of complex numbers. (In each case, let $C = a + jb$.)
(a) If $C = |C| e^{j\theta}$, then $C^* = |C| e^{-j\theta}$.
(b) $C + C^* = 2a$.
(c) $C - C^* = j2b$.
(d) $C^* C = CC^* = a^2 + b^2$.
(e) $(C_1 C_2)^* = C_1^* C_2^*$.

9-6 SINUSOIDS AND EXPONENTIALS

We have seen that Euler's identity in Eq. 9-49 relates the complex exponential to a complex sum of trigonometric quantities. The identity could be applied to functions of time as well and we can represent a complex exponential function as a complex sum of sinusoids,

$$e^{j\omega t} = \cos \omega t + j \sin \omega t \tag{9-73}$$

We can also incorporate the magnitude and phase of a sinusoid as

$$V_m e^{j(\omega t + \theta)} = V_m \cos (\omega t + \theta) + jV_m \sin (\omega t + \theta) \tag{9-74}$$

The sinusoid in Eq. 9-39 may now be written as

$$V_m \sin (\omega t + \theta) = \text{Im} \, [V_m e^{j(\omega t + \theta)}] \tag{9-75}$$

$$= \text{Im} \, [V_m e^{j\theta} \cdot e^{j\omega t}] \tag{9-76}$$

$$= \text{Im} \, [V e^{j\omega t}] \tag{9-77}$$

where

$$V = V_m e^{j\theta} = |V| e^{j\theta} = |V| /\underline{\theta} \tag{9-78}$$

The sinusoid is thus represented in terms of the corresponding phasor V and the complex exponential $e^{j\omega t}$. Equations 9-77 and 9-78 are the basis for the phasor representation of Section 9-4. It is generally easier to work with the exponential than with the sinusoid, and Eq. 9-77 represents the key in switching between the two. Purely for manipulations with sinusoids represented as complex exponentials, *the imaginary part designation Im may always be dispensed with and, unless derivatives or integrals are present, the exponential $e^{j\omega t}$ may also be dispensed with.* We illustrate these statements below.

Suppose that we wish to sum two sinusoids of the same frequency:

$$v_1 = |V_1| \sin(\omega t + \theta_1) = \text{Im}\,[V_1 e^{j\omega t}] \tag{9-79}$$

and

$$v_2 = |V_2| \sin(\omega t + \theta_2) = \text{Im}\,[V_2 e^{j\omega t}] \tag{9-80}$$

to yield the sum which must also be a sinusoid of the same frequency. We may write the unknown sum as

$$v = |V| \sin(\omega t + \phi) = \text{Im}\,[V e^{j\omega t}] \tag{9-81}$$

where the magnitude $|V|$ and phase ϕ of the phasor V are to be determined. Now

$$v_1 + v_2 = \text{Im}\,[V_1 e^{j\omega t}] + \text{Im}\,[V_2 e^{j\omega t}] \tag{9-82}$$

$$= \text{Im}\,[V_1 e^{j\omega t} + V_2 e^{j\omega t}] \tag{9-83}$$

The sum in Eq. 9-83 should equal the function v given in Eq. 9-81. That is,

$$\text{Im}\,[V e^{j\omega t}] = \text{Im}\,[V_1 e^{j\omega t} + V_2 e^{j\omega t}] \tag{9-84}$$

The phasors V_1 and V_2 are known in Eq. 9-84, and it is possible to determine the unknown phasor V. On the other hand, if we set the complex quantities in the parentheses on the two sides of the equation equal, we will also satisfy Eq. 9-84. Thus, it is sufficient to set

$$V e^{j\omega t} = V_1 e^{j\omega t} + V_2 e^{j\omega t} \tag{9-85}$$

in order to satisfy Eq. 9-84.

This is an important step and should be clearly understood. We have, in effect, replaced the *real sinusoidal functions* by *complex exponential functions* in the summation. That is, for ease in manipulation, we are *transforming* the real functions into complex functions,

$$v_1 \longrightarrow V_1 e^{j\omega t} \tag{9-86}$$

$$v_2 \longrightarrow V_2 e^{j\omega t} \tag{9-87}$$

Thus, the transformation is performed by dropping the Im designation of Eqs. 9-79 and 9-80. The summation is then performed as in Eq. 9-85, Finally, the sum in Eq. 9-85 must be *transformed back* to the real function v by putting the Im designation back as in Eq. 9-81 to obtain

$$v \longleftarrow V e^{j\omega t} \tag{9-88}$$

A further simplification is now possible in Eq. 9-85 by canceling out the common exponential. Then

$$V = V_1 + V_2 \tag{9-89}$$

The determination of the unknown phasor V is now direct and involves only the *sum of phasors*. This is the real foundation for the discussion in Section 9-4 on the phasor analog.

We thus have a simple way to *sum* sinusoids of the same frequency. Both the Im designation and $e^{j\omega t}$ are dropped and the sinusoids replaced by the corresponding phasors and summed together. The sum sinusoid, if needed, is obtained by putting $e^{j\omega t}$ and the Im designation back as in Eq. 9-81.

EXAMPLE 9-6

Find the sum of

$$v_1 = 10 \sin (300t + 90°)$$
$$v_2 = 10 \sin (300t + 0°)$$

Solution

By Eq. 9-89, we sum the corresponding phasors

$$V_1 = 10e^{j90°} = 0 + j10 \tag{9-90}$$

and

$$V_2 = 10e^{j0°} = 10 + j0 \tag{9-91}$$

to obtain the sum phasor

$$V = 0 + j10 + 10 + j0 = 10 + j10 \tag{9-92}$$
$$= 14.14e^{j45°} \tag{9-93}$$

The time function is then obtained by Eq. 9-81:

$$v(t) = 14.14 \sin (300t + 45°) \tag{9-94}$$

EXERCISES

9-6.1. Use the method of this section to solve Exercise 9-2.1.

9-6.2. Use the method of this section to solve Exercise 9-2.2.

9-6.3. Use the method of this section to solve Exercise 9-2.3.

PROBLEMS

9-1. Using the appropriate trigonometric identities, express the following summations of sinusoids in the general form $A \sin (\omega t + \theta)$:
  (a) $i(t) = 3 \sin t + 4 \cos (t + 30°)$
  (b) $i(t) = 4 \sin t + 3 \cos (t - 60°)$
  (c) $v(t) = 15 \cos (3t - 45°) + 7 \sin (3t + 22.5°)$

9-2. Express the following sums in the form of $A \sin (\omega t + \theta)$:
  (a) $2 \cos 3t + 3 \sin 3t$
  (b) $5 \cos (2t + 60°) + 7 \sin (2t - 90°)$

9-3. Given the phasors $V_1 = -3 - j2$, $V_2 = 2 + j4$, and $V_3 = 3 + j2$ and the following information on four additional phasors: $V_a = -V_1$, $V_b = V_2 + V_a$, $V_c = V_3^*$, and $V_d = V_b + V_c$, determine the magnitude and phase of the seven phasors and sketch their positions in the complex plane. Suppose that each of these phasors represents a sinusoid of frequency $\omega_0 = 75$ rad/s. Find the time function corresponding to each phasor.

9-4. Given the phasor in rectangular coordinates $V_1 = 5\sqrt{3} + j5$. Rotate this phasor by $-75°$ and express the result in rectangular form.

9-5. Repeat Prob. 9-4 for the phasor $5 + j5\sqrt{3}$ rotating through $-15°$ this time. Express the result in rectangular form.

9-6. Determine the polar form of the quantity

$$A = \frac{(8 + j6)(3.6 - j5)}{10\underline{/120°} - (-8.5 - j2)}$$

9-7. The sum of three phasors, A, B, and C, is equal to zero. Phasor A has a length of 5 and a phase angle of $0°$. Phasors B and C have angles of $-105°$ and $150°$, respectively. Compute the magnitude of the phasors B and C and sketch their locations graphically.

9-8. The steady-state voltage across a certain portion of a circuit is

$$v_1(t) = 45 \sin (377t + 20°) + 10 \cos (377t - 10°)$$

due to a source $i_0(t) = \cos 377t$ somewhere in the circuit. What would be the response at the portion of the circuit identified if the source is changed to $3 \sin (377t - 10°)$?

9-9. The signal observed in a circuit is equal to the sum

$$v(t) = 8.5 \sin (1000t - 30°) + 5 \sin (1000t - 90°)$$
$$+ 4 \cos (1000t - 45°) + 10 \sin (1000t + \theta°)$$

For what value of θ is the peak value of $v(t)$ the largest?

9-10. Given the phasor quantities $X = 3 - j5$, $Y = 7 + j3$, $U = 6\underline{/35°}$, and $V = 9\underline{/-75°}$, evaluate the following quantities and express the result in polar form.
(a) $(U + Y)X^*$
(b) X/Y
(c) $(X - Y)(UV)^*$

9-11. Let $C_1 = 6.5 + j1.3$, $C_2 = 8 + j5$, $C_3 = 12\underline{/35°}$, and $C_4 = 6\underline{/60°}$. Evaluate the following quantities and express the result in polar form.
(a) $C_1(C_3 + C_4)$
(b) $(C_2 + C_3)/C_4$
(c) $C_1C_2(C_1 + C_3)$

9-12. Verify the following identities involving the conjugate of the complex number $C = a + jb$, using rectangular and exponential forms.
(a) $C + C^* = 2a$
(b) $C - C^* = j2b$
(c) $CC^* = C^*C = a^2 + b^2$
(d) $(C_1C_2)^* = C_1^*C_2^*$

10

IMPEDANCE AND ADMITTANCE

In Chapter 9 we introduced the concept of the phasor to represent sinusoidal voltages and currents when operating in the steady state. In this chapter we show that elements or combinations of elements can also be characterized by complex numbers. With this, the rules for writing equations in circuits become as simple as they were for purely resistive circuits. We begin by stating the Kirchhoff laws in terms of phasor quantities.

10-1 KIRCHHOFF'S LAWS

We have seen that all voltages and currents in sinusoidal steady-state circuits are sinusoids of the *same frequency*. We have also seen that *sums of sinusoids of the same frequency* can be accomplished by *phasor sums*. We are now in a position to restate Kirchhoff's laws in terms of phasors.

> KVL: The algebraic sum of *phasor* voltages around a closed loop is zero; or the sum of phasor voltage rises equals the sum of phasor voltage drops around a closed loop.
>
> KCL: The algebraic sum of *phasor* currents at a node is zero; or the sum of phasor currents in at a node equals the sum of phasor currents out at that node.

A simple example will illustrate. Consider the circuit in Fig. 10-1, which is assumed to be in steady state. We can write KVL for this circuit in terms

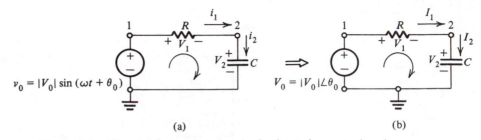

Figure 10-1.   Changing from a time-domain circuit to a frequency-domain circuit.

of the corresponding phasors,

$$V_0 = V_1 + V_2 \tag{10-1}$$

since all the voltages v_0, v_1, and v_2 are sinusoids of the same frequency; that is, by Eq. 9-81,

$$v_0 = \text{Im}\,[V_0 e^{j\omega t}], \qquad V_0 = |V_0|\, e^{j\theta_0} \tag{10-2}$$

$$v_1 = \text{Im}\,[V_1 e^{j\omega t}], \qquad V_1 = |V_1|\, e^{j\theta_1} \tag{10-3}$$

$$v_2 = \text{Im}\,[V_2 e^{j\omega t}], \qquad V_2 = |V_2|\, e^{j\theta_2} \tag{10-4}$$

The usual KVL for these voltages would lead to Eq. 10-1 in a manner analogous to that of Eq. 9-85. It is therefore convenient to replace the original circuit by that in (b) of the figure, and to write KVL directly in terms of phasor voltages. Since ω is implied in all phasor representations, it is sometimes shown at the bottom of the circuit.

Similarly, KCL at node 2 is written directly in terms of phasor currents,

$$I_1 = I_2 \tag{10-5}$$

where, it is, of course, understood that

$$i_1 = \text{Im}\,[I_1 e^{j\omega t}], \qquad I_1 = |I_1|\, e^{j\phi_1} \tag{10-6}$$

$$i_2 = \text{Im}\,[I_2 e^{j\omega t}], \qquad I_2 = |I_2|\, e^{j\phi_2} \tag{10-7}$$

To complete the analysis either by KVL or by KCL, we need to express either the phasor branch voltages in terms of the phasor loop currents or the phasor branch currents in terms of the phasor node voltages. Can we find phasor branch constraints for the circuit elements? If we can, we can perform either loop or node analysis directly in terms of phasors. The answer to our question is yes and leads to a simple sinusoidal steady-state analyses.

10-2 OHM'S LAW EXTENDED: IMPEDANCE

We now derive phasor current–voltage relationships for the circuit elements. Sinusoidal steady-state conditions are assumed throughout our discussion. We first concentrate on two-terminal or one-port elements (i.e., resistors, capacitors, and inductors). Standard terminal current–voltage polarities are assumed as

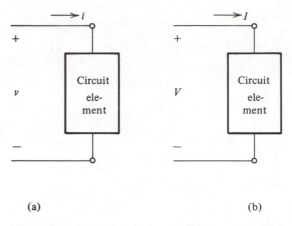

(a)   (b)

Figure 10-2. Time-domain conventions are identical with frequency-domain conventions.

shown in Fig. 10-2. The voltage and current are assumed to be

$$v = \text{Im}\,[Ve^{j\omega t}], \qquad V = |V|\,\underline{/\theta} \tag{10-8}$$

$$i = \text{Im}\,[Ie^{j\omega t}], \qquad I = |I|\,\underline{/\phi} \tag{10-9}$$

We are interested in determining the relationships between phasors V and I for the various elements as shown in (b) of the figure. In each case we will start with the known relationship between v and i, and derive that between V and I.

For the resistor, Ohm's law states that

$$v = Ri \tag{10-10}$$

or

$$\text{Im}\,(Ve^{j\omega t}) = R\,\text{Im}\,(Ie^{j\omega t}) \tag{10-11}$$

$$= \text{Im}\,(RIe^{j\omega t}) \tag{10-12}$$

since R is a *real* constant. The Im part on both sides will be equal if the complex quantities in the parentheses are equal. Then $e^{j\omega t}$ factors out to give

$$V = RI \tag{10-13}$$

We have been able to drop both Im and $e^{j\omega t}$ in this case. This equation represents Ohm's law in terms of phasors.

Next, consider the circuit element to be a capacitor. Then

$$i = C\frac{dv}{dt} \tag{10-14}$$

or

$$\text{Im}\,[Ie^{j\omega t}] = C\frac{d}{dt}\,\text{Im}\,[Ve^{j\omega t}] \tag{10-15}$$

$$= \text{Im}\left[C\frac{d}{dt}Ve^{j\omega t}\right] \tag{10-16}$$

where we have interchanged the Im operation with the derivative operation, and C is, of course, real. We can again drop Im designation and make the complex quantities equal. The right-hand side involves the time derivative and we cannot factor out the exponential. Then

$$Ie^{j\omega t} = C\frac{d}{dt}Ve^{j\omega t} \tag{10-17}$$

and we can carry out the derivative operation to give

$$Ie^{j\omega t} = j\omega C Ve^{j\omega t} \tag{10-18}$$

or, now, by factoring out the exponential,

$$I = j\omega CV \tag{10-19}$$

or

$$V = \frac{1}{j\omega C}I \tag{10-20}$$

The phasors V and I are again related by a constant, although a *complex* constant.

The equation above is reminiscent of Ohm's law except that we are dealing with phasors and complex constants. We therefore define an extended Ohm's law which expresses the phasor relation as

$$V = ZI \tag{10-21}$$

where the constant Z is called the *impedance*. The impedance Z may be complex, in general. By comparing with Eq. 10-13 we see that *the impedance of a resistor is R*, that is,

$$Z_R = R \tag{10-22}$$

Comparing Eq. 10-21 with the capacitor relation of Eq. 10-20, we see that *the impedance of a capacitor is $1/j\omega C$*, that is,

$$Z_C = \frac{1}{j\omega C} \tag{10-23}$$

The relation for an inductor involves a derivative. We are able to dispense with the Im designation but the exponential must be retained initially. The relation

$$v = L\frac{di}{dt} \tag{10-24}$$

becomes

$$Ve^{j\omega t} = L\frac{d}{dt}Ie^{j\omega t} \tag{10-25}$$

$$= j\omega LIe^{j\omega t} \tag{10-26}$$

Then

$$V = j\omega LI \tag{10-27}$$

and *the impedance of an inductor is $j\omega L$*, that is,

$$Z_L = j\omega L \qquad (10\text{-}28)$$

We thus see that all our two-terminal circuit elements are defined by impedances relating the voltage and current phasors. It is often important to determine the relative magnitudes and phases of the terminal phasors for each element. This is easily done using Eq. 10-21 and the appropriate impedance of each circuit dlement.

We use subscripts R, L, and C to designate the terminal phasors for the different circuit elements. Then, since for a resistor by Eq. 10-13,

$$V_R = RI_R \qquad (10\text{-}29)$$

we have, by using the voltage and current phasors defined in Eqs. 10-8 and 10-9,

$$|V_R| e^{j\theta} = R|I_R| e^{j\phi} \qquad (10\text{-}30)$$

Comparing the magnitudes and phases on the two sides,

$$|V_R| = R|I_R|, \qquad \theta = \phi \qquad (10\text{-}31)$$

The voltage and current phasors for a resistor have the same phase angle, that is, they are *in phase* with each other. The phasors are shown in their relative positions in a *phasor diagram* as in Fig. 10-3. The reference in a phasor

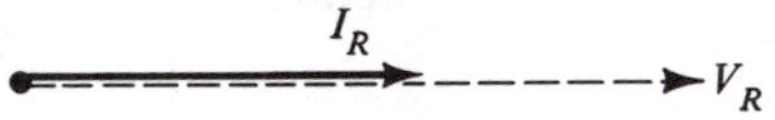

Figure 10-3. Phasor relationships in a resistor.

diagram is chosen arbitrarily. Usually, one of the phasors is selected as the reference with phase angle equal to zero. Since only relative angles between phasors are of interest, there is no problem with this practice. If the actual phase angle of a reference phasor is required to be different, we correspondingly change all the other phase angles. In the figure we have selected one of the phasors as the reference, and the other phasor coincides with it since they are in phase.

For the capacitor, by Eq. 10-19,

$$I_C = j\omega C V_C \qquad (10\text{-}32)$$

or, by using phasors defined in Eqs. 10-8 and 10-9, we obtain

$$|I_C| e^{j\phi} = \omega C e^{j90°} \cdot |V_C| e^{j\theta} \qquad (10\text{-}33)$$

Comparing magnitudes and phases again yields

$$|I_C| = \omega C |V_C|, \qquad \phi = \theta + 90° \qquad (10\text{-}34)$$

We thus see that the phase angle of the current phasor is 90° greater than that of the voltage phasor; that is, *the current in a capacitor leads the voltage by 90°*. This is shown in Fig. 10-4, in (a) with V_C as the reference, and in (b) with I_C as the reference.

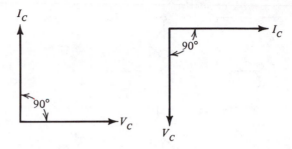

Figure 10-4. Phasor relationships in a capacitor.

Similarly, for the inductor we use Eq. 10-27 to give

$$V_L = j\omega L I_L \qquad (10\text{-}35)$$

from which we see that

$$|V_L| = \omega L |I_L|, \qquad \theta = \phi + 90° \qquad (10\text{-}36)$$

The current in an inductor lags the voltage by 90°. The phasor diagram for the inductor is given in Fig. 10-5. Compare these diagrams with those of Fig. 10-4. Our results for R, L, and C are shown in Table 10-1.

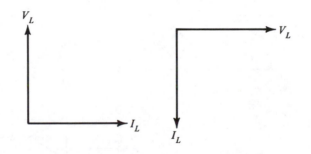

Figure 10-5. Phasor relationships in an inductor.

Table 10-1

	Time Equation	Extended Ohm's Law	Impedance	Phasor Diagram	
				V as reference	*I* as reference
R	$v_R = R i_R$	$V_R = R I_R$	$Z_R = R$	$I_R \rightarrow V_R$	$I_R \rightarrow V_R$
C	$i_C = C \dfrac{dv_C}{dt}$	$V_C = \dfrac{1}{j\omega C} I_C$	$Z_C = \dfrac{1}{j\omega C}$	$I_C \uparrow \rightarrow V_C$	$\downarrow V_C \rightarrow I_C$
L	$v_L = L \dfrac{di_L}{dt}$	$V_L = j\omega L I_L$	$Z_L = j\omega L$	$\downarrow I_L \rightarrow V_L$	$\uparrow V_L \rightarrow I_L$

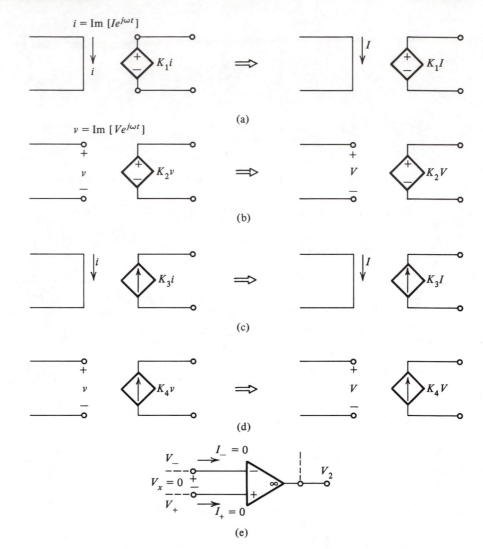

Figure 10-6. Phasor relationships for controlled sources.

Controlled sources present no special problems in sinusoidal steady state. The controlling signals will, of course, be sinusoidal, since all voltages and currents in the circuit are sinusoids of the same frequency. In writing Kirchhoff's laws with phasors, we use the appropriate phasor for the controlled signal value, which is just the (real) constant times the controlling signal phasor as in Fig. 10-6. Similarly, phasors replace the time functions in the op amp as in (e) of the figure.

Phasor current–voltage relationships for coupled coils of Fig. 10-7(a) are derived in a manner analogous to that for the inductor. Currents and voltages

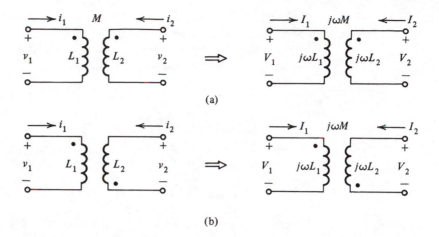

(a)

(b)

Figure 10-7.   Phasor relationships for coupled coils.

are expressed as in Eqs. 10-8 and 10-9 at the two ports with 1 and 2 used as subscripts to indicate the ports. As we have seen, we need not carry the Im designation in our discussion; just remember that it is implied whenever we need the sinusoid. Also, the exponential may be factored out only after all derivatives (or, integrals) have been taken.

The coupled-coil equations for our circuit in (a) are, from Chapter 7,

$$v_1 = L_1 \frac{di_1}{dt} + M \frac{di_2}{dt} \tag{10-37}$$

$$v_2 = M \frac{di_1}{dt} + L_2 \frac{di_2}{dt} \tag{10-38}$$

Replacing the voltages and currents by their exponetial form and (without the Im designations) yields

$$V_1 e^{j\omega t} = L_1 \frac{d}{dt} I_1 e^{j\omega t} + M \frac{d}{dt} I_2 e^{j\omega t} \tag{10-39}$$

$$V_2 e^{j\omega t} = M \frac{d}{dt} I_1 e^{j\omega t} + L_2 \frac{d}{dt} I_2 e^{j\omega t} \tag{10-40}$$

Performing the derivatives and canceling out the exponentials from both sides, we have

$$V_1 = j\omega L_1 I_1 + j\omega M I_2 \tag{10-41}$$

$$V_2 = j\omega M I_1 + j\omega L_2 I_2 \tag{10-42}$$

These are the phasor current–voltage relations for the coupled coils of our circuit. If the dot conventions are asymmetric as in (b) of the figure, the mutual inductance terms have negative signs.

If $I_2 = 0$ in the equations above, then $j\omega L_1 = V_1/I_1$ represents the *self-impedance* of coil 1 and $j\omega M = V_2/I_1$ is called the *mutual* (or *transfer*) impedance. Similarly, $j\omega L_2$ is the self-impedance of coil 2. The representations for phasor analysis are also shown in the figure.

Phasors at the ideal transformer terminals in Fig. 10-8 are obviously related by the turns ratio, as is the case for time functions:

$$\frac{V_1}{V_2} = -\frac{I_2}{I_1} = n \tag{10-43}$$

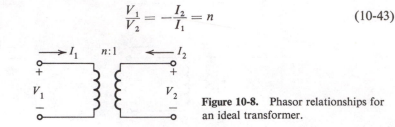

Figure 10-8. Phasor relationships for an ideal transformer.

We now have not only Kirchhoff's laws but also the branch constraints for the circuit elements in terms of phasors. We are thus in a position to perform sinusoidal steady-state analysis entirely in terms of phasors. The impedance concept, however, plays such an important role in analysis that we examine it in more detail before proceeding with circuit analysis using phasors.

EXERCISES

10-2.1. A new element has been found which is described by the equation

$$i(t) = k_1 \frac{d^2 v}{dt^2}$$

Draw a diagram similar to those shown in Table 10-1 which describes voltage and current relationships when the system containing the new element is in the sinusoidal steady state.

10-2.2. Repeat Exercise 10-2.1 for another element described by

$$i(t) = k_2 \frac{d^3 v}{dt^3}$$

10-3 IMPEDANCE AND ADMITTANCE

The concept of impedance, introduced in Section 10-2, is quite general and may be applied to any two-terminal or one-port circuit in sinusoidal steady state. Figure 10-9 shows a one-port circuit which may consist of any combination of elements, except that no independent sources are present inside the circuit. If the excitation at the terminals is a sinusoid, so would the steady-state response be. We have therefore only shown the corresponding phasors. We define the *impedance of the circuit* as a ratio of the phasors V and I,

$$Z = \frac{V}{I} \tag{10-44}$$

as in Section 10-2. The terminal behavior of the circuit is indicated by the

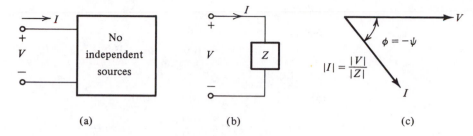

Figure 10-9. Generalization of phasor relationships in elements.

impedance as in (b) of the figure. The impedance is also known as *input imped-ance*. For a circuit containing independent sources, they are first turned off and the resulting impedance at the terminals is the input impedance.

For any circuit we are thus able to determine either phasor once we know the impedance and the other phasor. In general, the impedance is complex and may be expressed in polar form as

$$Z = |Z| e^{j\psi} = |Z| \underline{/\psi} \qquad (10\text{-}45)$$

where $|Z|$ is the magnitude and ψ the angle of Z. We can relate these to V and I from Eq. 10-44. Let us assume that the phase of V is θ and that of I is ϕ. Then

$$Z = \frac{V}{I} = \frac{|V| e^{j\theta}}{|I| e^{j\phi}} \frac{|V|}{|I|} e^{j(\theta - \phi)} \qquad (10\text{-}46)$$

$$= \frac{|V|}{|I|} \underline{/\theta - \phi} \qquad (10\text{-}47)$$

We see that

$$|Z| = \frac{|V|}{|I|}, \qquad \psi = \theta - \phi \qquad (10\text{-}48)$$

For phasor voltage and current magnitudes, the magnitude of the impedance plays a role similar to that of the resistance. For a given voltage magnitude, the current (magnitude) is smaller as the impedance (magnitude) is greater. These magnitudes are real numbers and are given the units volts, amperes, and ohms, respectively. With V as the reference (i.e., $\theta = 0$), the angle of the impedance, ψ, represents the angle by which I *lags* behind V as shown in (c) of the figure.

EXAMPLE 10-1

Find the steady-state current in each case below if $v(t) = 100 \sin 100t$.
(a) $R = 100\ \Omega$
(b) $L = 0.01$ H
(c) $C = 1\ \mu$F
(d) Circuit with impedance $Z = 1 + j1$

Solution

We have

$$|V| = 100, \qquad \omega = 100, \qquad \theta = 0° \tag{10-49}$$

Then, by Eq. 10-47,

(a)
$$Z_R = 100 = 100\underline{/0°} \tag{10-50}$$

$$|I| = \frac{100}{100} = 1 \text{ A}, \qquad \phi = 0° \tag{10-51}$$

$$i(t) = \sin 100t \tag{10-52}$$

(b)
$$Z_L = j\omega L = 1\underline{/90°} \tag{10-53}$$

$$|I| = \frac{100}{1} = 100, \qquad \phi = -90° \tag{10-54}$$

$$i(t) = 100 \sin (100t - 90°) \tag{10-55}$$

(c)
$$Z_C = \frac{1}{j\omega C} = 10^4\underline{/-90°} \tag{10-56}$$

$$|I| = \frac{100}{10^4} = 0.01, \qquad \phi = +90° \tag{10-57}$$

$$i(t) = 0.01 \sin (100t + 90°) \tag{10-58}$$

(d)
$$Z = 1 + j1 = \sqrt{2}\underline{/45°} \tag{10-59}$$

$$|I| = \frac{100}{\sqrt{2}} = 70.70, \qquad \phi = -45° \tag{10-60}$$

$$i(t) = 70.70 \sin (100t - 45°) \tag{10-61}$$

The impedance Z may also be represented in rectangular coordinates as

$$Z = R + jX \tag{10-62}$$

where

$$R = \operatorname{Re} Z \qquad \text{and} \qquad X = \operatorname{Im} Z \tag{10-63}$$

They are known as the *resistance* and *reactance*, respectively, and have the units ohms. Magnitude and angle of Z, in terms of R and X, are

$$|Z| = \sqrt{R^2 + X^2}, \qquad \psi = \tan^{-1} \frac{X}{R} \tag{10-64}$$

We frequently make use of the reciprocal of impedance, which is given the name *admittance* and the symbol Y. Then

$$Y = \frac{1}{Z} = G + jB \tag{10-65}$$

where G is the *conductance* and B the *susceptance*. The units of $|Y|$, G, and B are mhos (siemens). The relationship between R, X and G, B is easily established from our knowledge of complex numbers; thus,

$$R + jX = \frac{1}{G + jB} = \frac{G - jB}{G^2 + B^2} \tag{10-66}$$

and real and imaginary parts may be equated giving

$$R = \frac{G}{G^2 + B^2} \quad \text{and} \quad X = \frac{-B}{G^2 + B^2} \tag{10-67}$$

Similarly, we find that

$$G = \frac{R}{R^2 + X^2} \quad \text{and} \quad B = \frac{-X}{R^2 + X^2} \tag{10-68}$$

EXAMPLE 10-2

Measurements of V and I in the circuit of Fig. 10-9 establish that $|V| = (2)^{1/2}$ V and $|I| = 1$ A, with I lagging V by $45°$. For the element, determine R, X, G, and B.

Solution

From Eq. 10-44,

$$Z = \frac{V}{I} = (2)^{1/2}\underline{/+45°} = 1 + j1 \tag{10-69}$$

so $R = 1\,\Omega$ and $X = 1\,\Omega$. Using Eq. 10-68, we find that $G = \frac{1}{2}$ mho and $B = -\frac{1}{2}$ mho.

Table 10-2 summarizes these parameters for the simple circuit elements R, C, and L.

Table 10-2 Impedance and Admittance of R, C, and L

Element	Impedance, Z	$R = Re\ Z$	$X = Im\ Z$	Admittance, Y	$G = Re\ V$	$B = Im\ Y$
R	$Z_R = R$	R	0	$Y_R = \dfrac{1}{R}$	$\dfrac{1}{R}$	0
C	$Z_C = \dfrac{1}{j\omega C}$	0	$\dfrac{-1}{\omega C}$	$Y_C = j\omega C$	0	ωC
L	$Z_L = j\omega L$	0	ωL	$Y_L = \dfrac{1}{j\omega L}$	0	$\dfrac{-1}{\omega L}$

EXERCISES

10-3.1. For the voltage and current phasors identified in Fig. 10-9 it is found that $|V| = 10$ V, $|I| = 2$ A, and I leads V by $60°$. Determine the following quantities: R, X, G, and B.

10-3.2. For the voltage and current phasors identified in the figure of Exercise 10-3.1, it is known that $|V| = 0.1$ V and $|I| = 10$ A. It is found that I lags V by $30°$. Determine the following quantities: R, X, G, and B.

We are now equipped to use phasors to represent voltages and currents and impedances and admittances to represent voltage–current relationships in circuit elements. We understand, of course, that these concepts apply only to the sinusoidal steady state. We consider several examples of such circuits.

EXAMPLE 10-3

The circuit shown in Fig. 10-10 is operating in the sinusoidal steady state. We are given the output voltage, v_2, and we wish to find the input voltage, v_1.

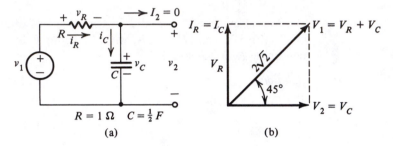

Figure 10-10. *RC* circuit driven by a sinusoidal source.

Solution

We first do this with sinusoids, and then employing phasors. We start with the output voltage, v_2, which is assumed given as

$$v_2 = 2 \sin 2t \qquad (10\text{-}70)$$

and since this is a ladder network, we determine quantities working toward the input. Thus,

$$v_C = v_2 = 2 \sin 2t \qquad (10\text{-}71)$$

$$i_C = C \frac{dv_C}{dt} = \tfrac{1}{2} \cdot 4 \cos 2t = 2 \cos 2t \qquad (10\text{-}72)$$

Since

$$i_R = i_C = 2 \cos 2t \qquad (10\text{-}73)$$

then

$$v_R = R i_R = 2 \cos 2t \qquad (10\text{-}74)$$

Finally, using KVL,

$$v_1 = v_R + v_C = 2 \cos 2t + 2 \sin 2t \qquad (10\text{-}75)$$

Combining these, as in Section 9-2,

$$v_1 = 2\sqrt{2} \sin (2t + 45°) \qquad (10\text{-}76)$$

We see that the magnitude of the input is $(2)^{1/2}$ times larger than the output and that the output lags the input by 45°.

Next, by the method of phasors, we write

$$V_2 = 2\underline{/0°} = V_C \tag{10-77}$$

$$I_C = Y_C V_C = 1\underline{/90°} \cdot 2\underline{/0°} = 2\underline{/90°} \tag{10-78}$$

$$I_R = I_C; \quad V_R = R I_R = 1\underline{/0°} \cdot 2\underline{/90°} = 2\underline{/90°} \tag{10-79}$$

Again, using KVL for phasors,

$$V_1 = V_C + V_R = 2\underline{/0°} + 2\underline{/90°}$$
$$= 2 + j2 \tag{10-80}$$

so that

$$V_1 = 2\sqrt{2}\underline{/45°} \tag{10-81}$$

From this phasor we may determine $v_1(t)$, which is

$$v_1 = 2\sqrt{2}\,\sin{(2t + 45°)} \tag{10-82}$$

the same as Eq. 10-76. Can you convince yourself that the method of phasors is more direct and easier?

It is generally more convenient to show phasor currents and voltages directly on the circuit and proceed with analysis. Frequently, it is not even necessary to determine the time function for the desired response. The phasor value of the response is all that is necessary.

EXAMPLE 10-4

Consider a different version of the circuit of Fig. 10-10 as in Fig. 10-11. Only phasor quantities are shown, together with the value of ω. This time V_1 is given and we wish to determine V_2.

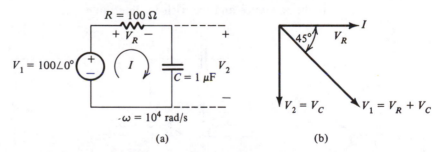

(a)   (b)

Figure 10-11.   *RC circuit analyzed beginning with Eq. 10-83.*

Solution

By KVL around the loop as shown,

$$V_1 = V_R + V_2 \tag{10-83}$$

We now express V_R and V_2 in terms of the loop current I by the impedance relations

$$V_1 = RI + \frac{1}{j\omega C}I \tag{10-84}$$

or

$$100 = 100I - j100I \tag{10-85}$$

or

$$I = \frac{100}{100 - j100} = \frac{100/0°}{100\sqrt{2}\,/-45°} \tag{10-86}$$

$$= 0.707/45° \tag{10-87}$$

Then

$$V_2 = \frac{1}{j\omega C}I \tag{10-88}$$

$$= 100/-90° \cdot 0.707/45° \tag{10-89}$$

$$= 70.70/-45° \tag{10-90}$$

We do not need the time function in this case. But the phasor V_2 has all the necessary information. It is of magnitude 70.70 and lags by 45° behind V_1. Usually, this is all that is desired. A phasor diagram is shown in (b) of the figure. Observe that this is the same as for the preceding case except that we have used I as the reference to draw the figure. This is merely a matter of convenience. To change the reference to V_1, we increase all angles by 45°.

EXAMPLE 10-5

The next circuit is an op-amp circuit of Fig. 10-12. Phasors are shown on the circuit and some value of ω is assumed but left unspecified. V_1 is the reference and we wish to determine V_2.

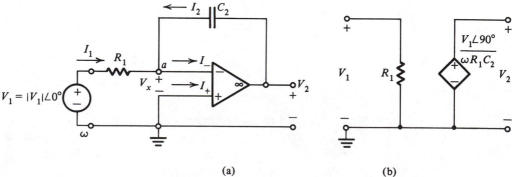

(a)   (b)

Figure 10-12. Op-amp integrator circuit.

Solution

We, of course, know that $I_- = I_+ = 0$ and $V_X = 0$ by the virtual short phenomenon. Then, by KCL at a,

$$I_1 + I_2 = 0 \tag{10-91}$$

Expressing these currents in terms of node voltages using impedances and admittances, we obtain

$$\frac{V_1 - V_a}{R_1} + (V_2 - V_a)(j\omega C_2) = 0 \tag{10-92}$$

But $V_a = 0$ by virtual short, since the $+$ terminal is grounded. Then

$$V_2 = -\frac{V_1}{\omega R_1 C_2}\underline{/-90°} = \frac{V_1}{\omega R_1 C_2}\underline{/+90°} \tag{10-93}$$

We can substitute the values of V_1, ω, R_1, and C_2 when we actually need the specific value of V_2. We may also express the op-amp circuit by a controlled source equivalent. We observe that since $V_a = 0$, the input impedance given by $V_1/I_1 = R_1$. The controlled source equivalent is then shown in (b) of the figure.

Let us find V_2 for two different frequency values, $\omega_1 = 1000$ rad/s and $\omega_2 = 10,000$ rads/s if $R_1 = 1000\ \Omega$ and $C = 1\ \mu F$. Then from Eq. 10-93, for ω_1

$$V_2 = V_1\underline{/90°} \tag{10-94}$$

and for ω_2

$$V_2 = 0.1V_1\underline{/90°} \tag{10-95}$$

EXAMPLE 10-6

The circuit of Fig. 10-13 includes a controlled source. We wish to determine the current I.

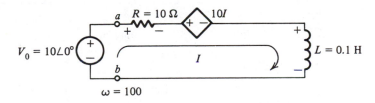

$\omega = 100$

Figure 10-13.   *RL circuit containing a controlled source.*

Solution

By KVL,

$$V_0 = RI + 10I + j\omega LI \tag{10-96}$$

or

$$10 = 20I + j10I \tag{10-97}$$

Then

$$I = \frac{10}{20 + j10} = 0.4 - j0.2 \tag{10-98}$$

and

$$|I| = \sqrt{0.2}, \qquad \phi = \tan^{-1}(-\tfrac{1}{2}) \tag{10-99}$$

EXAMPLE 10-7

What is the impedance of the circuit in Fig. 10-13 at terminals a–b?

Solution

By Eq. 10-44, the impedance is

$$Z_{ab} = \frac{V_0}{I} \tag{10-100}$$

which from Eq. 10-96 is

$$Z_{ab} = \frac{V_0}{I} = 20 + j10 \tag{10-101}$$

The resistance is 20 Ω and the reactance is 10 Ω. We may also write it in polar form as

$$Z_{ab} = \sqrt{20^2 + 10^2} \,\underline{/\tan^{-1} 10/20} \tag{10-102}$$

EXAMPLE 10-8

The coupled-coil circuit of Fig. 10-14 has two loops, as shown. We are interested in determining I_2.

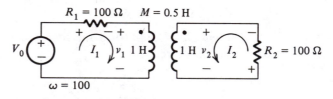

Figure 10-14.   Circuit containing coupled coils.

Solution

Using the coupled-coil equations, Eqs. 10-41 and 10-42, we write KVL for the two loops as

$$V_0 = R_1 I_1 + j\omega L_1 I_1 + j\omega M I_2 \tag{10-103}$$
$$0 = R_2 I_2 + j\omega L_2 I_2 + j\omega M I_1 \tag{10-104}$$

By substituting the values and rearranging the terms, we obtain

$$(100 + j100)I_1 + j50 I_2 = V_0 \tag{10-105}$$
$$j50 I_1 + (100 + j100)I_2 = 0 \tag{10-106}$$

These simultaneous equations are solved in the usual way except that we now have complex coefficients. From the second equation,

$$I_1 = (-2 + j2)I_2 \qquad (10\text{-}107)$$

Substituting in the first yields

$$-400I_2 + j50I_2 = V_0 \qquad (10\text{-}108)$$

Then

$$I_2 = \frac{V_0}{-400 + j50} \qquad (10\text{-}109)$$

which may be evaluated once V_0 is specified.

EXAMPLE 10-9

We are interested in determining the (input) impedance at a–b in the ideal transformer circuit of Fig. 10-15.

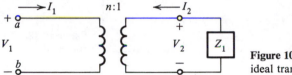

Figure 10-15. Circuit containing an ideal transformer.

Solution

The load is some impedance Z_L. The ideal transformer equation, specified by Eq. 10-43, is

$$\frac{V_1}{V_2} = -\frac{I_2}{I_1} = n \qquad (10\text{-}110)$$

Then

$$Z_{1n} = \frac{V_1}{I_1} = \frac{nV_2}{-I_2/n} \qquad (10\text{-}111)$$

$$= n^2 \frac{V_2}{-I_2} \qquad (10\text{-}112)$$

$$= n^2 Z_L \qquad (10\text{-}113)$$

EXERCISES

10-4.1. Use phasor analysis to find the steady-state response $v(t)$ of the circuit in Fig. 10-16. Assume the following value of the circuit parameters: $R = 5\ \Omega$, $C = 3$ mF, $f = 120$ Hz.

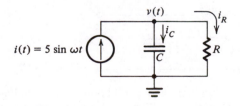

Figure 10-16. Circuit for Exercise 10-4.1.

10-4.2. Draw the phasor diagrams for I, V, V_L, and V_R for the circuit in Fig. 10-17, and express the impedance $Z = V/I$ in terms of R, L, and ω.

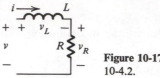

Figure 10-17.   Circuit for Exercise 10-4.2.

10-4.3. Suppose that the circuit in Fig. 10-18 is in steady state and $i(t) = 10 \sin (3t + \pi/6)$ and $v(t) = 270 \sin (3t + 3\pi/8)$. State the complex impedance of the circuit. Furthermore, suppose that the circuit within the box was a resistor in series with an inductor. Find the value of the resistance and the value of the inductance.

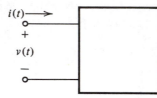

Figure 10-18.   Circuit for Exercise 10-4.3.

PROBLEMS

10-1. In the circuit given in the figure, suppose that it is known that the current is $I = 1/\underline{0°}$ A. (a) Find the phasors V_R, V_L, and V. (b) Give a phasor diagram showing V_R and V_L and their phasor addition to give V, all with respect to the given reference of I.

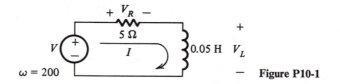

Figure P10-1

10-2. In the circuit of the figure, it is given that the output voltage is

$$v_2(t) = 10 \sin (2t + 45°)$$

(a) Find the current in the circuit, I. (b) Find the voltage driving the circuit, V_1. (c) Draw a phasor diagram showing the relationship of all voltages and the current.

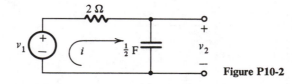

Figure P10-2

10-3. In the circuit given in the figure, it is known that the voltage across the capacitor has the phasor form $V_C = 10/\underline{45°}$. Solve for the phasor quantities in the circuit in the following sequence: I_C, I_R, I, V_L, and V. Draw a phasor diagram showing the phasor relationships that exist for all voltages and currents.

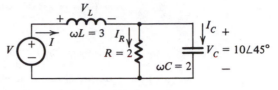

Figure P10-3

10-4. In the general circuit shown in the figure, it is intended that the two networks contain a single element or two elements. The units for these elements are ohms,

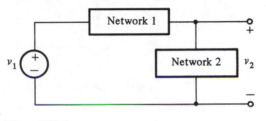

Figure P10-4

henries, and farads. In the table that follows a double entry in the first column implies a series connection, in the second column a parallel connection. For each problem you are given the output voltage

$$v_2(t) = V_m \sin(\omega t + \phi)$$

You are required to (a) solve for $v_1(t)$ by phasor methods, and (b) draw a complete phasor diagram showing all voltages and all currents and all relationships between voltages and currents due to the two Kirchhoff laws.

Problem Number	Network 1	Network 2	V_m	ω	ϕ
10-4.	$R = 1$	$C = 2$	2	$\frac{1}{2}$	$-30°$
10-5.	$R = 2$	$C = 1$	10	2	$45°$
10-6.	$R = 20$	$C = \frac{1}{2}$	1	0.1	$0°$
10-7.	$R = 2$	$L = 2$	100	$\frac{1}{2}$	$30°$
10-8.	$L = \frac{1}{2}$	$R = 1$	10	$\frac{1}{2}$	$0°$
10-9.	$C = 2$	$R = 2$	3	1	$45°$
10-10.	$L = 3$	$C = 1$	10	$\frac{1}{2}$	$-45°$
10-11.	$C = 1$	$L = \frac{1}{2}$	1	2	$0°$
10-12.	$R = 1, C = 1$	$L = 2$	2	1	$30°$
10-13.	$R = 1, L = 2$	$C = \frac{1}{2}$	2	1	$45°$
10-14.	$L = 1, C = 2$	$R = 1$	10	$\frac{1}{2}$	$0°$
10-15.	$R = 1, C = 1$	$R = 1, L = 2$	10	1	$90°$
10-16.	$R = 3, L = 2$	$R = 1, C = \frac{1}{2}$	1	$\frac{1}{2}$	$0°$
10-17.	$L = 1, C = 2$	$R = 1, C = 1$	100	1	$-90°$
10-18.	$L = 1, C = 2$	$R = 1, L = 2$	1	1	$0°$

10-19. A circuit within the box indicated in the figure is operating in the sinusoidal steady state. It is found that

$$i(t) = 10 \sin\left(3t + \frac{3\pi}{8}\right) \quad \text{A}$$

and

$$v(t) = 200 \sin\left(3t + \frac{\pi}{6}\right) \quad \text{V}$$

It is given that there are two circuit elements within the box connected in series. Find the element values.

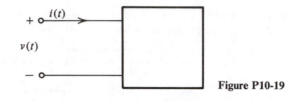

Figure P10-19

10-20. Given that the circuit shown in the figure is operating in the sinusoidal steady state. The voltage at the output terminals is

$$v_2(t) = 2 \sin(t + 30°) \quad \text{V}$$

For the element values given, find the input current

$$i_1(t) = A \sin(t + \theta) \quad \text{A}$$

with numerical values for A and θ.

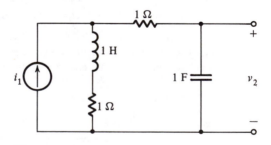

Figure P10-20

10-21. The circuit given in the figure is operating in the sinusoidal steady state. The values of the elements in the circuit are as follows: $R_1 = R_2 = 3\ \Omega$, $L =$

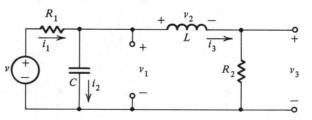

Figure P10-21

20×10^{-3} H, and $C = 2 \times 10^{-3}$ F. The voltage across the capacitor is given by

$$v_1(t) = 5 \sin (200t + 23.13°)$$

Determine numerical values for the phasor quantities V_1, V_2, V_3, I_1, I_2, and I_3. Sketch a phasor diagram with the various phasor additions corresponding to the Kirchhoff laws clearly indicated.

II

CIRCUIT FUNCTIONS AND ANALYSIS TECHNIQUES

We have seen how steady-state sinusoidal analysis is simplified by the use of the exponential form for the sinusoid. All signals are then represented in terms of the signal phasors and the exponentials. Except for the rare occasion when we need the sinusoids, the Im designation is dropped in our representation. That is, we have *replaced* the sinusoids by the exponentials.

This change to the exponential form leads to phasor analysis. The exponential is common to all signals and so factors out in Kirchhoff's laws, leaving only the phasor versions of the laws. The common exponential also cancels out in the current–voltage relations, leading to the phasor ratio, V/I, that is, a circuit constant called the impedance. The analysis is then carried out entirely with phasors.

In this chapter we extend all the resistive circuit analysis concepts to phasor analysis. Before doing this, we further generalize our signals so as to include all possible exponential forms. Thus, a new form called the *generalized* or *complex* exponential is introduced. We also introduce the concept of the frequency variable, which allows analysis in terms of an unknown frequency represented by a variable. The result of the analysis is then a *circuit function* (e.g., an impedance function) of the frequency variable. A specific value of the frequency is substituted as necessary to calculate the desired value of the impedance. Function values may thus be determined at different values of the frequency without having to analyze the circuit repeatedly. The generalized exponential further leads to many worthwhile features very important to sinusoidal analysis.

Throughout our study henceforth, unless otherwise stated, we assume steady-state conditions and exponential or sinusoidal excitations. We first introduce the generalized exponential and the frequency variable. We then formalize all our results in terms of the frequency variable.

11-1 GENERALIZED OR COMPLEX EXPONENTIALS

We begin with a review of the sinusoid and the corresponding exponential. The general form of the sinusoid is

$$v(t) = V_m \sin{(\omega t + \theta)} \qquad (11\text{-}1)$$

where V_m is the amplitude of the sinusoid, θ is the phase angle, and ω is the frequency in rad/s, or $f = \omega/2\pi$ is the frequency in Hz. The waveform of Eq. 11-1 is shown in Fig. 11-1. With respect to the reference shown as a dashed sine

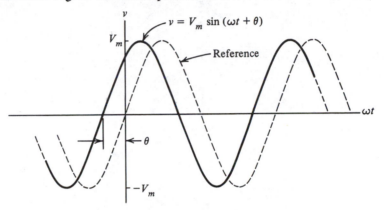

Figure 11-1. Sinusoidal waveform corresponding to Eq. 11-1.

wave, $v(t)$ is said to be leading the reference when θ is positive, lagging when θ is negative.

We have seen in Chapter 9 that there is an advantage in generalizing voltages and currents of the form of Eq. 11-1 to the exponential forms. To accomplish this, we make use of Euler's equation,

$$e^{j\omega t} = \cos{\omega t} + j \sin{\omega t} \qquad (11\text{-}2)$$

The exponential $e^{j\omega t}$ may be thought of as a *rotating vector or phasor* shown in Fig. 11-2, in which the real part may be projected as a cosine wave and the imaginary part as a sine wave. If $v(t)$ is in the form of Eq. 11-1, it may be written in exponential form as

$$v(t) = \text{Im}\,[V_m e^{j(\omega t + \theta)}] \qquad (11\text{-}3)$$

$$= \text{Im}\,[V_m e^{j\theta} e^{j\omega t}] = \text{Im}\,[V e^{j\omega t}] \qquad (11\text{-}4)$$

where Im indicates the imaginary part of the complex quantity. In this equation, $V_m e^{j\theta} = V$ is the phasor voltage, sometimes written in the short form $V_m\underline{/\theta}$, and $e^{j\omega t}$ is the rotating phasor which provides for the time variation of

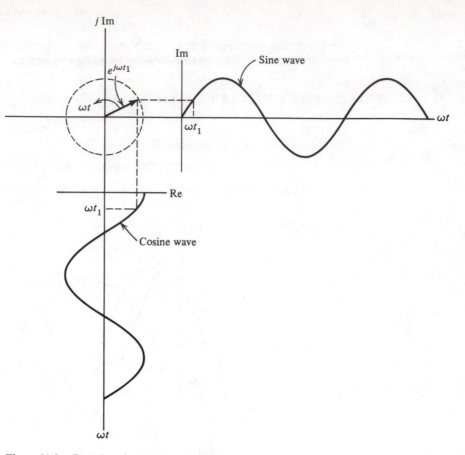

Figure 11-2. Rotating phasor generates a sine wave and a cosine wave.

the function. In our analysis we dispense with the designation Im, and assume the signal v to be of the form $Ve^{j\omega t}$ for the most part.

Other waveforms important in electrical engineering also have exponential form. The *decreasing and increasing exponentials* shown in Fig. 11-3 are described by the equation

$$v(t) = V_0 e^{\pm \alpha t} \qquad (\alpha > 0) \tag{11-5}$$

where the negative sign is used with an exponentially decreasing function and the positive sign for an exponentially increasing $v(t)$. The equation

$$v(t) = V_0 e^{-\sigma t}(\sin \omega t + \theta) \tag{11-6}$$

is known as *a damped sinusoid* for $\sigma > 0$ and has the appearance shown in Fig. 11-4. In the figure we have assumed that $\theta = 0$ for convenience.

The damped sinusoid may also be described by the exponential form as

$$v(t) = \text{Im} \, [V_0 e^{j\theta} \cdot e^{(-\sigma + j\omega)t}] \tag{11-7}$$

$$= \text{Im} \, [Ve^{(-\sigma + j\omega)t}] \tag{11-8}$$

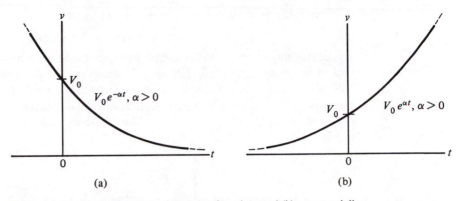

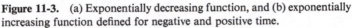

Figure 11-3. (a) Exponentially decreasing function, and (b) exponentially increasing function defined for negative and positive time.

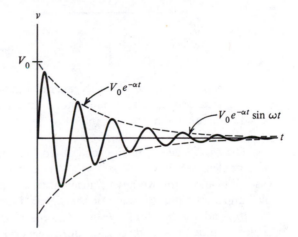

Figure 11-4. Exponentially decreasing function which is the envelope of a sinusoid.

We need not use the Im designation in our analysis, as discussed in Chapter 10. We can always use the complex exponential function itself. We can therefore replace v by the exponential form in parentheses,

$$v(t) \sim V e^{(-\sigma + j\omega)t} \tag{11-9}$$

All of these waveforms and others may be described by the *generalized complex exponential* form

$$v(t) = V e^{st} \tag{11-10}$$

where

$$s = \sigma + j\omega \tag{11-11}$$

and σ carries its sign. It is also convenient to assume that ω can be either positive or negative. A negative value of ω implies only that the sinusoid has a negative sign, since $\sin(-\omega t) = -\sin \omega t$. Then, since

$$e^{st} = e^{\sigma t}e^{j\omega t} = e^{\sigma t}(\cos \omega t + j \sin \omega t) \qquad (11\text{-}12)$$

we see that any of the waveforms that have been listed can be obtained by using *real or imaginary part or by letting σ or ω vanish*. Of particular interest to us will be *the case $s = j\omega$, which corresponds to the sinusoidal steady state*. The particular value of s under consideration will be displayed in the σ–ω coordinate system suggested by Eq. 11-11. This will be known as the *s-plane*, shown in Fig. 11-5.

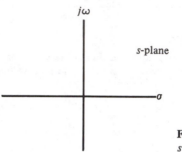

Figure 11-5.   Complex s-plane where $s = \sigma + j\omega$.

The exponent s is complex, in general, and has the dimensions of frequency. As discussed above, it is usually more convenient to leave the frequency unspecified as an unknown variable. For these reasons, s is called the *complex frequency variable*.

The *coefficient, V,* of the generalized exponential in Eq. 11-10 may be real or complex. When $s = 0$, $v(t) = V$; then V is real and we have just the dc case. When $s = j\omega$, we have sinusoidal steady state; then $V = |V| \underline{/\theta}$ is complex and is the phasor value of the signal. Our interest is only in this case of $s = j\omega$, and we therefore *refer to V as the phasor*. The dc case of $s = j\omega = 0$ is also included in our discussion of the general case of $s = j\omega$.

EXERCISES

11-1.1. For the waveform shown in Fig. 11-3(a), it is found that at $t = 1$, $v = 0.67$, and at $t = 3$, $v = 0.0333$. Determine the coefficients of Eq. 11-5.

Ans. $V_0 = 3$, $\alpha = 1.5$

11-1.2. For the waveform shown in Fig. 11-3(b), it is observed that $v(1) = 3.30$ while $v(2) = 5.4366$. Determine the coefficients of Eq. 11-5.

Ans. $V_0 = 2$, $\alpha = +0.5$

11-2 GENERALIZED IMPEDANCE

We next consider the voltage–current relationship in the circuit elements when the waveform is that of Eq. 11-10. A general two-terminal element is shown in Fig. 11-6(a) as a block, soon to be replaced by a resistor, capacitor, or inductor.

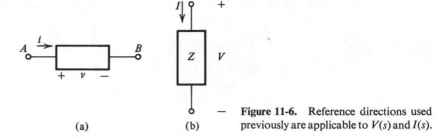

Figure 11-6. Reference directions used previously are applicable to $V(s)$ and $I(s)$.

The reference directions for the current i and the voltage v are as shown, with the $+$ sign of the voltage always placed at the tail end of the current arrow.

We assume steady-state conditions so that all voltages and currents are of the form in Eq. 11-10. Then the voltage

$$v(t) = Ve^{st} \tag{11-13}$$

and the current

$$i(t) = Ie^{st} \tag{11-14}$$

where V and I are complex constants, in general. The ratio of v to i will be a constant since the time dependence is the same in both and will be canceled out,

$$\frac{v}{i} = \frac{Ve^{st}}{Ie^{st}} = \frac{V}{I} = Z \tag{11-15}$$

The constant Z is called the *generalized impedance*. Note that the impedance is now the ratio of v and i. *This is true only when both are assumed in the expo-nential form.* More properly, with the signals in the sinusoidal or the expo-nential form, *the impedance is the ratio of phasors V and I.* The circuit element may then be represented as in Fig. 11-6(b).

Let the element in Fig. 11-6 be a resistor. Then, using the current of Eq. 11-14,

$$v(t) = Ri(t) = RIe^{st} \tag{11-16}$$

and the voltage phasor in Eq. 11-13 becomes $V = RI$. Then the impedance of the resistor is

$$Z_R = \frac{v}{i} = \frac{V}{I} = R \tag{11-17}$$

The reciprocal of Z_R is the *admittance* of the resistor, which is the conductance $G = 1/R$.

We will repeat this process for the capacitor by assuming a form of voltage like that of Eq. 11-13. Making use of the voltage–current relationship for the capacitor, we obtain

$$i = C\frac{dv}{dt} = CsVe^{st} \tag{11-18}$$

The current phasor is thus $I = CsV$. Then the impedance of the capacitor is

$$Z_c = \frac{v}{i} = \frac{V}{I} = \frac{1}{Cs} \qquad (11\text{-}19)$$

The admittance of the capacitor is $Y_c = 1/Z_c = Cs$.

Finally, for the inductor,

$$v = L\frac{di}{dt} = LsIe^{st} \qquad (11\text{-}20)$$

where LsI represents the voltage phasor V. Then the impedance of the inductor is

$$Z_L = \frac{v}{i} = \frac{V}{I} = Ls \qquad (11\text{-}21)$$

The admittance of the inductor is $Y_L = 1/Z_L = 1/Ls$.

Table 11-1 summarizes the information for the various elements.

Table 11-1

Symbol	Element	$Z(s)$	$Y(s)$
V + □ − / I		$Z = \dfrac{V}{I}$	$Y = \dfrac{1}{Z}$
R ⌇	R	R	$\dfrac{1}{R}$
C ‖	C	$\dfrac{1}{Cs}$	Cs
L ⌇	L	Ls	$\dfrac{1}{Ls}$

EXAMPLE 11-1

The current is given as $10 \sin(100t)$. Determine the voltage across each of the following elements: 100-Ω resistor, 0.01-F capacitor, 0.1-H inductor.

Solution

The respective impedances are

$$Z_R = R = 100 \qquad (11\text{-}22)$$

$$Z_c = \frac{1}{sC} = \frac{100}{s} \qquad (11\text{-}23)$$

$$Z_L = sL = \frac{s}{10} \tag{11-24}$$

The current phasor is

$$I = 10 \tag{11-25}$$

and

$$s = j\omega = j100 \tag{11-26}$$

The voltage phasors are

$$V_R = RI = 100 \times 10 = 1000 = 1000\underline{/0^\circ} \tag{11-27}$$

$$V_C = Z_C I = \frac{100 \times 10}{j100} = -j10 = 10\underline{/-90^\circ} \tag{11-28}$$

$$V_L = Z_L I = \frac{j100}{10} \times 10 = j100 = 100\underline{/90^\circ} \tag{11-29}$$

We can now write down the voltages using Eqs. 11-1 and 11-3.

We thus see that the generalized exponential leads to the same results as the sinusoid, except that s is used instead of $j\omega$. It is possible to go from one to the other by replacing s with $j\omega$, and vice versa. We therefore use these forms interchangeably, guided only by convenience. Equations 11-1 and 11-3 must, of course, be used whenever sinusoidal functions are desired.

Without further repetition then, we can write down the relationships for the two-port elements. There is, of course, no difference for controlled sources and op amps. All relations involve phasor voltages and currents. The same is true for ideal transformers. For coupled coils, the self-impedances of the coils are sL_1 and sL_2, while the mutual impedance is sM. These items are summarized in Table 11-2.

EXERCISES

11-2.1. A new element is described by the law

$$i = \frac{1}{D} \frac{d^2 v}{dt^2}$$

For this element, find $Z(s)$ and $Y(s)$.

$$\textit{Ans.} \ \ Z(s) = \frac{D}{s^2}$$

11-2.2. A series *RLC* circuit has the current $i = 2\sin 200t$ passing through it. If $R = 2\,\Omega$, $C = 0.1$ F, and $L = 0.2$ H, find the voltage across each of the three elements.

$$\textit{Ans.} \ \ V_R = 4 \text{ V}, \ V_C = 0.1\underline{/-90^\circ}, \ V_L = 80\underline{/90^\circ}$$

Table 11-2

Circuit Element	Circuit Symbol
Voltage-controlled voltage source	
Voltage-controlled current source	
Current-controlled voltage source	
Current-controlled current source	
Op amp with negative feedback	
Coupled coils	
Ideal transformer	

We now define the response-excitation ratios for circuits in terms of the complex frequency variable s. The ratios are determined *algebraically* in the unknown variable s. Specific values of s are then substituted, when necessary, to find specific values of the ratios. The ratios, which are functions of s, are called *circuit functions*. They are also known as *network* or *system functions*.

The circuit function *values* are of interest to us only in sinusoidal cases (i.e., when $s = j\omega$). These functions are therefore frequently written for $s = j\omega$ with ω (or $j\omega$) as a frequency variable. *It is easier algebra to work with s in derivations and then to use jω in computations.* The s-plane is also very useful for representations of functions of s and provides much insight, as we shall see in later chapters.

In all our discussion below, we assume that there are no independent sources within the circuit which is excited externally at its terminals.

Consider the representation of a circuit shown in Fig. 11-7. The current I

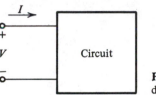
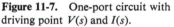

Figure 11-7. One-port circuit with driving point $V(s)$ and $I(s)$.

and the voltage V are phasors with the given reference directions. The phasors are defined for each value of s. Since s is an unknown variable, they may be written as functions of s [i.e., $I(s)$ and $V(s)$]. For each value of s, the current and voltage are as in Eqs. 11-13 and 11-14. Since both the current and the voltage have the same exponential, their ratio is just the ratio of the phasors as in Chapter 10. The *impedance function Z* is thus defined for any circuit as

$$Z(s) = \frac{V(s)}{I(s)} \tag{11-30}$$

In the above, we have shown the functional dependence on s explicitly. Frequently, this dependence is understood and we merely write Z or V or I. We may also write the impedance function for sinusoidal analysis by setting $s = j\omega$,

$$Z(j\omega) = \frac{V(j\omega)}{I(j\omega)} \tag{11-31}$$

where ω is now the sinusoidal frequency variable. The complex impedance function may be written in terms of its magnitude and phase as

$$Z(j\omega) = |Z(j\omega)| \underline{/\psi(j\omega)} \tag{11-32}$$

In this equation $|Z(j\omega)|$ is called the *magnitude function* and $\psi(j\omega)$ is, called the *phase function*. We will be interested in plotting these magnitude and phase functions for a range of values of ω.

Similarly, V_1 and V_2 shown in Fig. 11-8(a) are phasor voltages for the excitation and response. The voltages are again of the same exponential form and their ratio is just the ratio of the phasors. The *gain* or *voltage-ratio transfer function* is then

$$T(s) = \frac{V_2(s)}{V_i(s)} \tag{11-33}$$

which for sinusoidal analysis is

$$T(j\omega) = \frac{V_2(j\omega)}{V_1(j\omega)} \tag{11-34}$$

The complex function may be written as

$$T(j\omega) = |T(j\omega)| \underline{/\theta(j\omega)} \tag{11-35}$$

Again, we will be interested in plotting the magnitude and phase of $T(j\omega)$.

Although magnitude and phase are the usual way to describe $Z(j\omega)$ or $T(j\omega)$, the impedance and admittance of networks like that of Fig. 11-7 are sometimes expressed in rectangular coordinates. Thus,

$$Z(j\omega) = R(\omega) + jX(\omega) \tag{11-36}$$

where

$$R(\omega) = \text{Re } Z(j\omega) \tag{11-37}$$

is known as the *resistance function*, and

$$X(\omega) = \text{Im } Z(j\omega) \tag{11-38}$$

is known as the *reactance function*. Similarly, for the admittance

$$Y(j\omega) = G(\omega) + jB(\omega) \tag{11-39}$$

where

$$G(\omega) = \text{Re } Z(j\omega) \tag{11-40}$$

is known as the *conductance function*, and

$$B(\omega) = \text{Im } Y(j\omega) \tag{11-41}$$

is known as the *susceptance function*. The unit for Z, R, and X is the ohm (Ω), and for Y, G, and B is the mho ($\mho$). The values of these functions, at any value of the frequency ω are determined simply by a mere substitution. At any frequency, these quantities are constants, as in Chapter 10, and are easily manipulated.

EXAMPLE 11-2

Suppose that the impedance of a circuit is $Z(s) = s + 100$. Determine $|Z|$, R, X, $|Y|$, G, and B at frequency $\omega = 100$.

Solution

$$s = j\omega = j100 \tag{11-42}$$

Then

$$Z(s = j100) = j100 + 100 = 141.40\underline{/45°} \tag{11-43}$$

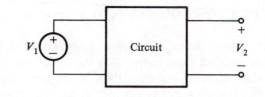

(a)

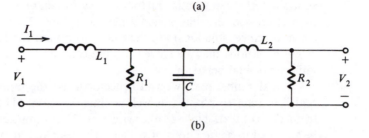

(b)

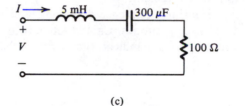

(c)

Figure 11-8. A two-port circuit with an open-circuit output is shown in (a), with specific examples given in (b) and (c).

and

$$Y = \frac{1}{Z} = 0.007\underline{/-45^\circ} \qquad (11\text{-}44)$$

Thus,

$$|Z| = 141.40, \qquad \theta_z = 45^\circ, \qquad R = 100, \qquad X = 100 \qquad (11\text{-}45)$$

and

$$|Y| = 0.007, \qquad \theta_Y = -45^\circ, \qquad G = 0.005, \qquad B = -0.005 \qquad (11\text{-}46)$$

EXERCISES

11-3.1. For the circuit shown in Fig. 11-8(b), (a) express the input impedance $Z_1 = V_1/I_1$ in terms of the complex frequency variable s; (b) express the voltage-ratio transfer function, V_2/V_1, in terms of the complex frequency variable s; and (c) find the complex value of the impedance Z_1 when $s = j200$, $R_1 = 100\ \Omega$, $R_2 = 15\ \Omega$, $L_1 = 5$ mH, $L_2 = 3$ mH, and $C = 200\ \mu$F.

11-3.2. For the series RLC combination given in Fig. 11-8(c), find the impedance $Z(j\omega)$ as a function of ω. At what frequency ω is the magnitude of the impedance at a minimum?

Analogous to the sinusoidal case, steady-state response to an exponential exci-tation is an exponential with the same exponent but with a different phasor. Also, sums of exponentials with the same exponent result in an exponential with that identical exponent. The sum can therefore be carried out by a phasor sum since the exponential factors out as in Chapter 10. Kirchhoff's laws are thus expressed in phasor quantities as before. The generalized impedances (or impedance functions) are then used to obtain the loop or node equations. These equations may be solved algebraically and $s = j\omega$ substituted to deter-mine sinusoidal behavior.

For the most part, we will henceforth use the terms "voltages" and "cur-rents" to refer to phasors as well as time functions. The meaning will be clear from the context. Unless otherwise stated, impedances and other ratios will be assumed to be functions of s. The variables s and $j\omega$ will be used interchange-ably as convenient.

Polarities, reference directions, and voltage rises and drops will all refer to phasor quantities. Capital letters will indicate phasors; impedances will be generally used to indicate the current–voltage relations for either simple elements

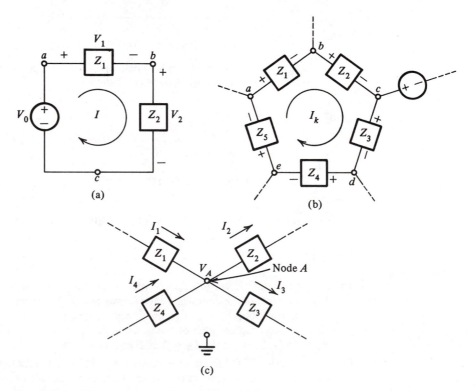

Figure 11-9 Circuits prepared for general analysis.

or for circuits. Loops and nodes, loop currents, and node voltages (with respect to ground or datum) are as defined before. Circuits are prepared for analysis in the same manner as before and as shown in Fig. 11-9 in general terms. A few simple examples will illustrate.

For example, the simple circuit of Fig. 11-10 gives us the KVL

$$V_1 = V_{z1} \tag{11-47}$$

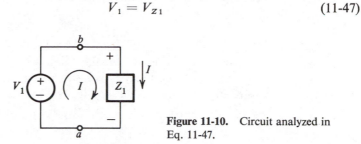

Figure 11-10. Circuit analyzed in Eq. 11-47.

where V_1 is the voltage rise phasor and V_{z1} is the voltage drop phasor across Z_1. But $V_{z1} = Z_1 I$, so that the KVL becomes

$$V_1 = Z_1 I \tag{11-48}$$

Similarly, the circuit of Fig. 11-11 gives us the KVL, in terms of the loop current,

$$V_1 = Z_1 I + Z_2 I \tag{11-49}$$

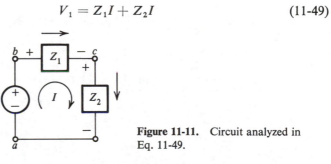

Figure 11-11. Circuit analyzed in Eq. 11-49.

For the two-loop circuit Fig of. 11-12, we choose window-pane-type loops. The loop currents are assumed to be I_1 and I_2. Then I_1 flows through Z_1 and I_2 flows through Z_3. The current in Z_2 is the composite current $(I_1 - I_2)$ in the direction from node c to a. The voltage drop in Z_2 is $Z_2(I_1 - I_2)$ from c to a.

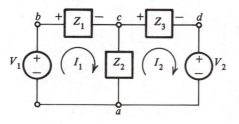

Figure 11-12. Two-loop circuit analyzed in Eqs. 11-50 and 11-51.

Equivalently, the drop in Z_2 from a to c is $Z_2(I_2 - I_1)$ since the composite current from a to c is $(I_2 - I_1)$. The loop equations are

$$Z_1 I_1 + Z_2(I_1 - I_2) = V_1 \tag{11-50}$$

$$Z_2(I_2 - I_1) + Z_3 I_2 = -V_2 \tag{11-51}$$

These equations are usually written as

$$(Z_1 + Z_2)I_1 - Z_2 I_2 = V_1 \tag{11-52}$$

$$-Z_2 I_1 + (Z_2 + Z_3)I_2 = -V_2 \tag{11-53}$$

These may now be solved for the unknowns.

Similarly, node analysis is also carried out just as before with a few details changed. In the circuit of Fig. 11-13, KCL gives

$$I_0 = I_1 \tag{11-54}$$

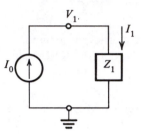

Figure 11-13.   Circuit analyzed in Eq. 11-55.

Since $I_1 = V_1/Z_1$, the equation becomes

$$I_0 = \frac{V_1}{Z_1} = Y_1 V_1 \tag{11-55}$$

For the circuit of Fig. 11-14, $I_2 = (V_1 - V_2)/Z_2$ and the node equations are

$$\frac{V_1}{Z_1} + \frac{V_1 - V_2}{Z_2} = I_0 \tag{11-56}$$

$$\frac{V_1 - V_2}{Z_2} = \frac{V_2}{Z_3} \tag{11-57}$$

Or by rearranging and using admittances, we obtain

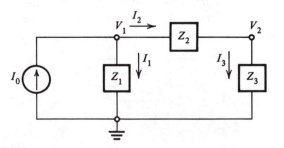

Figure 11-14.   Circuit analyzed in Eq. 11-56.

$$(Y_1 + Y_2)V_1 - Y_2V_2 = I_0 \qquad (11\text{-}58)$$
$$-Y_2V_1 + (Y_2 + Y_3)V_2 = 0 \qquad (11\text{-}59)$$

The equations are now ready to be solved.

We have illustrated the concepts using very simple examples, but in quite general terms, without knowing even the detailed circuits or the excitations. We use this approach next to derive several results for steady-state circuits analogous to those for resistive circuits. We should, of course, anticipate that the differences between the two will be that Z replaces R, and phasors replace time functions.

EXERCISES

11-4.1. Give the three circuit equations, in terms of the complex frequency variable s, that are necessary for the determination of the voltage v_C across the capacitor in the circuit given in Fig. 11-15.

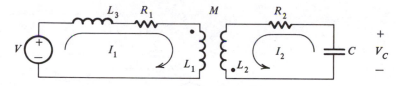

Figure 11-15. Circuit for Exercise 11-4.1.

11-4.2. For the circuit given in Fig. 11-16, find the values of the loop currents i_1 and i_2 and the voltage v_1.

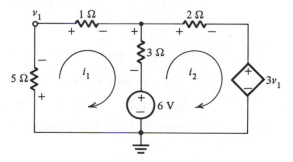

Figure 11-16. Circuit for Exercise 11-4.2.

$$\textit{Ans. } i_1 = \tfrac{4}{3}, \ i_2 = 6, \ v_1 = -\tfrac{20}{3}$$

11-5 SERIES AND PARALLEL IMPEDANCES

The equivalent circuit has been defined before as that circuit which replaces another but maintains the same current–voltage relationship. The current–voltage relation is now defined by an impedance. An *equivalent* impedance, Z_{eq}, is thus the impedance which, while replacing a combination of impedances, keeps invariant the ratio V/I at the circuit terminals.

Consider the series circuit of Fig. 11-17(a) consisting of a series connection of four circuit elements, each represented by impedance Z_i, $i = 1$ to 4.

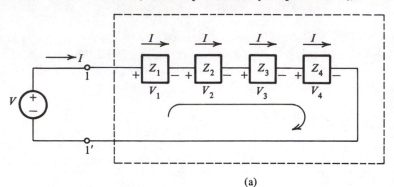

(a)

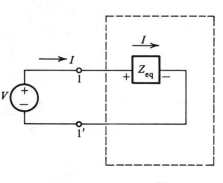

(b)

Figure 11-17.   Series circuit and equivalent.

The same current I is in each of the elements. Then KVL is

$$V = Z_1 I + Z_2 I + Z_3 I + Z_4 I \qquad (11\text{-}60)$$

If we divide this equation by I and then define the ratio of V and I as the equivalent impedance, Z_{eq}, we have

$$Z_{eq} = Z_1 + Z_2 + Z_3 + Z_4 \qquad (11\text{-}61)$$

We have thus replaced the circuit to the right of terminals or nodes marked 1–1′ in (a) with that in (b). In terms of measurements or calculations made at these nodes, the two circuits of Fig. 11-17 cannot be distinguished. They are *equivalent* since the same impedances are seen *at these nodes*.

The circuit of Fig. 11-18(a) consists of a parallel connection of elements. By KCL,

$$I = \frac{V}{Z_1} + \frac{V}{Z_2} + \frac{V}{Z_3} \qquad (11\text{-}62)$$

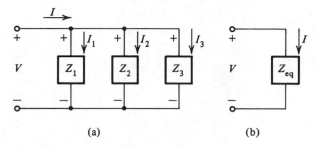

Figure 11-18. Parallel circuit and equivalent.

Then, for the equivalent impedance shown in (b),

$$I = \frac{V}{Z_{eq}} \tag{11-63}$$

so that

$$\frac{1}{Z_{eq}} = \frac{1}{Z_1} + \frac{1}{Z_2} + \frac{1}{Z_3} \tag{11-64}$$

It is more convenient to use admittances, in which case

$$Y_{eq} = Y_1 + Y_2 + Y_3 \tag{11-65}$$

We have illustrated series and parallel combinations using just a few impedances, but similar results are obtained for any number. The combination of two impedances in parallel is of sufficient importance that a special notation has been introduced for it:

$$Z_{eq} = Z_1 \| Z_2 = \frac{Z_1 Z_2}{Z_1 + Z_2} \tag{11-66}$$

which is easily remembered as the product over the sum of two impedances.

11-6 VOLTAGE- AND CURRENT-DIVIDER CIRCUITS

The circuit of Fig. 11-19 is the impedance version of the voltage-divider circuit. We will apply KVL to this circuit with the objective of finding the relationship of voltage V_2 to V_1. Around loop *abca*, KVL is

$$V_1 = Z_1 I + Z_2 I \tag{11-67}$$

But

$$V_2 = Z_2 I \tag{11-68}$$

Solving the first of these equations for I and substituting into the second equation, we have

$$\frac{V_2}{V_1} = \frac{Z_2}{Z_1 + Z_2} \tag{11-69}$$

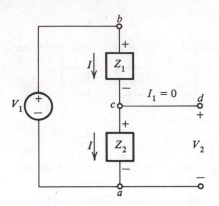

Figure 11-19. Generalized voltage-divider circuit.

This is the voltage-divider equation. In the special case that both elements are resistors and $Z_1 = R_1$ and $Z_2 = R_2$, then

$$\frac{V_2}{V_1} = \frac{R_2}{R_1 + R_2} \tag{11-70}$$

This equation indicates that the voltage V_2 is a fraction, $R_2/(R_1 + R_2)$, of the voltage V_1, and hence this circuit may be used when it is necessary to reduce a voltage. Writing the equation in the form

$$\frac{V_2}{V_1} = \frac{1}{1 + R_1/R_2} \tag{11-71}$$

which is displayed in Fig. 11-20, note that $R_1 = 0$ corresponds to no voltage reduction, so that $V_2 = V_1$. The larger the value of R_1 compared to R_2, the smaller the value of V_2 will be. Remember that V_2 and V_1 are *phasors* now.

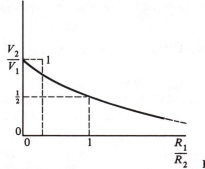

Figure 11-20. Plot of Eq. 11-71.

EXAMPLE 11-3

The input to an amplifier is required to be $\frac{1}{10}$ of that available from a source. Design a circuit to give the required voltage to the amplifier.

Solution

Substituting the requirement of $V_2/V_1 = \frac{1}{10}$ in Eq. 11-71, we find that $R_1/R_2 = 9$. We may make any selection of element values as long as the ratio is maintained. One example that meets the requirement is $R_2 = 1\text{K}\Omega$ and $R_1 = 9\text{K}\Omega$.

EXAMPLE 11-4

Find V_2/V_1 for the circuit shown in Fig. 11-21; in this case use Eq. 11-69.

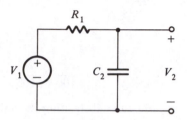
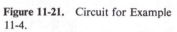

Figure 11-21. Circuit for Example 11-4.

Solution

Observe that $Z_1 = R_1$ and $Z_2 = 1/C_2 s$. Then we find that

$$\frac{V_2}{V_1} = \frac{1/C_2 s}{R_1 + 1/C_2 s} = \frac{1/R_1 C_2}{s + 1/R_1 C_2} \tag{11-72}$$

This may be written in a different form by recalling that $R_1 C_2$ is the time constant, T_0, of the RC circuit defined in Chapter 6. Then

$$\frac{V_2}{V_1} = \frac{1/T_0}{s + 1/T_0} = \frac{1}{T_0 s + 1} \tag{11-73}$$

This equation will express the relationship that exists between V_1 and V_2 for a sinusoidal source when $s = j\omega$. Then

$$\frac{V_2}{V_1} = \frac{1}{1 + j\omega T_0} \tag{11-74}$$

The current-divider circuit of Fig. 11-22 is analyzed in a similar manner, as follows:

$$V = (Z_1 \| Z_2)I \tag{11-75}$$

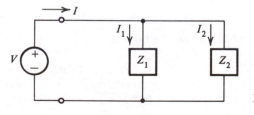

Figure 11-22. Current-divider circuit.

so that

$$I_1 = \frac{V}{Z_1} = \frac{Z_2}{Z_1 + Z_2} I \qquad (11\text{-}76)$$

and

$$I_2 = \frac{V}{Z_2} = \frac{Z_1}{Z_1 + Z_2} I \qquad (11\text{-}77)$$

EXAMPLE 11-5

What is the current I_1 in Fig. 11-22 if $Z_1 = 100$, $Z_2 = 10{,}000/s$, and $I = 10\underline{/90^\circ}$ at $\omega = 100$?

Solution

By Eq. 11-76,

$$I_1 = \frac{10^4/s}{100 + 10^4/s} I \qquad (11\text{-}78)$$

Setting $s = j\omega = j100$ and substituting the value of I, we obtain

$$I_1 = \frac{100/j}{100 + 100/j} j10 = \frac{10}{1 - j1} \qquad (11\text{-}79)$$

$$= \frac{10}{\sqrt{2}}\underline{/45^\circ} \qquad (11\text{-}80)$$

EXERCISE

11-6.1. The circuit shown in Fig. 11-23 is operating in the sinusoidal steady state. The following is given:

$$R_1 = R_2 = 3\,\Omega, \quad L = 20\,\text{mH}, \quad C = 2\,\text{mF}, \quad v = 5\sin(200t + 113.13^\circ)$$

Find the phasor quantities V_1, V_2, V_3, I_1, I_2, and I_3 and sketch these phasors with phasor additions clearly indicated.

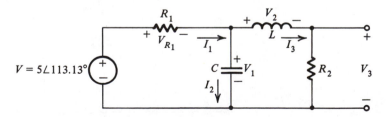

Figure 11-23. Circuit for Exercise 11-6.1.

Ans. $V_1 = 3.249\underline{/85.23^\circ}$, $I_1 = 0.872\underline{/148.67^\circ}$; $V_2 = 2.6\underline{/122.09^\circ}$, $I_2 = 1.300\underline{/175.23^\circ}$; $V_3 = 1.95\underline{/32.09^\circ}$, $I_3 = 0.650\underline{/32.09^\circ}$

Next we show that a voltage source may be transformed into an equivalent current source, and vice versa, provided only that the voltage source has a series impedance and the current source has a parallel impedance. The procedure for doing this is found by starting from the circuit representation of Fig. 11-24(a). For the loop shown, KVL is

$$V_0 = ZI_t + V_t \tag{11-81}$$

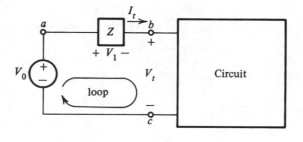

(a)

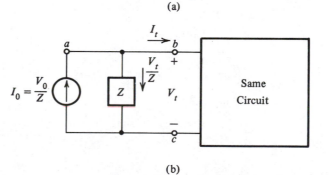

(b)

Figure 11-24. Equivalent circuits found through source transformation.

where V_t is the phasor voltage and I_t the phasor current at the input to the remainder of the circuit. We require, of course, that V_t and I_t not be changed after the source transformation. That is, no change may take place in the rest of the circuit.

After the source transformation, the circuit of (b) in the figure will result. We require the same circuit behavior as before at terminals b–c. By KCL at node b,

$$I_0 = \frac{V_t}{Z} + I_t \tag{11-82}$$

By comparison with Eq. 11-81, we see that the equations are identical if

$$V_0 = ZI_0 \tag{11-83}$$

in which case the same terminal behavior at b–c will result.

The source transformation to current source is carried out by making $I_0 = V_0/Z$ and placing Z in parallel. For transformation to the voltage source we use Eq. 11-83 and place Z in series. Note that the transformation is not defined for either $Z = 0$ or $Z = \infty$.

EXAMPLE 11-6

The source shown in Fig. 11-25(a) is described by the phasor voltage $V_1 = 1\underline{/45°}$ and is in series with an impedance of $Z_1 = 1.732 + j1$ ohms. It is required to find the current-source equivalent.

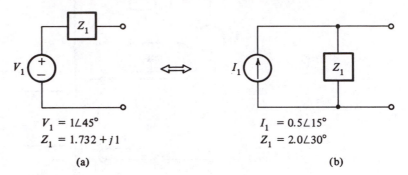

$V_1 = 1\angle 45°$
$Z_1 = 1.732 + j1$

$I_1 = 0.5\angle 15°$
$Z_1 = 2.0\angle 30°$

(a)   (b)

Figure 11-25.   Circuit for Example 11-6.

Solution

The current source has the value V_1/Z_1, which is

$$I_2 = \frac{1\underline{/45°}}{2\underline{/30°}} = 0.5\underline{/15°} \text{ A} \tag{11-84}$$

as shown in the figure.

The transformation may be used for successive simplification, as illustrated by the steps shown in Fig. 11-26.

EXERCISE

11-7.1. Find the values $|V|, \theta, |Z|$, and ψ that are the result of the application of source transformation on the circuit shown in Fig. 11-27 at $\omega = 200$.

11-8 THÉVENIN AND NORTON EQUIVALENT CIRCUITS

We start with the representation shown in Fig. 11-28, in which a circuit A is enclosed in a box with terminals marked a and b. The circuit may contain any elements, independent sources, and controlled sources. The *load* connected at terminals a–b is quite arbitrary. We wish to replace the circuit to the left of

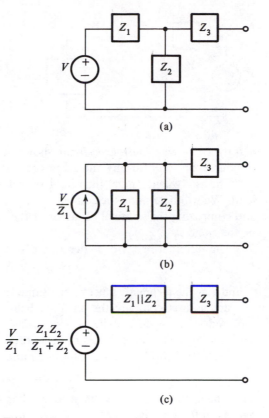

(a)

(b)

(c)

Figure 11-26. Successive steps of circuit reduction.

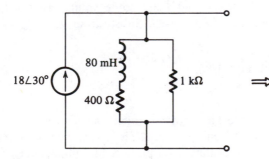

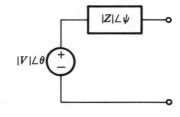

Figure 11-27. Circuit for Exercise 11-7.1.

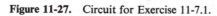

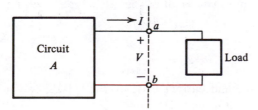

Figure 11-28. Circuit used in explaining Thévenin's theorem.

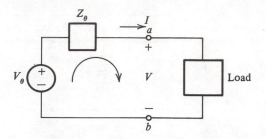

Figure 11-29. The Thévenin equivalent circuit.

a–b by a *Thévenin equivalent circuit* shown in Fig. 11-29. The equivalent must, of course, be valid for any and all possible loads.

The equivalence must thus hold true if the load is replaced by any other load. We will determine the equivalent circuit by considering two distinct loads and comparing the terminal behavior of the two circuits, which must, of course, be the same.

The terminal behavior is described by KVL, for the equivalent circuit, as

$$V_\theta = Z_\theta I + V \tag{11-85}$$

V and I at the terminals must be the same for the two equivalent circuits.

Let us first compare the terminal behavior when the load is temporarily replaced by an open circuit. In Fig. 11-28, $I = I_{oc} = 0$ and V is some value, say $V = V_{oc}$. The same values must be present in Fig. 11-29 if the equivalence is to hold. Substituting these values of I and V in Eq. 11-85,

$$V_\theta = V_{oc} \tag{11-86}$$

The Thévenin equivalent voltage source has the value of the open-circuit voltage as in the resistive case.

Next, let us replace the load by a short circuit. Then in the original circuit of Fig. 11-28, $V = V_{sc} = 0$ and I has some value, say $I = I_{sc}$. These same values must be present in Fig. 11-29. Substituting $V = V_{sc} = 0$ and $I = I_{sc}$ in Eq. 11-85,

$$V_\theta = Z_\theta I_{sc} \tag{11-87}$$

or, combining with Eq. 11-86,

$$Z_\theta = \frac{V_{oc}}{I_{sc}} \tag{11-88}$$

gives us the value of the Thévenin equivalent impedance.

We see that the Thévenin equivalent is determined pretty much as before except that we are now dealing with phasors and impedances. In addition, we may determine Z_θ in an alternative manner by setting all independent sources to zero in the two equivalent circuits. Then, for Fig. 11-29, Eq. 11-85 becomes

$$0 = Z_\theta I + V \tag{11-89}$$

or

$$Z_\theta = \frac{V}{-I}\bigg|_{\text{ind.sources}\equiv 0} \tag{11-90}$$

The same voltage and current exist in Fig. 11-28, in which all independent sources are also set to zero. Here

$$Z_{in} = \frac{V}{-I}\bigg|_{ind.sources \equiv 0} \tag{11-91}$$

where $-I$ represent the current into the circuit A, and Z_{in} is, of course, the input impedance of A. Then, by Eqs. 11-90 and 11-91,

$$Z_\theta = Z_{in} \tag{11-92}$$

or the Thévenin impedance is just the input impedance of the original circuit. This may be obtained in many cases by inspection.

Either Eq. 11-88 or 11-92 may be used to obtain Z_θ. Together with Eq. 11-86, these equations represent the Thévenin equivalent. The Norton equivalent of Fig. 11-30 is obtained by a source transformation,

$$I_\theta = \frac{V_\theta}{Z_\theta} \tag{11-93}$$

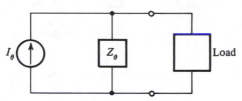

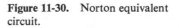

Figure 11-30.   Norton equivalent circuit.

which from Eqs. 11-86 and 11-88 gives us

$$I_\theta = I_{sc} \tag{11-94}$$

The Norton equivalent may be obtained directly by using Eqs. 11-94 and 11-92 or 11-88.

EXAMPLE 11-7

We wish to determine both the Thévenin and Norton equivalent circuits for the voltage divider studied in Section 11-5, specialized to the form shown in Fig. 11-31(a).

Solution

We first short out the voltage source so that $V_1 = 0$ and then determine the input impedance of the R and C in parallel. It is the same as Thévenin impedance Z_θ:

$$Z_\theta = Z_{in} = R \bigg|\bigg| \frac{1}{Cs} = \frac{R(1/Cs)}{R + 1/Cs} = \frac{1/C}{s + 1/RC} \tag{11-95}$$

The Thévenin source is given by the open-circuit voltage, determined by the voltage-divider relationship, which is

$$V_\theta = V_{oc} = \frac{1/Cs}{R + 1/Cs} V_1 = \frac{1/RC}{s + 1/RC} V_1 \tag{11-96}$$

The Thévenin equivalent circuit is shown in Fig. 11-31(b), and the Norton equivalent circuit in Fig. 11-31(c).

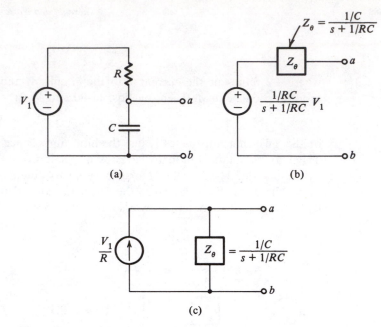

(a)   (b)

(c)

Figure 11-31. Thévenin (b) and Norton (c) equivalent of the circuit of (a).

EXAMPLE 11-8

Determine the Thévenin equivalent for the circuit of Fig. 11-32.

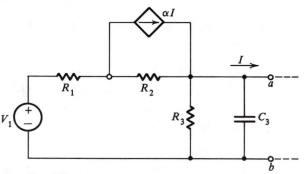

Figure 11-32. Circuit for Example 11-7.

Solution

We first determine the open-circuit voltage at a–b. Since for an open circuit $I = 0$, the controlled source is also open. By voltage-divider action,

$$V_\theta = V_{\text{OC}} = \frac{\dfrac{R_3}{R_3 C_3 s + 1}}{R_1 + R_2 + \dfrac{R_3}{R_3 C_3 s + 1}} V_1 \tag{11-97}$$

$$= \frac{R_3}{R_3 + (R_1 + R_2)(R_3 C_3 s + 1)} V_1 \tag{11-98}$$

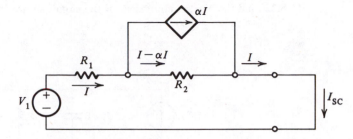

Figure 11-33. Circuit of Fig. 11-32 modified to determine I_{SC}.

Next, we use Eq. 11-88 to determine Z_θ. Figure 11-33 shows the short-circuited version of Fig. 11-32. R_3 and C_3 are not shown since they are shorted out. By KVL,

$$V_1 = R_1 I + R_2 (I - \alpha I) \tag{11-99}$$

so that

$$I_{SC} = I = \frac{V_1}{R_1 + R_2 (1 - \alpha)} \tag{11-100}$$

Then by Eq. 11-88,

$$Z_\theta = \frac{V_{OC}}{I_{SC}} = \frac{R_3 [R_1 + R_2 (1 - \alpha)]}{R_3 + (R_1 + R_2)(R_3 C_3 s + 1)} \tag{11-101}$$

The two equations, Eqs. 11-98 and 11-101, represent the Thévenin equivalent.

EXERCISES

11-8.1. Find the Thévenin equivalent of the circuit shown in Fig. 11-34 at terminal pair *a–b*.

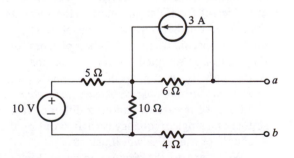

Figure 11-34. Circuit of Exercise 11-8.1.

Ans. $V_{eq} = -11.33$, $R_{eq} = 13.33$

11-8.2. Find the Norton equivalent of the circuit shown in Fig. 11-35 at terminal pair
a–b for $\omega = 300$ rad/s.

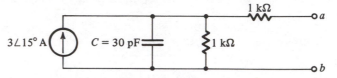

Figure 11-35.   Circuit of Exercise 11-8.2.

Ans. $I_{eq} = 0.474\underline{/-56.57°}$, $Z_{eq} = 1039.7\underline{/-8.97°}$

11-9 SUPERPOSITION

The principle of superposition applies equally well to the sinusoidal analysis.
All our circuits must, of course, be made up of only linear elements. Two different possibilities, each with multiple sources, may occur. One possibility is
that all sources are sinusoidal of the *same* frequency. The other is that the sources
are of *different* frequencies.

The general principle of superposition applies to all situations, including
the two mentioned above. Let us summarize the principle:

1. Select one independent source in the circuit.
2. Turn off all other independent sources.
3. Determine the desired response component.
4. Repeat the steps above for each source in the circuit.
5. Sum the (response) components to give the total response.

Step 3 is carried out in sinusoidal analysis using phasors and impedances
determined for the *specific source frequency* selected. If the source frequencies
are *different* for different sources, the *impedances must be reevaluated for each
frequency*. Furthermore, the phasor description of the response component
implies a specific frequency. When phasors for different frequencies are present,
they *cannot* be summed, in step 5, as a phasor sum. Remember, the phasor sum
is used only to sum signals of the same frequency. *Step 5 must use time functions
or sinusoids in summing responses corresponding to different frequencies.*

Except for care in handling different frequencies, the analysis for a single
source or a single frequency is carried out by the usual phasor analysis methods.
A couple of examples will illustrate.

We return to the circuit shown in Fig. 11-31(a). Suppose that the source
is now made up of two sinusiodal components, given by the time equation

$$v_1(t) = 100 \sin 1000t + 10 \sin 2000t \qquad (11\text{-}102)$$

The second term is described as a 10% second harmonic. We first represent each of the components as a separate generator and then connect the two in series as shown in Fig. 11-36. To this circuit we will apply superposition.

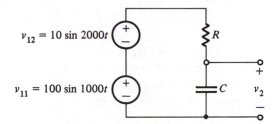

Figure 11-36. Circuit with two different sources.

We must determine the output phasor for each frequency input. It is best to use the transfer function for the circuit already determined for V_2 as output and some V_1 as input in Eq. 11-96. The function is

$$T(s) = \frac{V_2}{V_1} = \frac{1/RC}{s + 1/RC} \qquad (11\text{-}103)$$

Suppose that we are given the values $R = 1 \ \text{K}\Omega$ and $C = 1 \ \mu\text{f}$. Then $RC = 10^{-3}$. For sinusoidal analysis, we let $s = j\omega$ in Eq. 11-103:

$$\frac{V_2(j\omega)}{V_1(j\omega)} = \frac{1000}{1000 + j\omega} \qquad (11\text{-}104)$$

The response v_2 in Fig. 11-36 is determined by first selecting v_{11} and setting $v_{12} = 0$. Then $\omega_1 = 1000$, and the input phasor is $V_1(j1000) = V_{11} = 100\underline{/0°}$. The response component for ω_1 is found from Eq. 11-104,

$$V_{21} = V_2(j1000) = \frac{1000}{1000 + j\omega_1} V_1(j1000) \qquad (11\text{-}105)$$

$$= \frac{1000}{1000 + j1000} 100 \qquad (11\text{-}106)$$

$$= 70.7\underline{/-45°} \qquad (11\text{-}107)$$

Similarly, we next select v_{12} for excitation and set $v_{11} = 0$. Then $\omega_2 = 2000$ and input is $V_1(j2000) = V_{12} = 10\underline{/0°}$. The response component for ω_2 is

$$V_{22} = V_2(j2000) = \frac{1000}{1000 + j2000} 10 \qquad (11\text{-}108)$$

$$= 4.47\underline{/-63.43°} \qquad (11\text{-}109)$$

The total response is found by summing the responses at the two frequencies. That is, they must be first transformed to the time functions, since they are phasors at two different frequencies. Thus,

$$v_2(t) = 70.7 \sin (1000t - 45°) + 4.47 \sin (2000t - 63.43°) \qquad (11\text{-}110)$$

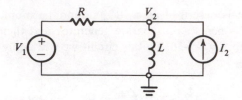

Figure 11-37. Circuit with voltage source and current source to which superposition is applied.

The circuit in Fig. 11-37 is shown with phasor voltages and currents, since a single sinusoidal frequency ω_0 is specified. The impedances are R and $j\omega_0 L$ for the two elements. Since both sources are of the same frequency, the phasor response components may be added by phasor sum. Suppose that we wish to find V_2.

We first select V_1 to drive the circuit with $I_2 = 0$. Then, by voltage-divider action,

$$V_{21} = \frac{sL}{R + sL} V_1 \tag{11-111}$$

$$= \frac{j\omega_0 L}{R + j\omega_0 L} V_1 \tag{11-112}$$

Next we let $V_1 = 0$ and use I_2 to drive the circuit. In this case

$$V_{22} = \frac{sLR}{R + sL} I_2 \tag{11-113}$$

$$= \frac{j\omega_0 LR}{R + j\omega_0 L} I_2 \tag{11-114}$$

The total response is the phasor sum

$$V_2 = V_{21} + V_{22} \tag{11-115}$$

$$= \frac{j\omega_0 L}{R + j\omega_0 L}(V_1 + RI_2) \tag{11-116}$$

The response is of course, a sinusoid of frequency ω_0. Its phase and magnitude for specific values of the parameters may be determined from Eq. 11-116.

EXERCISE

11-9.1. Find V_1, V_2, I_1, and I_2 at $\omega = 20$ for the circuit shown in Fig. 11-38 by use of the principle of superposition.

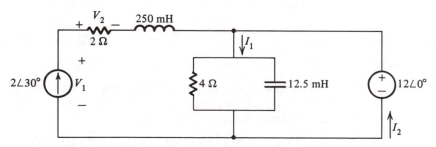

Figure 11-38. Circuit for Exercise 11-9.1.

PROBLEMS

11-1. For the circuit given in the figure, find $Z(j\omega)$ for the frequency $\omega = 10^4$ rad/s. Repeat solving for $Z(j\omega)$ for all frequencies.

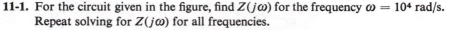

Figure P11-1

11-2. For the circuit given in the figure, show that

$$Z(s) = \frac{(s+2)(s+5)}{(s+1)(s+3)}$$

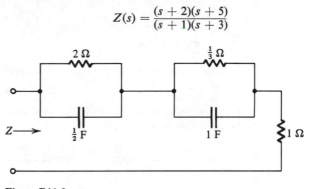

Figure P11-2

11-3. For the circuit given in the figure, find expressions for the following ratios: V_2/V_1, I_2/I_1, V_2/I_1, and I_2/V_1.

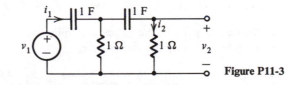

Figure P11-3

11-4. For the circuit given in the figure, determine a value for $Z = V_1/I_1$ and for V_2/V_1.

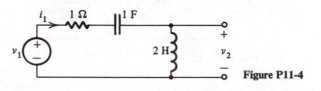

Figure P11-4

287

11-5. For the circuit given in the figure, determine a value for $Z = V_1/I_1$ and for V_2/V_1.

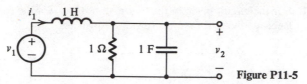

Figure P11-5

11-6. For the circuit given in the figure, show that

$$\left|\frac{V_2(j\omega)}{V_1(j\omega)}\right|^2 = \frac{1}{1 + \omega^6}$$

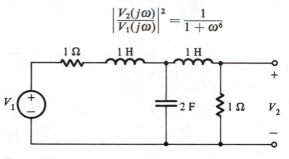

Figure P11-6

11-7. For the circuit shown in the figure, it is given that

$$v_1 = 2\sqrt{2}\,\sin 2t$$

Find the value of C that will give maximum power in R_2 and determine the value of this maximum power.

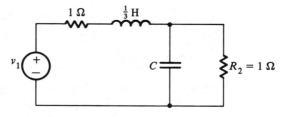

Figure P11-7

11-8. For the circuit given in the figure, determine the ratio of i_4/i if the circuit is operating in the sinusoidal steady state.

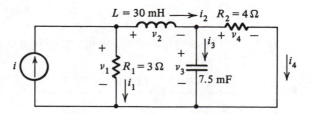

Figure P11-8

11-9. In the given circuit, the voltage sources are

$$v_1(t) = 1250 \sin 377t \quad \text{and} \quad v_2 = 417 \sin 754t \qquad \text{for all } t$$

For the element values given, determine the output voltage $v_3(t)$.

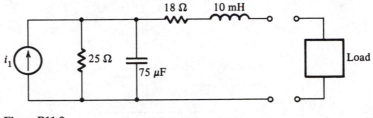

Figure P11-9

11-10. In the circuit given in the figure, it is given that

$$v_L(t) = 20 \sin (2000t + 90°)$$

For the element values given on the figure, determine V_2/i.

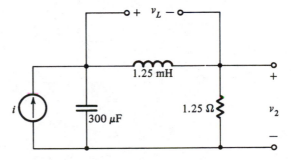

Figure P11-10

11-11. For the circuit given in the figure, the current source has the value

$$i_1 = 13\sqrt{2} \sin (400t + 30°)$$

Find the Thévenin equivalent circuit with respect to the load.

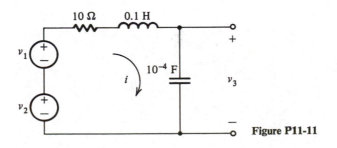

Figure P11-11

11-12. The circuit shown in the figure is known to contain independent sinusoidal sources operating at one frequency plus passive elements. An experiment was performed to determine the Thévenin equivalent circuit at terminals a–b with the following results:

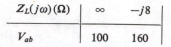

$Z_L(j\omega)\,(\Omega)$	∞	$-j8$
V_{ab}	100	160

Determine the Thévenin equivalent network

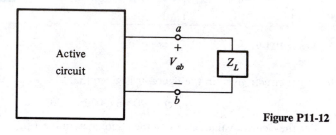

Figure P11-12

12

AVERAGE POWER

In this chapter we develop the concept of average power as distinguished from instantaneous power. We will also be concerned with power transfer, or the adjustment of circuit parameters such that the maximum amount of power is transferred from the source to the load.

12-1 POWER AND ENERGY

The power delivered to a circuit is given by

$$p(t) = v(t)i(t) \qquad \text{W} \tag{12-1}$$

where the polarities are as in Fig. 12-1. A positive value of $p(t)$ represents an actual power delivered to the circuit and a negative value represents power supplied by the circuit. The power in Eq. 12-1 is a function of time and in general changes with time. It is constant only in the dc case when v and i are constants, in which case the circuit is purely resistive. In the dc case discussed in Chapter 4, the power delivered to an input resistance R at the terminal is

$$P = VI = \frac{V^2}{R} = I^2R \qquad \text{W} \tag{12-2}$$

This power is dissipated in a resistor as heat and the corresponding energy dissipated in T_0 seconds is given by

$$W = \int_0^{T_0} P \, dt = PT_0 \qquad \text{W-s} \tag{12-3}$$

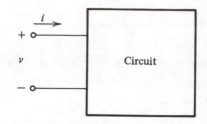

Figure 12-1. Voltage and current reference directions for the definition of power.

We are concerned here with the ac case (i.e., sinusoidal steady state), where

$$v(t) = V_m \sin(\omega t + \theta) \qquad (12\text{-}4)$$

and

$$i(t) = I_m \sin(\omega t + \phi) \qquad (12\text{-}5)$$

The phase angle between the voltage and the current is $\psi = \theta - \phi$. The period of the sinusoids is $T = 2\pi/\omega$. It follows that $v(t + T) = v(t)$ and $i(t + T) = i(t)$. We now observe that the power is also *periodic* with period T since, by Eq. 12-1,

$$
\begin{aligned}
p(t + T) &= v(t + T)i(t + T) \\
&= v(t)i(t) \qquad (12\text{-}6) \\
&= p(t)
\end{aligned}
$$

The value of power thus depends on time but *repeats itself with the same period T*. It is generally more useful to know the *average power* delivered or supplied over a given interval of time than to know the actual time variation. Since power is periodic, we need only determine the average power over one period. The same value of average power is valid over every period.

The average value of power delivered over one period T is given by

$$P_{av} = \frac{1}{T} \int_0^T v(t)i(t)\, dt \qquad (12\text{-}7)$$

For simplicity, let the current phasor be the reference so that $\phi = 0$ and $\psi = \theta$. Then substituting Eqs. 12-4 and 12-5 into Eq. 12-7 yields

$$P_{av} = \frac{1}{T} \int_0^T V_m \sin(\omega t + \psi) I_m \sin \omega t\, dt \qquad (12\text{-}8)$$

$$= \frac{V_m I_m}{T} \int_0^T \sin^2 \omega t \cos \psi\, dt$$

$$+ \frac{V_m I_m}{T} \int_0^T \cos \omega t \sin \omega t \sin \psi\, dt \qquad (12\text{-}9)$$

$$= \frac{V_m I_m}{T} \int_0^T \sin^2 \omega t \cos \psi\, dt + 0 \qquad (12\text{-}10)$$

$$= \frac{V_m I_m}{T} \int_0^T (\tfrac{1}{2} - \tfrac{1}{2} \cos 2\omega t) \cos \psi\, dt \qquad (12\text{-}11)$$

$$= \frac{V_m I_m}{T} \int_0^T \tfrac{1}{2} \cos \psi\, dt \qquad (12\text{-}12)$$

so that

$$P_{av} = \frac{V_m I_m}{2} \cos \psi \qquad\qquad (12\text{-}13)$$

Note that we must use the *cosine of the angle between the voltage and the current phasor*. If $\phi \neq 0$, then $\psi = \theta - \phi$. Note also that $\cos(-\psi) = \cos \psi$, so that we need only the absolute value of ψ.

Let us now take some specific cases and examine the average power in each case. We assume that there are no independent sources present in the circuits considered below.

1. *Resistor:* The voltage and current are in phase, so that $\psi = 0$ and $\cos \psi = 1$.

$$P_{av} = \frac{V_m I_m}{2} \qquad\qquad (12\text{-}14)$$

Since $V_m = R I_m$, we can also write

$$P_{av} = \frac{V_m^2}{2R} = \frac{I_m^2 R}{2} \qquad\qquad (12\text{-}15)$$

2. *Capacitor:* The voltage and current are 90° apart, so that $\psi = 90°$ and $\cos \psi = 0$.

$$P_{av} = 0 \qquad\qquad (12\text{-}16)$$

3. *Inductor:* Again, $\psi = 90°$ and $\cos \psi = 0$.

$$P_{av} = 0 \qquad\qquad (12\text{-}17)$$

The average power delivered to the reactive elements is zero. This does not mean that the instantaneous power in Eq. 12-1 is zero for all t. What really happens is that the power is delivered to and stored in the reactive element for part of the cycle and, for an equal duration of the cycle, it is returned or supplied by the reactive element to the rest of the circuit. This is shown in Fig. 12-2.

4. Impedance: Let the circuit be defined by an input impedance Z. Let

$$Z = |Z| \underline{/\psi} = R + jX \qquad\qquad (12\text{-}18)$$

If

$$I = I_m \underline{/\phi} \qquad\qquad (12\text{-}19)$$

and

$$V = V_m \underline{/\theta} \qquad\qquad (12\text{-}20)$$

then

$$V_m = |Z| I_m \qquad\qquad (12\text{-}21)$$

and

$$\psi = \theta - \phi \qquad\qquad (12\text{-}22)$$

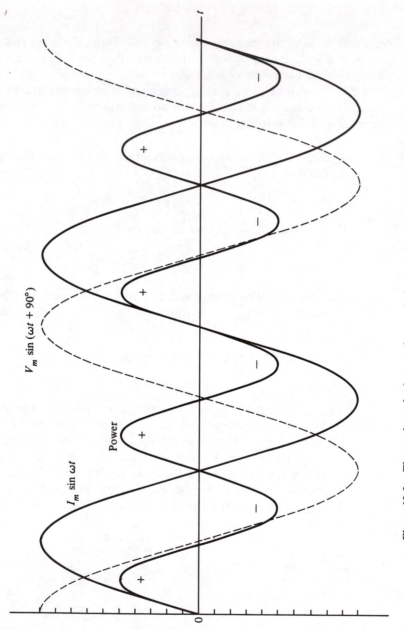

$I_m \sin \omega t$

$V_m \sin (\omega t + 90°)$

Power

Figure 12-2. The product of voltage and power is shown to be a quantity having an average value of zero.

Also,

$$|Z| \cos \psi = R \qquad (12\text{-}23)$$

or

$$\cos \psi = \frac{R}{|Z|} \qquad (12\text{-}24)$$

Substituting in Eq. 12-13, we have

$$P_{av} = \frac{V_m I_m}{2} \cos \psi \qquad (12\text{-}25)$$

$$= \frac{V_m I_m R}{2|Z|} \qquad (12\text{-}26)$$

$$= \frac{V_m^2 R}{2|Z|^2} \qquad (12\text{-}27)$$

$$= \frac{I_m^2 R}{2} \qquad (12\text{-}28)$$

Here R is the resistive part of the general impedance Z.

In all other circuits for which the impedance is not known or there are independent sources present, we must resort to Eq. 12-13.

EXERCISES

12-1.1. In the circuit shown as Fig. 12-1, it is given that

$$v = 180 \cos \omega t$$

$$i = 2.5 \cos (\omega t - 45°)$$

Find P_{av}, R, and $|Z|$.

Ans. 159, 50.9, 72

12-1.2. In the first exercise, the voltage is the same, but the current is changed to

$$i = 0.30 \cos (\omega t + 60°)$$

Find P_{av}, R, and $|Z|$.

Ans. 112.5, 300, 600

12-2 RMS VALUES

If we compare Eq. 12-15 with Eq. 12-2, we see that they are identical except for a factor of 2. It is possible to make them identical by defining *the effective values* of the voltage and the current as follows:

$$V_{eff} = \frac{V_m}{\sqrt{2}} \qquad (12\text{-}29)$$

$$I_{eff} = \frac{I_m}{\sqrt{2}} \qquad (12\text{-}30)$$

Then the average power in the resistive case is

$$P_{av} = \frac{V_{eff}^2}{R} = I_{eff}^2 R \tag{12-31}$$

Similarly, for the impedance case,

$$P_{av} = V_{eff}^2 \frac{R}{|Z|^2} \tag{12-32}$$

$$= I_{eff}^2 R \tag{12-33}$$

The effective values of the voltage and the current are also called the *rms values*. rms stands for *root mean square*, which is the square root of the average of the square of the function. We show this by equating the expression of Eq. 12-31 with that of Eq. 12-8. For the resistive case $\psi = 0$ and

$$V_m = RI_m \tag{12-34}$$

Then

$$\frac{V_{eff}^2}{R} = \frac{1}{T} \int_0^T V_m I_m \sin^2 \omega t \, dt \tag{12-35}$$

$$= \frac{1}{T} \int_0^T \frac{V_m^2}{R} \sin^2 \omega t \, dt \tag{12-36}$$

Thus,

$$V_{eff} = \sqrt{\frac{1}{T} \int_0^T V_m^2 \sin^2 \omega t \, dt} \tag{12-37}$$

$$= \sqrt{\frac{1}{T} \int_0^T v^2(t) \, dt} \tag{12-38}$$

$$= V_{rms} \tag{12-39}$$

In the same way,

$$I_{eff} = \sqrt{\frac{1}{T} \int_0^T i^2(t) \, dt} \tag{12-40}$$

$$= I_{rms} \tag{12-41}$$

The rms values are the ones used in measurements of voltages and currents. If we are told that the ac voltage is 110 V, it is implied that the rms value of the sinusoidal voltage $V_{rms} = 110$. The magnitude is $V_m = \sqrt{2} V_{rms}$.

Since rms quantities are the measured quantities, most ac relations are assumed to be written in terms of rms quantities. Then the subscript rms is dropped and average power is simply $|V||I| \cos \Psi$. Also, $V = ZI$ with $|V| = |Z||I|$ and so forth. Here phasors are assumed to have rms magnitudes. The impedance relation looks the same because both voltage and current phasors are scaled by a factor of $\sqrt{2}$. We will, however, use the definition of phasors as we have introduced it and indicate explicitly if rms values are to be used.

Several of the terms used in the last paragraph are identified by names. The product of magnitudes $|V||I|$ is known as the *apparent power* and is measured in volt-amperes (VA) or kilovolt-amperes (KVA). The quantity $\cos \psi$ in Eq. 12-25 is known as the *power factor* (pf), and ψ as the *power factor angle*. Power factor is further specified by whether the current leads or lags the voltage. Thus an inductive load results in a lagging power factor while a capacitive load results in a leading power factor. The power factor multiplies the apparent power to give the average power. The apparent power will equal the average power when voltage and current are in phase such that $\psi = 0$ and pf $= 1$.

In electric power systems, an objective is to deliver maximum power to the load (sometimes called the plant.) Most power systems are inductive. To obtain a power factor as close to unity as possible, it is common practice to add special large capacitors in parallel with the load. These capacitors achieve power factor correction.

The next section discusses the maximum power delivered to a load when it is possible to vary the load but the source impedance is fixed.

Finally, the energy delivered to a circuit is easily calculated from the average power. Since the average power is constant, the energy delivered in any time interval T_0 is

$$W = P_{av}T_0 \qquad \text{W-s} \qquad (12\text{-}42)$$

The common unit in electrical power systems is the kilowatt-hour, where P_{av} is in kilowatts and T_0 is in hours.

EXERCISES

12-2.1. It is known that a given capacitor will fail if the terminal voltage exceeds 180 V. What is the maximum rms voltage that may be applied to the capacitor, assuming that the circuit is operating in the sinusoidal steady state?

Ans. 127.3 V

12-2.2. Find the rms value of a periodic sawtooth voltage that is described as

$$v(t) = t \qquad \text{from 0 to } T$$

Ans. $V_{rms} = 1/\sqrt{3}$

12-3   MAXIMUM POWER TRANSFER

In Chapter 4 we determined that for resistive circuits maximum power delivered to a load R_L, by a source with fixed source resistance R_S, occurs when $R_L = R_S$. We now wish to determine the conditions for maximum power transfer in the ac case.

Let us consider the circuit of Fig. 12-3, where the Thévenin source network has an impedance Z_S and the load impedance is Z_L. Let

$$Z_S = R_S + jX_S \qquad (12\text{-}43)$$

and

$$Z_L = R_L + jX_L \qquad (12\text{-}44)$$

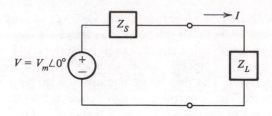

Figure 12-3. Thévenin equivalent circuit to which a load is connected.

The average power delivered to the load is, by Eq. 12-28,

$$P_L = \frac{I_m^2 R_L}{2} = I_{rms}^2 R_L \tag{12-45}$$

$$= \frac{V_{rms}^2}{|Z_S + Z_L|^2} R_L \tag{12-46}$$

$$= V_{rms}^2 \frac{R_L}{(R_S + R_L)^2 + (X_S + X_L)^2} \tag{12-47}$$

Reactance can take on either positive or negative value and we see that the best that X_L can do to maximize P_L is to make $X_S + X_L = 0$, or

$$X_L = -X_S \tag{12-48}$$

Then the power in the load becomes

$$P_L = V_{rms}^2 \frac{R_L}{(R_S + R_L)^2} \tag{12-49}$$

which is the same expression as in the resistive case of Chapter 4. This expression is maximized by making

$$R_L = R_S \tag{12-50}$$

giving us

$$P_{max} = \frac{V_{rms}^2}{4R_S} \tag{12-51}$$

The condition the load impedance must satisfy for maximum power transfer from a fixed source impedance is given by Eqs. 12-48 and 12-50,

$$Z_L = R_L + jX_L = R_S - jX_S \tag{12-52}$$

$$= Z_S^* \tag{12-53}$$

The load impedance is thus repuired to be the *complex conjugate* of the source impedance. For these reasons the term *conjugate match* or *conjugate impedance match* is frequently used. Effort devoted to impedance matching is fully rewarded by the increased power in the load.

EXAMPLE 12-1

Determine the load in Fig. 12-4 for maximum power transfer. Find the energy delivered to the load in 1 h.

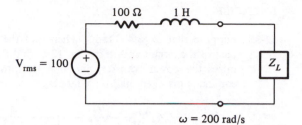

$\omega = 200$ rad/s

Figure 12-4. Circuit for Example 12-1.

Solution

$$Z_S = 100 + j200 \tag{12-54}$$

and

$$Z_S^* = 100 - j200 \tag{12-55}$$

For maximum power transfer we need

$$Z_L = Z_S^* = 100 - j200 \tag{12-56}$$

$$= R_L - j\frac{1}{\omega C_L} \tag{12-57}$$

Then

$$R_L = 100, \qquad X_L = -200 \tag{12-58}$$

Negative reactance is provided by a capacitance, so that

$$-\frac{1}{\omega C_L} = -200 \tag{12-59}$$

or

$$C_L = \frac{1}{200 \times 200} = 0.25 \times 10^{-4} \text{ F} \tag{12-60}$$

Average power delivered to the load is

$$P_L = \frac{V_{rms}^2}{4R_S} = \frac{100^2}{400} = 25 \text{ W} \tag{12-61}$$

Energy delivered in 1 h is

$$W = 0.025 \times 1 = 0.025 \text{kWh} \tag{12-62}$$

We conclude this chapter by once again emphasizing that sinusoidal steady-state analysis is quite analogous to resistive analysis. We now use phasors and impedances for a single frequency. More generally, we use functions of frequency for greater versatility. The complex frequency variable s is convenient to express the circuit functions. For the sinusoidal case, we let $s = j\omega$. In later chapter we will see that it is important to cultivate the habit of first deriving all circuit functions in terms of s *even if* our interest is in the sinusoidal case. The algebra is much easier, the functions always lead to ratios of polynomials in s, and there are other benefits besides as we shall see.

EXERCISE

12-3.1. Suppose that $\omega = 400$ rad/s. Then find the Thévenin equivalent of the left part of the circuit shown in Fig. 12-5, find the load impedance Z_L that maximizes the power transfer, and find the power that is transferred with the conjugate matched load impedance.

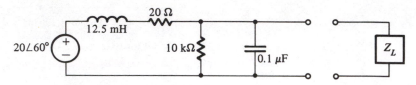

Figure 12-5. Circuit for Exercise 12-3.1.

PROBLEMS

12-1. The figure shows three different waveforms which are made up of all or part of the same sinusoid. If v_3 has an rms value of 0.76, what is the rms value of v_2 and v_1?

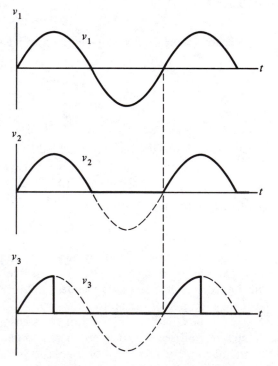

Figure P12-1

12-2. The figure shows a sine wave in which some half-cycles are removed; the waveform is periodic. It is known that the period is 0.001 s and that the rms value of the function is 0.93. What is the equation of the function when it is not zero?

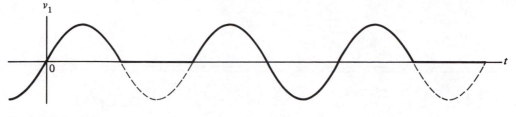

Figure P12-2

12-3. Determine the rms value of the signal $v(t) = 5 + 2 \sin 3t$.

12-4. Calculate the rms value of the signal given by the equation

$$v(t) = 5 \sin \omega t + 4 \sin \left(\omega t + \frac{\pi}{6}\right)$$

12-5. The average power for the plant shown in the figure has a rating of 1000 kW and a power factor of 0.8 lagging. The voltage of the source is 2300 V and the frequency is $\omega_0 = 377$ rad/s (60 Hz). A capacitor C is to be placed in parallel with the plant as shown in the figure to correct the power factor. Determine the value of C such that the following power factor is obtained: (a) 0.9 lagging, (b) 1.0, (c) 0.9 leading.

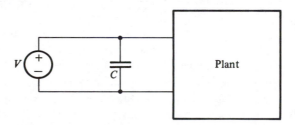

Figure P12-5

12-6. Consider the same system represent in the figure for Prob. 12-5. Let the rating of the plant be 2000 kVA and the power factor be 0.8 lagging with the voltage of the source being 1000 V. The system is operating in the sinusoidal steady state at the frequency given in Prob. 12-5. Determine the value of C such that (a) power factor = 0.95 lagging, and (b) power factor = 0.95 leading.

12-7. The current source shown by its phasor value in the circuit of the figure generates a sinusoidal current of frequency 1000 Hz. Find (a) the value of the load that maximizes the average power delivered to the load, and (b) the value of this power.

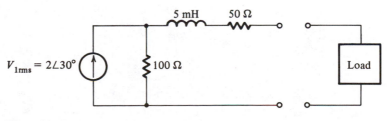

Figure P12-7

13

THE FREQUENCY SPECTRUM

So far we have considered sinusoids of a single frequency and determined the steady-state response of circuits excited by them. We have also used superposition to determine the steady-state response when there are several sinusoidal sources present, each of a different frequency. We now extend our consideration of *periodic functions*. A function $f(t)$ is periodic with period T_0 if $f(t) = f(t \pm T_0)$. Thus a periodic function repeats itself every T_0 seconds.

13-1 HARMONICS

We first show that if two or more sinusoids are added together, the resulting waveform is periodic if the frequencies of the individual sinusoids are multiples of frequency called the fundamental. For example, if ω_0 is the fundamental frequency with period $T_0 = 2\pi/\omega_0$, then the signal

$$v(t) = A_1 \sin(\omega_0 t + \theta_1) + A_k \sin(k\omega_0 t + \theta_k) \tag{13-1}$$

with k an integer is also periodic with the same period T_0. This can be seen from

$$
\begin{aligned}
v(t + T_0) &= A_1 \sin(\omega_0(t + T_0) + \theta_1) + A_k \sin(k\omega_0(t + T_0) + \theta_k) \tag{13-2} \\
&= A_1 \sin(\omega_0 t + \theta_1 + \omega_0 T_0) + A_k \sin(k\omega_0 t + \theta_k + k\omega_0 T_0) \\
&= A_1 \sin(\omega_0 t + \theta_1 + 2\pi) + A_k \sin(k\omega_0 t + \theta_k + k2\pi) \\
&= A_1 \sin(\omega_0 t + \theta_1) + A_k \sin(k\omega_0 t + \theta_k) \\
&= v(t)
\end{aligned}
$$

An example is shown in Fig. 13-1 for the case when $k = 2$ and both the phase angles are zero. The sinusoid of frequency $k\omega_0$ is called the kth harmonic.

302

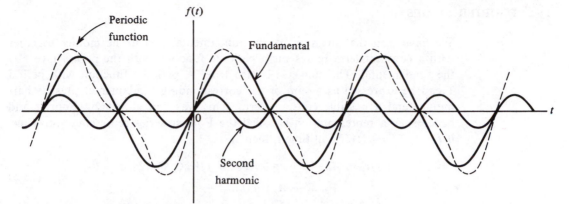

Figure 13-1. Three waveforms are shown: the fundamental, the second harmonic, and their sum, which is also periodic.

Of course, the sum of any number of harmonics added to the fundamental will also be periodic with the same fundamental period. Thus,

$$v(t) = A_1 \sin(\omega_0 t + \theta_1) + A_2 \sin(2\omega_0 t + \theta_2) + \ldots$$
$$+ A_k \sin(k\omega_0 t + \theta_k) + \ldots \tag{13-3}$$

will be a periodic function with period T_0. Since a constant added to a periodic function does not change its periodic nature, we could also have

$$v(t) = A_0 + A_1 \sin(\omega_0 t + \theta_1) + \ldots + A_k \sin(k\omega_0 t + \theta_k) + \ldots \tag{13-4}$$

as a periodic function with period T_0.

EXERCISE

13-1.1. The half-wave rectified sine wave shown in Fig. 13-2 has the Fourier series

$$f(t) = \frac{A}{\pi} + \frac{A}{2} \sin \omega_0 t - \frac{2A}{\pi} \left(\frac{1}{1 \times 3} \cos 2\omega_0 t + \frac{1}{3 \times 5} \cos 4\omega_0 t \right.$$
$$\left. + \frac{1}{5 \times 7} \cos 6\omega_0 t + \ldots \right)$$

where $\omega_0 = 2\pi/T$. Show that the fourth harmonic is the highest harmonic with a magnitude of at least 5% of the magnitude of the fundamental.

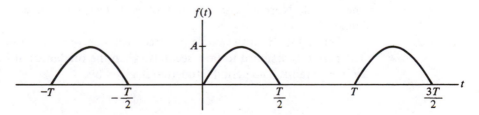

Figure 13-2. Half-wave rectified sine wave.

We have seen that the sum of any fundamental and its harmonics together with a constant term represents a periodic function with the period given by the fundamental. The converse is also true! A periodic function with period T can be expressed as a sum of the corresponding fundamental plus its harmonics and a possible constant term. This was first shown by Fourier and hence such a representation is called the Fourier series representation. Thus, if $f(t + T) = f(t)$ for all time t, then

$$f(t) = A_0 + A_1 \sin (\omega_0 t + \theta_1) + A_2 \sin (2\omega_0 t + \theta_2)$$
$$+ \ldots + A_k \sin (k\omega_0 t + \theta_k) + \ldots \tag{13-5}$$

where $\omega_0 = 2\pi/T$. Notice that this is an infinite sum in general and for a specific function $f(t)$ some or many coefficients may of course be zero. In other cases, none of them may be zero.

Given a periodic function it is possible to find the coefficients A_k and the phases θ_k of Eq. 13-5. However, it is easier to work with a somewhat simpler expression where each sinusoid is assumed to be expanded in its sine and cosine terms. Thus,

$$f(t) = \frac{a_0}{2} + a_1 \cos \omega_0 t + a_2 \cos 2\omega_0 t + \ldots$$
$$+ b_1 \sin \omega_0 t + b_2 \sin 2\omega_0 t + \ldots \tag{13-6}$$

In the above we have expressed A_0 as $a_0/2$ for later convenience and $A_k \sin \theta_k$ as a_k and $A_k \cos \theta_k$ as b_k. The coefficients in Eq. 13-6 are then given by

$$a_0 = \frac{2}{T} \int_a^{a+T} f(t)\, dt \tag{13-7}$$

$$a_k = \frac{2}{T} \int_a^{a+T} f(t) \cos k\omega_0 t\, dt, \qquad k = 1, 2, \ldots \tag{13-8}$$

and

$$b_k = \frac{2}{T} \int_a^{a+T} f(t) \sin k\omega_0 t\, dt, \qquad k = 1, 2, \ldots \tag{13-9}$$

The integration may be taken starting at any point a and extending over one period. Normally, $a = 0$, but other values may be chosen for ease of integration, such as $a = -T/2$.

Let us look at an example of a rectified sine wave as shown in Fig. 13-3. The period is assumed to be π seconds. Then the fundamental frequency $\omega_0 = 2\pi/T = 2$ rad/s. The periodic function is given by

$$f(t) = A\,|\sin t| \tag{13-10}$$

and from Eqs. 13-7 to 13-9, we have

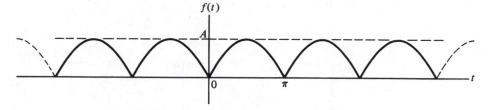

f(t)

A

0   π   t

Figure 13-3. Rectified sine wave.

$$a_0 = \frac{2A}{\pi} \int_0^\pi \sin t\, dt = \frac{4A}{\pi} \tag{13-11}$$

$$a_k = \frac{2A}{\pi} \int_0^\pi \sin t \cos 2kt\, dt = \frac{4A}{\pi} \cdot \frac{1}{1 - 4k^2} \tag{13-12}$$

$$b_k = \frac{2A}{\pi} \int_0^\pi \sin t \sin 2kt\, dt = 0 \tag{13-13}$$

The Fourier series is then given by

$$f(t) = \frac{2A}{\pi}\left[1 + \sum_{k=1}^\infty \frac{2}{1 - 4k^2} \cos 2kt\right] \tag{13-14}$$

If we let $A = \pi$ the series will simplify to

$$f(t) = 2 + \sum_{k=1}^\infty \frac{4}{1 - 4k^2} \cos 2kt \tag{13-15}$$

It should be observed that the Fourier coefficients become small as $k \rightarrow \infty$. In fact, for $k = 10$, the coefficient is of the order of 10^{-2} and rapidly decreases with increasing k. Thus, a fairly good approximation is obtained by taking only those terms with significant amplitudes. In any case, only a finite number of terms significantly contribute to $f(t)$.

It is also possible to write the series as in Eq. 13-4:

$$f(t) = 2 + \sum_{k=1}^\infty \frac{4}{1 - 4k^2} \sin\left(2kt + \frac{\pi}{2}\right) \tag{13-16}$$

giving us the amplitudes and phases of all the harmonics.

Let us consider one more example of a periodic function. Figure 13-4 shows a pulse train with the pulse width given by τ and the period by T. Then the Fourier coefficients are obtained easily using $\omega_0 = 2\pi/T$, $a = -T/2$, and $f(t) = A$ for $-\tau/2 \leq t \leq \tau/2$ and $f(t) = 0$ otherwise.

$$a_0 = \frac{2}{T} \int_{-T/2}^{T/2} f(t)\, dt = \frac{2}{T} \int_{-\tau/2}^{\tau/2} A\, dt = \frac{2A\tau}{T} \tag{13-17}$$

$$a_k = \frac{2}{T} \int_{-\tau/2}^{\tau/2} A \cos \frac{2\pi}{T} kt\, dt = \frac{2A\tau}{T} \cdot \frac{\sin (k\pi/T)\tau}{(k\pi/T)\tau}; \qquad b_k = 0 \tag{13-18}$$

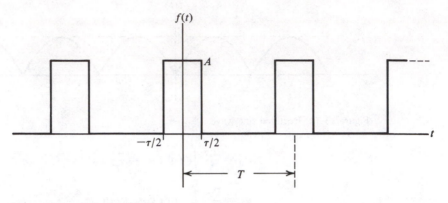

Figure 13-4.   Pulse of duration τ and period T.

The Fourier series is given by

$$f(t) = \frac{2A\tau}{T}\left[\frac{1}{2} + \sum_{k=1}^{\infty} \frac{1}{k\pi\tau/T} \sin\left(\frac{k\pi\tau}{T}\right) \cos k\omega t\right] \qquad (13\text{-}19)$$

Once again we see that the amplitudes become small with increasing k and the phase of every harmonic is $\pm\pi/2$.

EXERCISES

13-2.1. Find the values of the Fourier coefficients a_0, a_k, and b_k in Fig. 13-5 for the function given below by evaluating the appropriate integral formula.

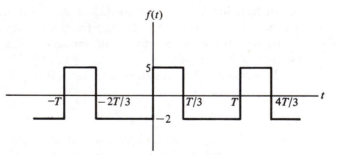

Figure 13-5.   Pulse waveform.

$$\textbf{\textit{Ans.}}\ a_0 = \frac{2}{3},\ a_k = \frac{7}{k\pi}\sin k\frac{2\pi}{3},$$

$$b_k = \frac{7}{k\pi}\left(1 - \cos k\frac{2\pi}{3}\right)$$

13-2.2. Find the coefficients of the sawtooth function given in Fig. 13-6. Show the actual calculus manipulations required to obtain the coefficients. (*Hint:* Find a_0, then find a_k and b_k without replacing the symbol k with a specific value.)

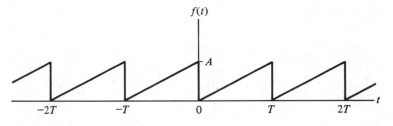

Figure 13-6. Sawtooth waveform.

$$\textbf{\textit{Ans.}}\ \ a_0 = A/2,\ a_k = 0,\ b_k = -\frac{A}{k\pi}$$

13-3   SPECTRUM

The Fourier series gives the frequency content or makeup of a periodic waveform. The constant term gives its dc content, the other terms give its harmonic contents. To find the appropriate content we must combine the cosine and sine terms of kth harmonic into a single sinusoid as in Eq. 13-4 using the fact that $a_k = A_k \sin \theta_k$ and $b_k = A_k \cos \theta_k$.

The amplitude (or magnitude) and the phase of the kth harmonic is thus given by

$$A_k = \sqrt{a_k^2 + b_k^2}$$

and

$$\theta_k = \tan^{-1} \frac{a_k}{b_k} \qquad \text{for all } k = 1, 2, \ldots \tag{13-20}$$

The fundamental is the first harmonic and the constant may be called the zeroth harmonic with $A_0 = a_0/2$. All the relevant information of a periodic function may then be summarized in terms of the values of these amplitudes and phases at the appropriate frequencies, as shown in Figs. 13-7 and 13-8. Such a frequency description of a function is called the *spectrum* or the *frequency spectrum*. A periodic function has a spectrum specified at the discrete frequencies of the harmonics as discrete lines of appropriate height, so it is called a *discrete* or *line* spectrum.

If the fundamental period becomes very large, then the fundamental frequency ω_0 is very small and the space between each successive harmonic (which is of course ω_0) is also very small. An aperiodic signal or waveform function may be considered approximately as being repeated with a very large period. The assumed period should, of course, be so large that successive repetitions do not interfere materially with the signal of interest or the circuit behavior.

Another view is that we have taken an aperiodic signal and, over a given interval of interest, we have represented it as an infinite sum of harmonic sinusoids. This sum yields, over the desired time interval, the same aperiodic

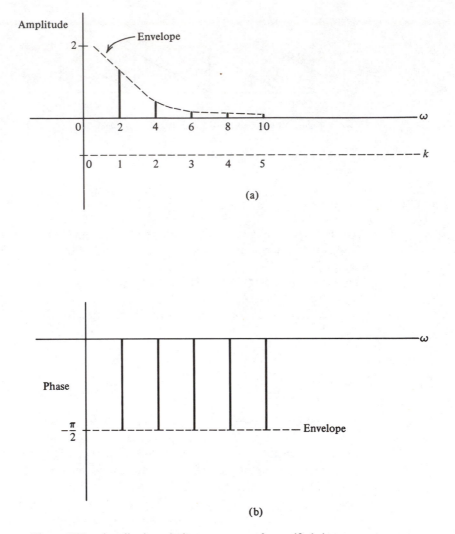

Figure 13-7.   Amplitude and phase spectrum of a rectified sine wave.

time-domain signal. Outside the interval the sum of sinusoids yields a time-domain signal different from the original. We are, however, interested only in representing the original time-domain signal over the desired time interval.

An example of such a situation is shown in Fig. 13-9. The aperiodic signal is a single pulse of width τ. The approximation to the signal is the same pulse repeated with a period T, where $T = 100\tau$. In particular situations we might need to assume T to be $10^3\tau$ or even greater. But in any case, we can always make such an assumption and obtain the corresponding line spectrum for the appropriate approximating period function.

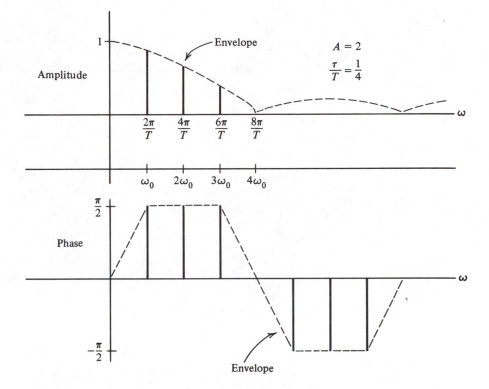

Figure 13-8. Amplitude and phase spectrum for a pulse train.

A true spectrum for the aperiodic signal may be obtained by assuming the period $T \rightarrow \infty$ and taking appropriate limits. This leads to a *continuous* spectrum obtained from a *Fourier integral* representation. We discuss the Fourier integral in later chapters, but for the present take the approach that all signals are either periodic or made periodic as an approximation in the manner discussed above.

Thus, all the signals will be assumed periodic and will have a line spectrum. Frequently, the detailed specifics of a line spectrum may be suppressed and only the *envelope* of the corresponding spectrum may be shown as in Fig. 13-9. The envelope gives a general picture of the nature of a spectrum and may be sufficient information in many situations. Details of the line spectrum are, of course, necessary when specific information is needed.

There are several useful properties of the signal spectrum. It should come as no surprise that if two signals are almost the same in the time-domain, then so are their frequency spectrums. The converse is also true. Thus, it would be reasonable to use idealized signals, instead of actual signals, to determine the frequency spectrum.

Another important concept is that of relative power in the signal spectrum

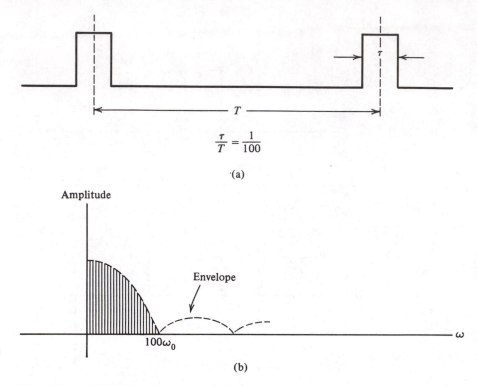

$$\frac{\tau}{T} = \frac{1}{100}$$

(a)

(b)

Figure 13-9. Pulse train with a large time between pulses.

Let the signal be replaced by its frequency spectrum (*i.e.*, a sum of sinusoids). The hypothetical power that would be dissipated, in a unit resistor, by each component sinusoid is proportional to the square of its magnitude which is A_k^2 or $a_k^2 + b_k^2$. Thus, the relative power corresponding to each frequency gives a good idea of its contribution to the overall signal composition.

EXERCISES

13-3.1. A function $f(x)$ is called an *even* function if $f(x) = f(-x)$. It is an *odd* function if $f(x) = -f(-x)$. Show that any function $f(x)$ can be written as the sum of an even part $f_e(x)$ and an odd part $f_o(x)$,

$$f(x) = f_e(x) + f_o(x)$$

by finding expressions for $f_e(x)$ and $f_o(x)$ in terms of $f(x)$ and showing that their sum is equal to $f(x)$.

Ans. $f_e = \frac{1}{2}[f(t) + f(-t)]$,
$f_o = \frac{1}{2}[f(t) - f(-t)]$.

13-3.2. Show that if $f_e(x) = f_o(x)$ for all values of x, then f is identically equal to zero [i.e., $f(x) = 0$ for all values of x].

13-3.3. Find the even and odd parts of the functions $1/(1 + x)$, $e^{j\omega x}$, $a_0 + a_1 x + a_2 x^2$, and $\cos(2x + \pi/3)$.

Since the Fourier series is an infinite series, it is not uncommon to find that many signals have a frequency content or spectrum that is infinite in width. The pulse train and a rectified sine wave both have infinite terms and hence have infinite widths. However, we have also seen that the higher harmonics become small quite fast and are negligibly small for very high k. In other situations we may find that both the lower and the higher harmonics are quite small compared to some intermediate harmonics. It is generally convenient to ignore terms or harmonics beyond certain points and consider only the finite number that are within those points. These *critical* points are generally located on the spectrum envelope at 0.707 (or $1/\sqrt{2}$) of the maximum magnitude and the width between such points is called the *bandwidth* of the signal and is expressed in Hz.

In our examples shown in Fig. 13-10 the bandwidth is 400 Hz for the first case (a) and 550 Hz for the second (b).

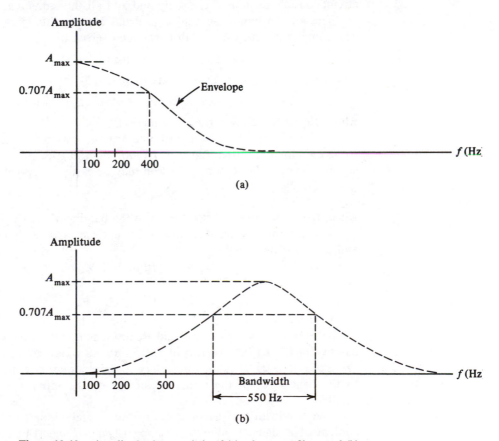

(a)

(b)

Figure 13-10. Amplitude characteristic of (a) a low-pass filter, and (b) a band-pass filter.

EXERCISE

13-4.1. Suppose that the functions given below specify the envelope of the line spectrum of a periodic function. Find the critical points and the bandwidth for each case.

(a) $10e^{-f^2/500}$ **Ans.** Bandwidth = 13.16.

(b) $\dfrac{5}{|1 + 0.003f|}$ **Ans.** Bandwidth = 138.07.

(c) $\dfrac{1}{1 + 0.02(|f| - 100)^2}$ **Ans.** Bandwidth = 9.1.

13-5 SIGNAL PROCESSING

We have seen how steady-state response of circuits to sinusoids of a single frequency can be easily determined using phasors, impedances, transfer ratios, and so on. We have also expressed impedances and transfer ratios as functions of the frequency variable ω. It is easy to obtain the response of a circuit to any periodic signal using superposition. We merely find the response of the circuit to each harmonic separately and add all the individual responses.

The kth harmonic phasor of a signal has amplitude A_k and phase θ_k. The complex phasor for the kth harmonic is given by

$$A_k e^{j\theta_k} \qquad \text{for all } k \qquad (13\text{-}21)$$

and the response phasor at the same frequency is

$$Be^{j\phi_k} = H(jk\omega_0)(A_k \cos \theta_k + jA_k \sin \theta_k) \qquad \text{for all } k \qquad (13\text{-}22)$$

where $H(j\omega)$ is the given transfer function.

In this manner it is possible to obtain the spectrum for the response signal from the circuit transfer function $H(j\omega)$ and the spectrum of the excitation signal. The periodic response function may be easily expressed as

$$B_0 + \sum_k B_k \sin (k\omega_0 t + \phi_k) \qquad (13\text{-}23)$$

where the summation is extended up to the bandwidth of the input signal.

The response spectrum could also be expressed separately for the amplitude and the phase as

$$B_k = |H(jk\omega_0)| A_k \qquad (13\text{-}24)$$

and

$$\phi_k = \underline{/H(jk\omega_0)} + \theta_k \qquad (13\text{-}25)$$

We thus see that B_k depends not only on A_k but also on $|H(jk\omega_0)|$, and if the latter were very small so would B_k be. It is therefore convenient to define bandwidth for $H(j\omega)$ in terms of critical points which are $0.707 \times$ (maximum value of $|H(j\omega)|$). It is then necessary to determine the response spectrum up to the bandwidth of the transfer function or the input signal, whichever is smaller.

The bandwidth of a transfer function merely indicates what frequency content of a signal is allowed to appear (*pass through*) in the response and what frequencies are so reduced (*attenuated or stopped*) in amplitude that they are nonexistent in the response. *Signal processing* is performed by circuits by selectively affecting appropriate frequency content of the input signals to yield output signals of desired characteristics. Of course, circuits may also affect signals in undesirable ways.

Let us consider some examples of transfer functions of certain classes of circuits called *filters*. Both realistic as well as idealized versions are shown in Fig. 13-11. In (a) we have the *low-pass* transfer function that allows only

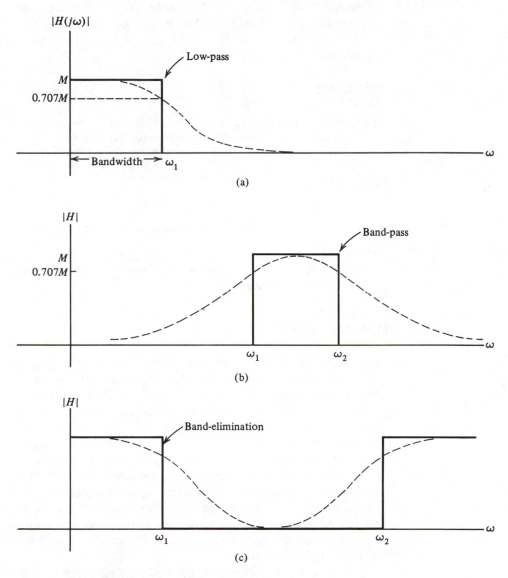

Figure 13-11. Amplitude characteristic of a low-pass, band-pass, and band-elimination filter.

the low-frequency content of signals to pass through and attenuates the high-frequency content. The *band-pass* filter allows a band of frequencies between ω_1 and ω_2 to pass through and attenuates the rest (b). The *band-elimination* filter attenuates the band between ω_1 and ω_2 and lets the rest pass through (c).

Modern communication systems send many distinct items of information by arranging each message in an appropriate frequency range. A composite message is then sent from which each distinct message is filtered out by means of bandpass or other filters. Telephone, radio, television, and so on, all use these techniques and the process of *filtering* or selecting the proper information is performed at the receiving end. In radio and television this process is known as tuning.

EXERCISES

13-5.1. Find the transfer function $T(s) = V_2/V_1$ for the circuit given in Fig. 13-12 and evaluate the magnitude of the transfer function for $s = j\omega$ and $\omega = 0$, ω_0, $2\omega_0$, $4\omega_0$, $6\omega_0$, assuming that $\omega_0 RC = 1$.

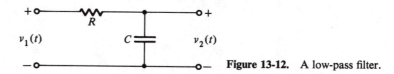

Figure 13-12.   A low-pass filter.

13-5.2. Suppose that the half-wave rectified sine function is the input to the two-port circuit given in Fig. 13-12. Find the magnitude of the dc component and the first six harmonics of the output.

13-6 DISTORTION AND SHAPING

Circuits usually affect signals in some undesirable ways, leading to distortions of signals. Let us illustrate this with a simple example. Suppose that our useful signal consists of a fundamental and the second harmonic but with other higher harmonics which are to be filtered out. Let us also say that our filter stops all higher harmonics and allows only the fundamental and the second harmonic to pass through but changes their amplitudes and phases. Figure 13-13 shows the situation, with part (a) showing the original undistorted case. Part (b) shows the case when amplitudes are affected differently but the phases are unaltered. This leads to *amplitude distortion* due to changes in relative amplitudes. Part (c) shows the case when amplitudes are unchanged but the relative phases are affected, leading to *phase distortion*. Distortion is usually a combination of both amplitude and phase distortion but sometimes one of the two might dominate.

Shaping or waveform shaping is correcting the effects of distortion by appropriate amplitude and phase *equalization* or correction. Suppose that our

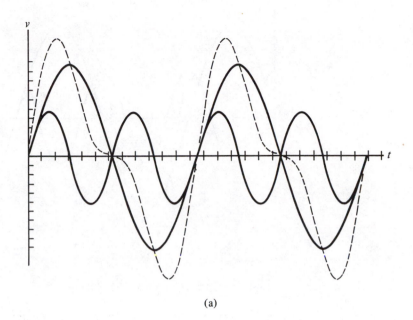

(a)

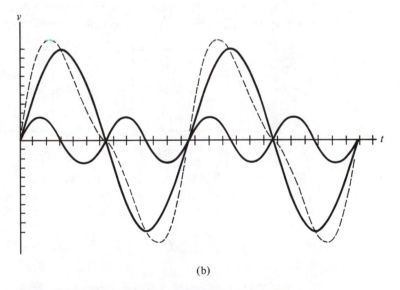

(b)

Figure 13-13. (a) No distortion, (b) amplitude distortion.

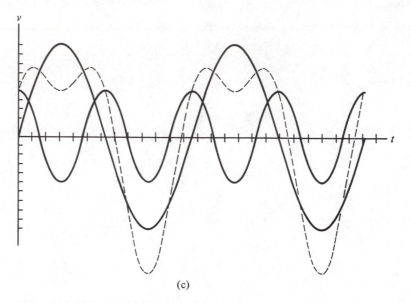

(c)

Figure 13-13. (*Cont.*)   (c) Phase distortion.

signal was distorted by the effects of a transfer function H_1, as shown in Fig. 13-14. We wish to shape it by means of H_2, which corrects the distortion. If the distortion was predominantly amplitude type, all we need to do is make $|H_1||H_2|$ flat over the frequency range of interest. The resulting transfer function will give an output signal with no amplitude distortion.

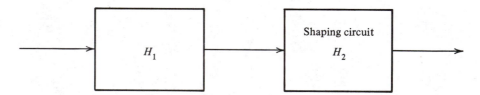

Figure 13-14.   Block diagram representation of cascaded circuits.

If the phase is a linear function of ω as shown in Fig. 13-15(b), then the signal is merely *delayed* but free of phase distortion. To see this let us change the origin of a signal from $t = 0$ to $t = t_0$. Then let the Fourier series of the original signal be

$$f(t) = A_0 + A_1 \sin(\omega_0 t + \theta_1)$$
$$+ A_k \sin(k\omega_0 t + \theta_k) \qquad (13\text{-}26)$$

We have only added a kth harmonic, where k is arbitrary, to illustrate the point. Now let us change the origin to t_0 by replacing t by $(t - t_0)$. Then we have

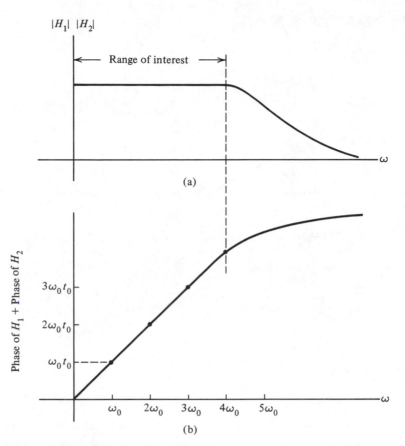

Figure 13-15. In many applications, there is a range of interest in both the amplitude and phase characteristics.

$$g(t) = A_0 + A_1 \sin(\omega_0(t - t_0) + \theta_1) + A_k \sin(k\omega_0(t - t_0) + \theta_k) \qquad (13\text{-}27)$$

$$= A_0 + A_1 \sin(\omega_0 t + \theta_1 - \omega_0 t_0) + A_k \sin(k\omega_0 t + \theta_k - k\omega_0 t_0) \qquad (13\text{-}28)$$

From Eqs. 13-26 and 13-28 we see that if the phase of the first harmonic is changed by $\omega_0 t_0$ and that of each kth harmonic by $k\omega_0 t_0$, the shape of the signal remains unchanged. Only a delay of t_0 seconds takes place in the signal. A plot of such a phase function is shown in Fig. 13-15(b). Phase distortion is thus corrected by making the angle of $H_1 H_2$ linear with ω. That is,

$$\underline{/H_1} + \underline{/H_2}$$

is made linear through the origin.

EXERCISE

13-6.1. Find a reasonable approximation to the voltage $v_2(t)$ in the circuit shown in Fig. 13-16, when $v_1(t)$ is a triangular function as shown in Fig. 13-17, with

317

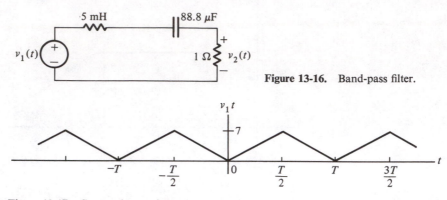

Figure 13-16.   Band-pass filter.

Figure 13-17.   Sawtooth waveform.

Fourier series

$$v_1(t) = \frac{7}{2} - \frac{28}{\pi^2}\left(\cos \omega t + \frac{1}{3^2}\cos 3\omega t + \frac{1}{5^2}\cos 5\omega t + \ldots\right)$$

where $\omega = 500$.

PROBLEMS

13-1. Using the identities

$$\sin x \sin y = \tfrac{1}{2}[\cos (x - y) - \cos (x + y)]$$
$$\cos x \cos y = \tfrac{1}{2}[\cos (x - y) + \cos (x + y)]$$
$$\sin x \cos y = \tfrac{1}{2}[\sin (x + y) + \sin (x - y)]$$
$$\cos x \sin y = \tfrac{1}{2}[\sin (x + y) - \sin (x - y)]$$

show that for $\omega_0 = 2\pi/T$,

$$\int_0^T \cos k\omega_0 t \cos m\omega_0 t \, dt = \begin{cases} 0, & k \neq m \\ \dfrac{T}{2}, & k = m \end{cases}$$

$$\int_0^T \sin k\omega_0 t \sin m\omega_0 t \, dt = \begin{cases} 0, & k \neq m \\ \dfrac{T}{2}, & k = m \end{cases}$$

$$\int_0^T \sin k\omega_0 t \cos m\omega_0 t \, dt = 0$$

13-2. Show that the same results as in Prob. 13-1 obtain if the limits of integration are changed from 0 to T to δ to $\delta + T$. Explain the implications of this conclusion.

13-3. Assuming the results of Exercise 13-3.1, find the even and odd parts of the waveform shown in Fig. 13-6.

13-4. Repeat Problem 13-3 for the waveform shown in Fig. 13.5.

13-5. The figure shows a periodic function which is sinusoidal and of amplitude V_m from $t = 0$ to $t = T/2$. It is also sinusoidal from $t = T/2$ to $t = T$ but of half-amplitude, $V_m/2$.
   (a) Determine the rms value of $v(t)$.
   (b) Determine a_0 of the Fourier series.
   (c) What can you say about the spectrum of $v(t)$ without actually determining a_n and b_n?

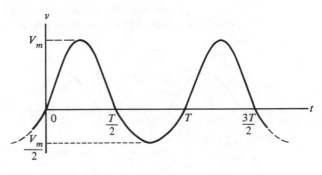

Figure P13-5

13-6. Find the Fourier coefficients for the sawtooth function that is given in the figure. Show the actual manipulations required to obtain the coefficients, first finding a_0 and then a_k and b_k without replacing the symbol k with a specific value.

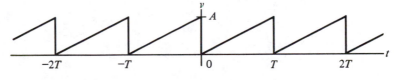

Figure P13-6

13-7. Repeat Prob. 13-6 for the pulsed waveform shown in the figure.

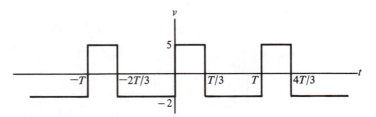

Figure P13-7

13-8. Consider the half-wave rectified sinusoid shown in the figure.

(a) Show that this waveform has the following Fourier series:

$$v(t) = \frac{A}{\pi} + \frac{A}{2} \sin \omega_0 t - \frac{2A}{\pi}\left(\frac{1}{1 \times 3} \cos 2\omega_0 t + \frac{1}{3 \times 5}\cos 4\omega_0 t \right.$$

$$\left. + \frac{1}{5 \times 7} \cos 6\omega_0 t + \ldots\right)$$

(b) We are always interested in the rate at which the coefficients of the Fourier series approach zero. Show that the fourth harmonic is the last with a magnitude of at least 5% of the magnitude of the fundamental.

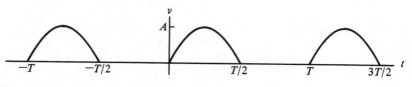

Figure P13-8

14

FOURIER ANALYSIS AND THE FREQUENCY SPECTRUM

In this chapter we continue the study of topics first introduced in Chapter 13. There we showed that any periodic function could be expanded in a Fourier series containing terms that were harmonically related. The amplitude and phase of each term in the Fourier series when plotted as a function of frequency was defined as the frequency spectrum. We did this as motivation for our study of signal filtering. Now that we have completed our study of various kinds of filters, we will bring the two topics together once more.

14-1 CIRCUITS AS SIGNAL PROCESSORS

All circuits may be thought of as signal processors. Even purely resistive circuits process a signal in the sense that the amplitude may be changed. If the circuit contains capacitors or inductors or equivalent elements, the signal will be changed considerably by the circuit. This may be visualized with the aid of Fig. 14-1. The signal input to the circuit is processed by the circuit, and this results in an output signal. Given the nature of the input signal and the signal processor, we would like to determine the properties of the output signal. In particular,

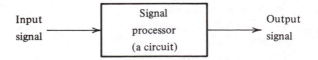

Figure 14-1. Signal processor viewed as a block diagram.

we are interested in the case in which the signal processor is a filter with prescribed amplitude and phase characterisrics.

As an example, consider the signal shown in Fig. 14-2. There we show a

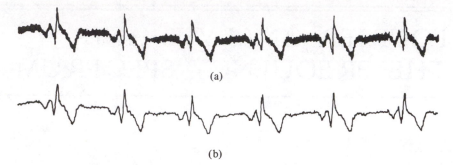

(a)

(b)

Figure 14-2. Waveform known as an electrocardiogram (ECG), showing the voltage produced by contractions of the heart. The waveform of (a) contains 60-Hz noise, while that of (b) results when the 60-Hz noise is removed by a filter.

tracing of a waveform showing the changes in electric potential produced by contractions of the heart known as an electrocardiogram or ECG. The signal waveform shown in (a) of the figure is contaminated by 60-Hz noise, which must be removed. This is accomplished with a band-reject or *notch* filter tuned to have a null at 60 Hz. The result is shown in (b) of the figure, where the noise has been eliminated without distortion of the signal. If we think of this process in terms of the block diagram of Fig. 14-1, the ECG plus noise is the input signal, which varies with time. This signal is processed by a filter with prescribed variation with frequency. The resulting signal has the 60-Hz "noise" removed and is identified as the output signal. In this discussion we have mixed the time domain and the frequency domain. If the input signal can be described in terms of frequency-domain concepts, our reasoning can be entirely in the frequency domain. This is made possible by characterizing the input signal as a Fourier series, which, in turn, is interpreted in terms of a frequency spectrum. This frequency spectrum for the signal consists of both magnitude and phase spectra.

A concept that has been introduced by means of the ECG signal is described further in Fig. 14-3, which shows how various time signals are processed by passing through specific kinds of filters: low-pass, high-pass, band-pass, and band-reject or notch. The figure mixes concepts from the time domain and the frequency domain. Our objective is to express all three quantities in terms of the frequency domain.

To review the concept of the Fourier series representation of a signal, we turn to Fig. 14-4, which shows how a square wave, shown in (e) of the figure, is constructed from harmonically related periodic sinusoidal functions. The Fourier series for the square wave may be written

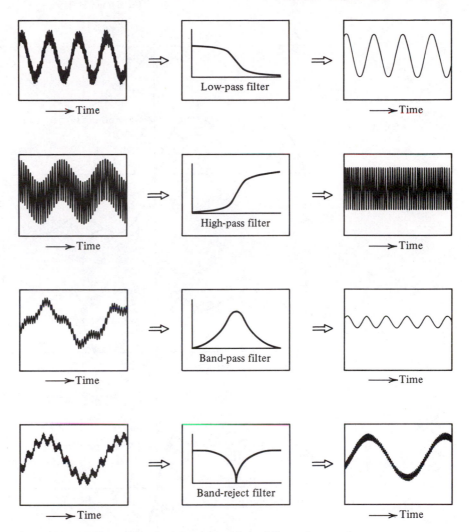

Figure 14-3.   Effect on signals of the various kinds of filters, shown as low-pass, high-pass, band-pass, and band-reject filters.

$$v(t) = V \sin t + \frac{V}{3} \sin 3t + \frac{V}{5} \sin 5t + \frac{V}{7} \sin 7t + \ldots \qquad (14\text{-}1)$$

if $\omega_0 = 1$. The first term in this series is known as the fundamental and this is shown in Fig. 14-4(a). In (b), the solid line shows the waveform that results when the first and third harmonics are added. In (c), the fifth harmonic is added to the waveform of (b), and this process is continued in (d) of the figure, which shows the addition of the seventh harmonic to the waveform of (c). The solid line in (d) of the figure is the periodic function made up of the four terms of an infinite series shown in Eq. 14-1, and this sum is seen to approximate the square

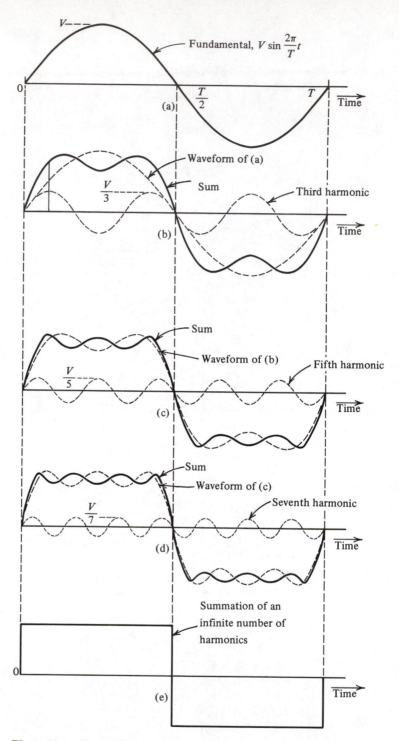

Figure 14-4. How a square wave is constituted from harmonically related periodic functions.

wave. If an infinite number of the terms shown in Eq. 14-1 are added together, the result will be the true square wave shown in Fig. 14-4(e). The spectrum corresponding to Eq. 14-1 is that shown in Fig. 14-5, from which we may make a number of observations:

1. Only odd terms are present in the series.
2. The terms that are present fall off with increasing frequency as $1/n$, where n is the number of the harmonic.
3. The phase of each term in the series of Eq. 14-1 is $0°$.

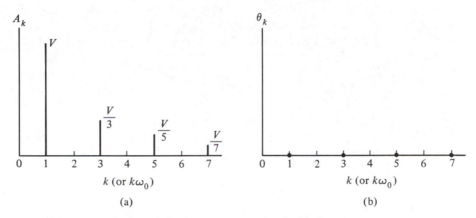

Figure 14-5. Magnitude and phase spectrum associated with the square-wave signal of Fig. 14-4.

Now let us imagine that we have a filter with the ideal brickwall characteristic shown in Fig. 14-6. This is a low-pass filter, which in the ideal case will pass all frequencies below ω_0, but stop all frequencies larger than ω_0. Suppose further that the frequency ω_0 may be adjusted or *tuned*, perhaps by adjusting some resistor in an RC op-amp circuit. First, we adjust ω_0 so that it is just larger than the seventh harmonic frequency. The result is that the ninth harmonic, the eleventh harmonic, and all higher harmonics are rejected. In terms of Eq. 14-1, we say that the Fourier series has been *truncated*. Now if the input signal

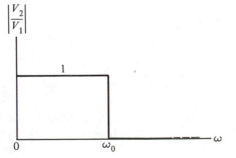

Figure 14-6. Magnitude characteristics of an ideal low-pass filter with cutoff frequency at ω_0.

is the square wave shown in Fig. 14-4(e), the output signal will be that shown by the solid line in Fig. 14-4(d). By the same reasoning, we see that if the filter is adjusted so that cutoff frequency, ω_0, is just larger than the third harmonic, then with the square-wave input, the output will be that shown by the solid line of Fig. 14-4(b).

This example illustrates the process by which the spectrum of the output signal is determined. The line spectrum of the input signal is multiplied by the continuous spectrum of the filter. This product is the spectrum of the output signal. This multiplication for magnitude spectra of the input and the filter is replaced by addition in the case of the phase. In other words, the phase of the output signal spectrum is the sum of the phase of the input signal and that contributed by the filter. This process is illustrated by Fig. 14-7.

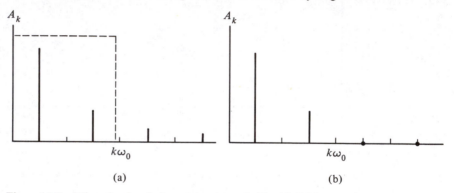

(a)   (b)

Figure 14-7.   When the signal of spectrum shown in Fig. 14-5(a) is passed through the ideal filter characterized in Fig. 14-6, the result is that higher harmonic components of the signal are removed.

14-2   FOURIER COEFFICIENTS

In Chapter 13 we found that the Fourier coefficients are determined from integral equations derived from the properties of the Fourier series. To review, these equations for a periodic function $v(t)$ were

$$a_0 = \frac{2}{T} \int_{t_0}^{t_0+T} v(t)\, dt \tag{14-2}$$

$$a_k = \frac{2}{T} \int_{t_0}^{t_0+T} v(t) \cos k\omega_0 t\, dt, \qquad k = 1, 2, \ldots \tag{14-3}$$

$$b_k = \frac{2}{T} \int_{t_0}^{t_0+T} v(t) \sin k\omega_0 t, \qquad k = 1, 2, \ldots \tag{14-4}$$

Here T is the period of the waveform, t_0 is an arbitrary time during that period normally selected to be $t_0 = 0$, and ω_0 is the frequency of the fundamental, which is equal to $2\pi/T$. We recall also that, the amplitude and phase of the kth harmonic of $v(t)$ are given by

$$A_k = \sqrt{a_k^2 + b_k^2} \tag{14-5}$$

and

$$\theta_k = \tan^{-1}\frac{a_k}{b_k} \qquad \text{for all } k = 1, 2, \ldots \tag{14-6}$$

The dc term is given by $A_0 = a_0/2$. The plot of A_k as a function of k or $k\omega_0$ is known as the line or discrete spectrum, and the plot of θ_k as a function of either k or $k\omega_0$ is known as the phase spectrum.

    The number of waveforms of special interest to us—the square wave, the pulse, the triangular wave—is relatively small. Once these have been determined, they may be tabulated in a table and stored for later reference. If a new waveform should appear, we may return to the equations of Eqs. 14-2 through 14-6 and carry out the integrations to determine A_k and θ_k, which may then be added to the tables. Some common waveforms are shown in Table 14-1 together with a few terms of their Fourier series.

    In applications of signal processing, we will be interested in both the waveforms shown in the table and in random waveforms such as those representing speech. One waveform of special interest is a train of pulses representing digital signals. A tool commonly employed by the engineer to study the spectrum of signals is the *spectrum analyzer*. In this instrument, the amplitude of the magnitude spectrum is displayed as a function of frequency. The instrument is designed to sweep through the frequency range of interest at a high rate so that the spectrum may be displayed on the face of a cathode-ray tube. A block diagram representation of the system by which this is accomplished is shown in Fig. 14-8,

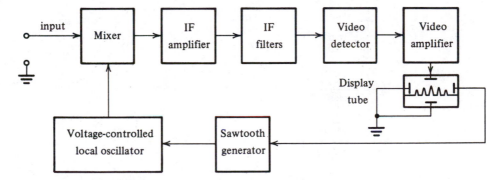

Figure 14-8. Block diagram representation of a spectrum analyzer.

and two typical spectra are shown in Fig. 14-9. The point of this discussion is that the determination of spectra can be automated, and engineers seldom have to carry out the integrations of Eqs. 14-2 through 14-4. However, there are certain properties of signals which follow from these equations that are of great importance to the engineer, and these will be reviewed next.

Table 14-1

Plot of $v(t)$	Equation	Fourier Series

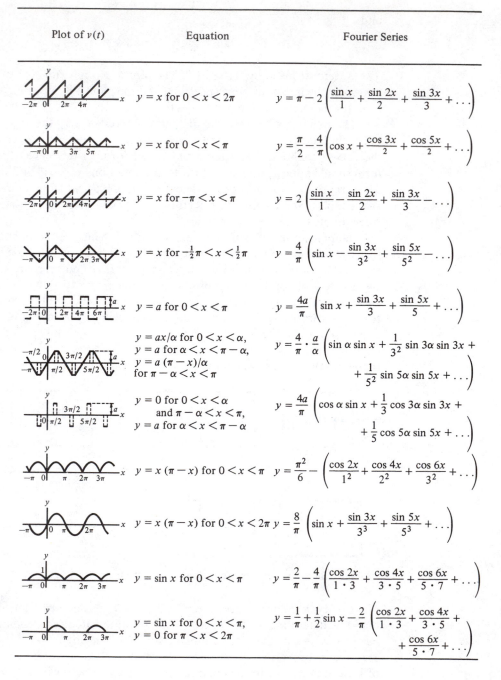

$y = x$ for $0 < x < 2\pi$ → $y = \pi - 2\left(\dfrac{\sin x}{1} + \dfrac{\sin 2x}{2} + \dfrac{\sin 3x}{3} + \ldots\right)$

$y = x$ for $0 < x < \pi$ → $y = \dfrac{\pi}{2} - \dfrac{4}{\pi}\left(\cos x + \dfrac{\cos 3x}{2} + \dfrac{\cos 5x}{2} + \ldots\right)$

$y = x$ for $-\pi < x < \pi$ → $y = 2\left(\dfrac{\sin x}{1} - \dfrac{\sin 2x}{2} + \dfrac{\sin 3x}{3} - \ldots\right)$

$y = x$ for $-\frac{1}{2}\pi < x < \frac{1}{2}\pi$ → $y = \dfrac{4}{\pi}\left(\sin x - \dfrac{\sin 3x}{3^2} + \dfrac{\sin 5x}{5^2} - \ldots\right)$

$y = a$ for $0 < x < \pi$ → $y = \dfrac{4a}{\pi}\left(\sin x + \dfrac{\sin 3x}{3} + \dfrac{\sin 5x}{5} + \ldots\right)$

$y = ax/\alpha$ for $0 < x < \alpha$,
$y = a$ for $\alpha < x < \pi - \alpha$,
$y = a(\pi - x)/\alpha$
for $\pi - \alpha < x < \pi$
→ $y = \dfrac{4}{\pi} \cdot \dfrac{a}{\alpha}\left(\sin \alpha \sin x + \dfrac{1}{3^2}\sin 3\alpha \sin 3x + \dfrac{1}{5^2}\sin 5\alpha \sin 5x + \ldots\right)$

$y = 0$ for $0 < x < \alpha$
and $\pi - \alpha < x < \pi$,
$y = a$ for $\alpha < x < \pi - \alpha$
→ $y = \dfrac{4a}{\pi}\left(\cos \alpha \sin x + \dfrac{1}{3}\cos 3\alpha \sin 3x + \dfrac{1}{5}\cos 5\alpha \sin 5x + \ldots\right)$

$y = x(\pi - x)$ for $0 < x < \pi$ → $y = \dfrac{\pi^2}{6} - \left(\dfrac{\cos 2x}{1^2} + \dfrac{\cos 4x}{2^2} + \dfrac{\cos 6x}{3^2} + \ldots\right)$

$y = x(\pi - x)$ for $0 < x < 2\pi$ → $y = \dfrac{8}{\pi}\left(\sin x + \dfrac{\sin 3x}{3^3} + \dfrac{\sin 5x}{5^3} + \ldots\right)$

$y = \sin x$ for $0 < x < \pi$ → $y = \dfrac{2}{\pi} - \dfrac{4}{\pi}\left(\dfrac{\cos 2x}{1 \cdot 3} + \dfrac{\cos 4x}{3 \cdot 5} + \dfrac{\cos 6x}{5 \cdot 7} + \ldots\right)$

$y = \sin x$ for $0 < x < \pi$,
$y = 0$ for $\pi < x < 2\pi$
→ $y = \dfrac{1}{\pi} + \dfrac{1}{2}\sin x - \dfrac{2}{\pi}\left(\dfrac{\cos 2x}{1 \cdot 3} + \dfrac{\cos 4x}{3 \cdot 5} + \dfrac{\cos 6x}{5 \cdot 7} + \ldots\right)$

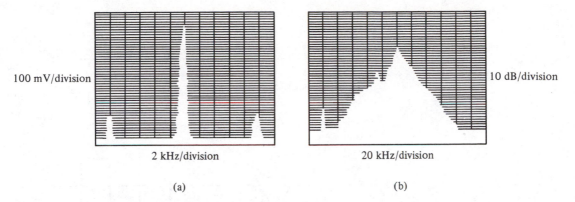

100 mV/division

10 dB/division

2 kHz/division

20 kHz/division

(a)   (b)

Figure 14-9. Typical spectrum recorded by a spectrum analyzer.

14-3 SIGNALS WITH EVEN AND ODD SYMMETRY

To describe a signal waveform, voltage as a function of time for example, it is necessary to select some time reference, which will be designated by $t = 0$. Once this is done, observations may be made concerning the symmetry or lack of symmetry of waveforms. With respect to the time reference, $t = 0$, a signal may be said to be *even* with respect to $t = 0$ when

$$v(-t) = v(t) \tag{14-7}$$

and *odd* when

$$v(-t) = -v(t) \tag{14-8}$$

Another symmetry of interest is known as *half-wave* symmetry, which occurs when

$$v(t) = -v\left(t \pm \frac{\tau}{2}\right) \tag{14-9}$$

Examples of the three symmetries are shown in Fig. 14-10. We observe that even symmetry might be called mirror symmetry if we regard the $t = 0$ axis as a mirror; thus, as we move away from the $t = 0$ axis, $v(t)$ has the same numerical value for equal distances in the positive and negative directions. In the case of odd symmetry, the sign of $v(t)$ is reversed but the magnitude of the values are maintained. Thus, as we move away from the $t = 0$ axis in the positive and negative directions, the signal values have equal amplitude but opposite signs. Half-wave symmetry is best understood by the fact that it is known as *rotational* symmetry being a form of symmetry produced by rotating electrical machinery. These waveforms are equal in value but reversed in sign on alternate half-cycles, as seen in the figure.

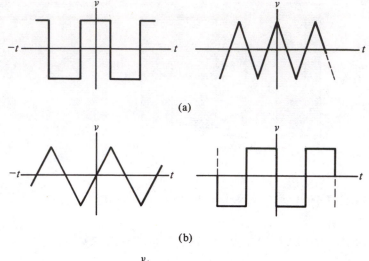

(a)

(b)

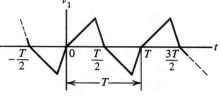

(c)

Figure 14-10. Waveforms used to illustrate even and odd symmetry.

For waveforms with these symmetries, the Fourier series representations simplify as follows:

- a. For a waveform with even symmetry, $v(t)$ contains only cosine terms.
- b. For a waveform with odd symmetry, $v(t)$ contains only sine terms.
- c. For a waveform with half-wave symmetry, $v(t)$ contains odd k terms only.

Conclusions (a) and (b) follow from the fact that cosine functions are even and sine functions are odd. Further, the sum of any number of even functions remains even, while the sum of any number of odd functions remains odd. Hence, the presence of a sine term in a series otherwise made up solely of cosine terms would destroy the even symmetry.

There is another kind of symmetry in waveforms that comes from the definition of a_0 in Eq. 14-2 with $t_0 = 0$:

$$a_0 = \frac{2}{T} \int_0^T v(t)\, dt \qquad (14\text{-}10)$$

Now the integral part of this equation simply represents the area under the curve for $v(t)$ over one cycle from $t = 0$ to $t = T$. When this area is divided by T, the result becomes the average value of $v(t)$ over one period, T. The multiplier 2 appears in the equation to provide a common form for the a coefficients, and results in the first term in the Fourier series being $a_0/2$. Hence, we see that Eq. 14-10 for a_0 is twice the average value of $v(t)$ taken over one cycle. In many cases, the average value is zero since the positive area is equal to the negative area over one period. Nothing specific can be said about a_0 for even functions; however, it is clear that both odd functions and functions with half-wave symmetry always satisfy the condition that $a_0 = 0$. This information is summarized in Table 14-2.

Table 14-2

Name of Symmetry	Condition	Property	a_0	a_n	b_n
Even	$v(t) = v(-t)$	Cosine terms only	General	General	0
Odd	$v(t) = -v(-t)$	Sine terms only	0	0	General
Half-wave	$v(t) = -v\left(t \pm \dfrac{T}{2}\right)$	Odd n only	0	General	General

14-4 THE RATE AT WHICH FOURIER COEFFICIENTS DECREASE

In Table 14-1 we observe that some Fourier series contain terms that decrease as $1/k$, others as $1/k^2$, and still others as $1/k^3$. How can we tell from the waveform which law will apply to a particular waveform? To answer this equation, we first discuss *jump discontinuities* in $v(t)$. Such discontinuities are permitted in $v(t)$ and have the appearance shown in Fig. 14-11 for an arbitrary waveform, $v(t)$. In fact, these have been encountered earlier in this chapter when we discussed the square wave, which has such a discontinuity each time it changes from a positive to a negative value. The law that determines how rapidly the Fourier coefficients decrease in value with k is determined by the number of times that $v(t)$ must be differentiated to produce a jump discontinuity. The law can be stated simply in terms of a few statements:

1. If $v(t)$ contains jump discontinuties, the upper bound for the decrease in $|a_k|$ and $|b_k|$ with k is $1/k$. An example of such a waveform is the square wave; another is the pulse.
2. If $v(t)$ contains no jump discontinuities, but dv/dt does, the upper

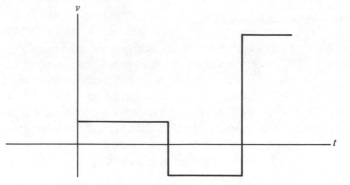

Figure 14-11. Signal with jump discontinuities.

bound for the decrease of $|a_k|$ and $|b_k|$ with k is $1/k^2$. An example of such a waveform is the triangular wave.

3. If neither $v(t)$ nor dv/dt contain jump discontinuties, but d^2v/dt^2 does, the decrease in $|a_k|$ and $|b_k|$ is $1/k^3$. An example of such a waveform is one that is parabolic.

Clearly, the rules can be extended to other functions.

The value of the concept we have just discussed is that it permits us to anticipate the spectrum of a given waveform in terms of its smoothness. A

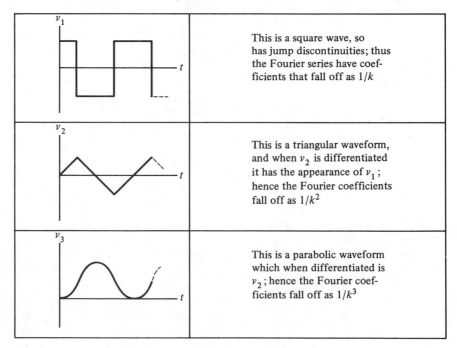

Figure 14-12. Waveforms with jump discontinuities as such or in their derivatives.

waveform with jumps in it will require more lines of significant amplitude in the spectrum than for a wave form that is relatively smooth. Several such waveforms are shown in Fig. 14-12.

14-5 SPECTRUM SHAPING

We are now perpared to discuss spectrum shaping, one of the foundation concepts of electrical engineering. This concept was discussed earlier in Chapter 13 before we had considered filters in detail. Now we are prepared to consider spectrum shaping in greater detail.

Let us first return to Fig. 14-3, which for four different cases shows the input and output waveforms as a function of time. But the signal processor itself is characterized by the magnitude of the transfer function as it varies with frequency. Thus, the descriptions of Fig. 14-3 mix the concepts of the time and the frequency domains. This is no longer necessary since with the aid of the Fourier series representation of a periodic function, we may characterize the input and output in the frequency domain. Later in this chapter, we will extend the concepts to signals that are not repeated. The relationship we have been discussing involves a product:

$$\begin{bmatrix} \text{line spectrum} \\ \text{of the input} \\ \text{signal} \end{bmatrix} \times \begin{bmatrix} \text{continuous} \\ \text{spectrum of} \\ \text{the transfer} \\ \text{function,} \\ T = V_2/V_1 \end{bmatrix} = \begin{bmatrix} \text{line spectrum} \\ \text{of the output} \\ \text{signal} \end{bmatrix} \qquad (14\text{-}11)$$

In the form of symbols, for the magnitude

$$|V_2(j\omega)| = |T(j\omega)| \cdot |V_1(j\omega)| \qquad (14\text{-}12)$$

and for the phase

$$\theta_2(j\omega) = \theta_T(j\omega) + \theta_1(j\omega) \qquad (14\text{-}13)$$

This is shown in the form of a signal flow chart in Fig. 14-13.

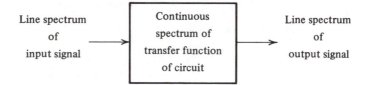

Line spectrum of input signal → Continuous spectrum of transfer function of circuit → Line spectrum of output signal

Figure 14-13. Block diagram showing how the line spectrum of the input signal is modified or processed to give the line spectrum of the output signal.

Referring once more to the illustration of signal processing shown in Fig. 14-3, we now present a unified description of each example of signal processing, using only frequency-domain concepts. So in Fig. 14-14(a) we have shown the spectrum of the signal shown in Fig. 14-3(a). We see that the spectrum has

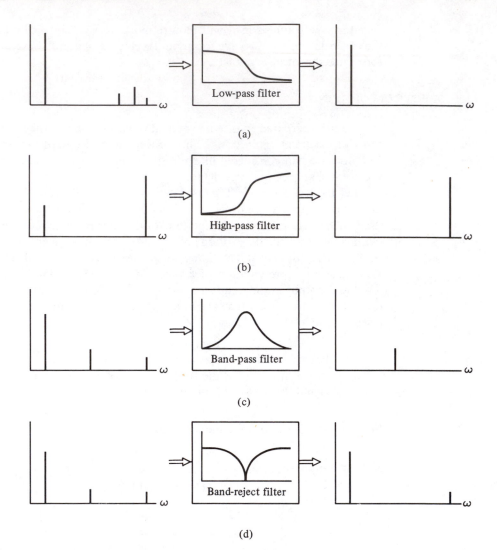

Figure 14-14.   Application of the concept of Fig. 14-13 to four kinds of filters.

a strong low-frequency line and low-amplitude lines at higher frequencies. The function of the low-pass filter is to pass the strong low-frequency line but reject or attentuate the high-frequency lines. The pure signal which is a sinusoid in Fig. 14-3(a) is represented by one line in the output spectrum of Fig. 14-14(a).

Turning now to Fig. 14-14(b), we see the input spectrum has a low-amplitude line at low frequency and a higher-amplitude line at high frequency. Passing this signal through a high-pass filter removes the low-frequency line, so that the output is a high-frequency sinusoid.

The input signal shown in Fig. 14-14(c) has three frequency lines. The function of the bandpass filter is to remove both the low- and high-frequency lines, leaving only the intermediate-frequency signal as shown for the output. The inverse of this spectrum shaping is shown in (d) of the figure, where a notch (or band reject) filter is used to remove the intermediate-frequency line from the spectrum.

Another example that will illustrate the engineering application of the concept of frequency spectrum may be described in term of Fig. 14-15. A UHF

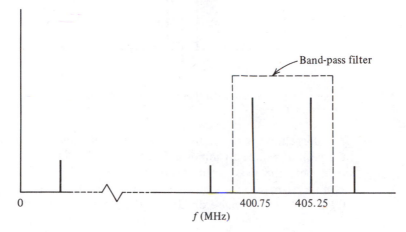

Figure 14-15. Function of a band-pass filter to pass some lines in the spectrum and attenuate all other lines.

television station is assigned the band of frequencies from 400 to 406 MHz. As shown by the figure, the picture carrier is assigned the frequency 400.75 MHz and the sound carrier 405.25 MHz, the two frequencies being separated by 4.5 MHz. To receive this station, the receiver must have a bandwidth of at least 4.5 MHz, but must be limited in bandwidth in order that other frequencies, both lower and higher, will be rejected to avoid interference. The situation is made perfectly clear in terms of the spectrum of Fig. 14-15. Using the frequency spectrum, we may consider the possibility of interference from signals in adjacent channels or other stations with different systems of signal modulation.

14-6 SPECTRUM OF A PULSE TRAIN

An important waveform in electrical engineering is the train of pulses depicted in Fig. 14-16. This may represent the transimssion or reception of a radar or sonar system, it may represent a timing system such as the "clock" in a digital

Figure 14-16. Train of pulses.

computer. Since the signal itself is so important, so the Fourier series representation is also of importance.

To determine the spectrum for the pulse train, we label the pulse to have width a and amplitude V, as shown in Fig. 14-17. We select our time reference

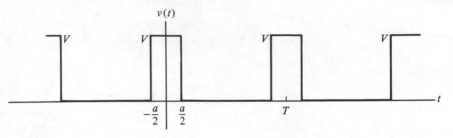

Figure 14-17.   Specification that the pulse be of width a, of height V, and of period T.

$t = 0$ as shown in the figure, and observe that the ratio of a to T, where T is the period, will be an important parameter. Having made this choice of time reference, the pulse train becomes an even function and because of this symmetry, $b_k = 0$ for all k. The value of the dc component, a_0, is seen to be the area under the pulse divided by the period, T, so that

$$a_0 = \frac{Va}{T} \tag{14-14}$$

Then there remains only the determination of a_k, making use of Eq. 14-3. Thus,

$$a_k = \frac{2}{T} \int_{-a/2}^{a/2} V \cos k\omega_0 t \, dt \tag{14-15}$$

This equation is easily integrated to give

$$a_k = \frac{2V}{\pi k} \sin \frac{k\omega_0 a}{2}, \qquad k = 1, 2, 3, \ldots \tag{14-16}$$

Then the Fourier series representation of the train of pulses becomes

$$v(t) = \frac{Va}{T} + \sum_{k=1}^{\infty} \left(\frac{2V}{\pi k} \sin \frac{k\omega_0 a}{2} \right) \cos k\omega_0 t \tag{14-17}$$

and the line spectrum is given by Eq. 14-16.

To study the equation for a_k further, we consider a specific example when the pulse width is one-fourth of the period, or $a = T/4$. When this value for a is substituted into Eq. 14-16 together with replacing ω_0 by $2\pi/T$ since $f_0 = 1/T$, we have

$$a_k = \frac{V}{2} \left(\frac{\sin k\pi/4}{k\pi/4} \right) \tag{14-18}$$

The form of this equation may be related to the mathematical function $(\sin x)/x$ if $x = k\pi/4$. The $(\sin x)/x$ function may be thought of as the *envelope* of the values of a_k. Actually, a_k has a value only for integer values of K. For example, when $k = 1$, we find that

$$a_1 = \frac{V}{2}\left(\frac{\sin \pi/4}{\pi/4}\right) = 0.45 \text{ V} \qquad (14\text{-}19)$$

and so on. The envelope of the values of a_k, the $(\sin x)/x$ function, is shown in Fig. 14-18. The line magnitude spectrum A_k defined by Eqs. 14-5 and 14-18

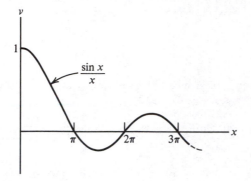

Figure 14-18. Plot of the function $v = (\sin x)/x$.

is shown in Fig. 14-19 together with the envelope as a dashed line. From the envelope of Fig. 14-18 we see that θ_k is either positive or negative, and this corresponds to values of θ_k in Eq. 14-6 of either $0°$ or $180°$. The phase spectrum is also shown in Fig. 14-19.

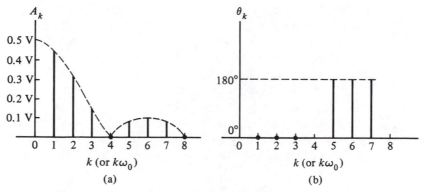

Figure 14-19. Magnitude and phase spectrum of a pulse.

We may repeat this analysis by taking a second example of a train of pulses in which the pulse width is one-eighth of the period, or when $a = T/8$. Substituting this value of a into the spectrum of Eq. 14-16, we see first that $a_0 = V/8$, and that

$$a_k = \frac{V}{4}\left(\frac{\sin k\pi/8}{k\pi/8}\right) \qquad (14\text{-}20)$$

Letting k have values starting with 1 and increasing as integers, we obtain the

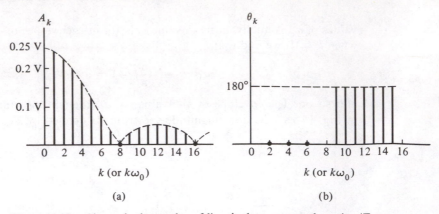

(a)   (b)

Figure 14-20. Change in the number of lines in the spectra as the ratio a/T changes.

spectrum of Fig. 14-20. Comparing the two examples, we can observe a number of trends.

a. As the ratio of the pulse width to the period decreases, the number of lines in both amplitude and phase spectra increases.

b. At the same time that the number of lines is increasing, the amplitude of the magnitude spectrum is decreasing.

This trend seems intuitively reasonable: the pulse width to period ratio, a/T, determines the relative time that the pulse is on. For the periodic sinusoids to add together to give the pulse in the time in which the pulse is on, and at the same time to cancel while the pulse is off is a difficult requirement. As the duration of the pulse decreases, the number of frequency components, which is the number of lines in the frequency spectrum, must increase.

We next consider what will happen if the pulse train is passed through a filter. For example, suppose that the pulse train is the input to a low-pass filter that has an abrupt cutoff (the so-called "brick wall"); such a system is known as a band-limited system. This will cause lines at a frequency greater than the cutoff frequency of the filter to be removed. This is illustrated in Fig. 14-21.

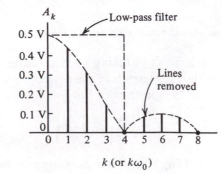

Figure 14-21. If a pulse train passes through a low-pass filter, the higher-frequency lines are removed.

A rule of thumb in this case is that if the cutoff frequency of the band-limiting low-pass filter is equal to or greater than the first null in the $(\sin x)/x$ form of response, the distortion of the output signal will not be so great that the pulse cannot be recognized. This is an approximation, of course. If an exact analysis is required, the Fourier series which has been truncated can be summed. Or the response can be found using a simulation program that is available making use of the digital computer.

14-7 CONTINUOUS SPECTRA: THE FOURIER INTEGRAL

A bolt of lightning produces a voltage that has the te^{-kt} waveform shown in Fig. 14-22. This lightning occurs only once, or at least it is a long time before it occurs again. Our common experience is that this lightning bolt will cause

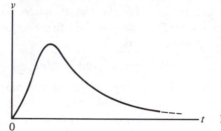

Figure 14-22.   Nonrecurring waveform.

interference no matter what the frequency range might be for the receiver we are using. Thus, we will hear a crackling noise whether we are listening to the broadcast band, short waves, or television stations, and it does not matter which station we are tuned to. This experience leads us to believe that a non-recurring signal such as that produced by lightning has a frequency spectrum which contains *all frequencies* (i. e. , a *continuous* spectrum rather than discrete).

This is a surprising result: It tells us that the a/T ratio for the pulse train studied in the last section is permitted to have the limiting value of 0, corresponding to a pulse that occurs only once. This single pulse also has a frequency spectrum which will turn out to be continuous rather than discrete.

Thus far in this chapter, we have associated a particular waveform with a particular spectrum. This concept might be written in the form of an equation:

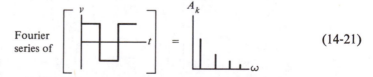

$$\text{Fourier series of} \left[\quad \right] = \qquad\qquad (14\text{-}21)$$

We interpret this equation in the following way: the Fourier series is the *transform* by which a function of time is converted into a function of frequency. For a nonrecurring waveform such as a single pulse, the means by which the frequency representation is found is known as the *Fourier integral*, which is

$$V(j\omega) = \int_{-\infty}^{\infty} v(t)e^{-j\omega t}\, dt \tag{14-22}$$

and the associated *inverse transform*

$$v(t) = \frac{1}{2\pi} \int_{-\infty}^{\infty} V(j\omega)e^{j\omega t}\, d\omega \tag{14-23}$$

A notation that is used to denote the operations of these two equations is the script letter $\mathcal{F}$:

$$V(j\omega) = \mathcal{F}v(t) \tag{14-24}$$

and

$$v(t) = \mathcal{F}^{-1}V(j\omega) \tag{14-25}$$

The equations involving the integral, Eqs. 14-22 and 14-23, are known as the *Fourier integrals*, and in the shorthand form of Eqs. 14-24 and 14-25 are known as *Fourier transforms*. Further, the *continuous spectrum* of a nonrecurring waveform is the Fourier transform of the time function, $v(t)$. It is written in the usual magnitude and phase forms,

$$|V(j\omega)| \quad \text{and} \quad \underline{/V(j\omega)} \tag{14-26}$$

Next we calculate the continuous spectrum, magnitude, and phase for several familiar waveforms. We should remind ourselves in advance that the concept applies to all waveforms, and not only those that can easily be described by an equation. So if the waveform shown in Fig. 14-23 occurs only once, it will have a spectrum representation. This spectrum may be determined

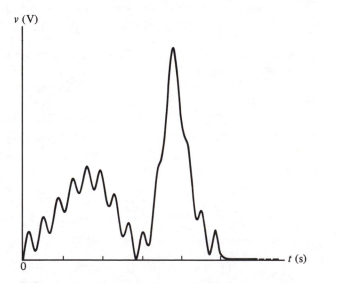

Figure 14-23. Waveform that cannot easily be described by equations.

by computer using a method, well known as the fast Fourier transform, or it may be determined by a spectrum analyzer represented in Fig. 14-8.

It is appropriate that our first example be the pulse shown in Fig. 14-24,

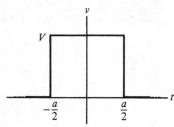

Figure 14-24. Nonrecurring pulse of amplitude V and width a.

which occurs once and then never again. The continuous spectrum is found using Eq. 14-22.

$$V(j\omega) = \int_{-a/2}^{a/2} Ve^{-j\omega t}\, dt = \frac{Ve^{-j\omega t}}{-j\omega}\bigg|_{-a/2}^{a/2}$$

$$= aV\left[\frac{\sin\,(\omega a/2)}{\omega a/2}\right]$$

(14-27)

which is seen to be the familiar $(\sin x)/x$ form but now defined for all values of ω. The continuous spectrum corresponding to the single pulse is that shown in Fig. 14-25 for both magnitude and phase. As before, this spectrum may be

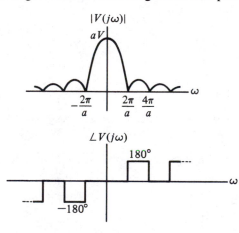

Figure 14-25. Continuous spectrum of a nonrecurring pulse.

multiplied by the continuous spectrum of the transfer function to obtain the output spectrum.

$$V_1(j\omega) \times T(j\omega) = V_2(j\omega)$$

(14-28)

so all concepts developed in connection with Fourier series apply to the Fourier integral.

As our second example, consider the time function that represents an exponential decay,

$$v(t) = Ae^{-\alpha t} \quad (t > 0)$$

(14-29)

342

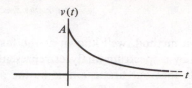

Figure 14-26. Decaying exponential described by Eq. 14-29.

which is shown in Fig. 14-26. Again, we make use of Eq. 14-22 to write

$$\mathscr{F}v(t) = \int_0^\infty Ae^{-\alpha t}e^{-j\omega t}\,dt$$

$$= \left| \frac{Ae^{-\alpha t}e^{-j\omega t}}{-(\alpha + j\omega)} \right|_0^\infty \tag{14-30}$$

Now if $\alpha > 0$, the function has 0 value at the upper limit, and we obtain

$$V(j\omega) = \frac{A}{\alpha + j\omega} \tag{14-31}$$

From this result, we see that the magnitude and phase functions are

$$|V(j\omega)| = \frac{A}{\sqrt{\alpha^2 + \omega^2}} \tag{14-32}$$

and

$$\underline{/V(j\omega)} = -\tan^{-1}\frac{\omega}{\alpha} \tag{14-33}$$

These two functions are shown plotted in Fig. 14-27.

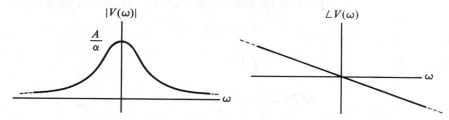

Figure 14-27. Magnitude and phase functions representing the continuous spectrum of the decaying exponential function.

The two-sided decaying exponential of Fig. 14-28 is defined by the equation

$$v(t) = Ae^{-\alpha|t|} \qquad \text{for all } t \tag{14-34}$$

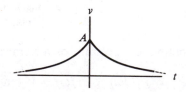

Figure 14-28. Two-sided decaying exponential function given by Eq. 14-34.

The corresponding continuous spectrum, $V(j\omega)$, is found from Eq. 14-22 and is

$$V(j\omega) = A\int_{-\infty}^0 e^{\alpha t}e^{-j\omega t}\,dt + A\int_0^\infty e^{-\alpha t}e^{-j\omega t}\,dt$$

$$= \frac{A}{\alpha - j\omega} + \frac{A}{\alpha + j\omega} = \frac{2\alpha A}{\alpha^2 + \omega^2} \tag{14-35}$$

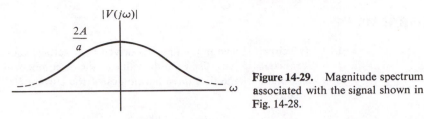

Figure 14-29.   Magnitude spectrum associated with the signal shown in Fig. 14-28.

The magnitude spectrum is that shown in Fig. 14-29. We see that this spectrum is different in that the phase spectrum is zero for all ω. The three examples given are representative of a large class of signal waveforms for which equations can be written. In every case the amplitude of the magnitude continuous spectrum falls off with increasing frequency, at rates that are determined by the abruptness of changes in the signal, $v(t)$.

As a final example, let us consider spectra associated with amplitude modulation such as that used on AM broadcast stations. Let the signal representing voice or music or a combination be $v(t)$ and let this signal be used to modulate the amplitude of a carrier signal

$$v_c(t) = \cos \omega_c t \tag{14-36}$$

Then the AM signal is

$$v_{AM}(t) = v(t)v_c(t) = v(t) \cos \omega_c t \tag{14-37}$$

We may find the Fourier transform of $v_{AM}(t)$ using Eq. 14-22. Substituting results in the equation

$$\mathcal{F}v_{AM}(t) = \int_\infty^\infty v(t)\left(\frac{e^{j\omega_c t} + e^{-j\omega_c t}}{2}\right)e^{-j\omega t}\,dt \tag{14-38}$$

Carrying out the individual steps in integration, we obtain the result

$$\mathcal{F}v_{AM}(t) = \tfrac{1}{2}V[j(\omega - \omega_c)] + \tfrac{1}{2}V[j(\omega + \omega_c)] \tag{14-39}$$

where $V(j\omega)$ is the transform of the signal $v(t)$. Thus, we see that, modulating a signal by the scheme of Eq. 14-37 results in a shift of the spectrum of $v(t)$ by $\pm\omega_c$. If we let $v(t)$ be a sine wave of frequency ω, the spectrum of the modulated signal is that shown in Fig. 14-30. This is only one of many forms of signal modulation which are considered in courses on communication systerms.

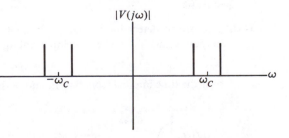

Figure 14-30.   Spectrum of an amplitude-modulated signal.

PROBLEMS

14-1. The circuit shown in the figure is driven by a voltage source having the waveform $v(t) = 7 |\sin 10t|$. Find the value of the dc component and the magnitude and phase of the first three harmonics of the current $i(t)$ as identified in the circuit.

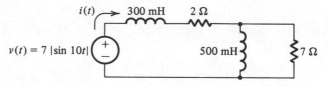

Figure P14-1

14-2. The voltage source in the circuit shown in the figure is that shown, which has a Fourier series

$$v_1(t) = \frac{7}{2} - \frac{28}{\pi^2}\left(\cos \omega_0 t + \frac{1}{3^2}\cos 3\omega_0 t + \frac{1}{5^2}\cos 5\omega_0 t + \ldots\right)$$

The source is adjusted such that $\omega_0 = 500$. Sketch the spectrum of the output voltage $v_2(t)$

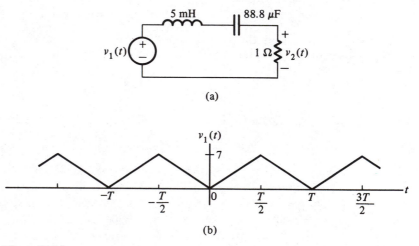

(a)

(b)

Figure P14-2

14-3. You are to determine the dc component and the first two harmonics of the output voltage $v_2(t)$ when the input voltage is

$$v_1 = t(\pi - t)$$

over the time interval from 0 to π, and periodic in other such intervals.

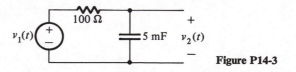

Figure P14-3

14-4. The input voltage $v_1(t)$ is a full-wave rectified sine wave as shown in the figure. This input drives a circuit having the transfer function

$$\frac{V_2}{V_1} = \frac{4s}{s^2 + s + 100}$$

Determine and sketch the amplitude spectrum for the output, $V_2(j\omega)$.

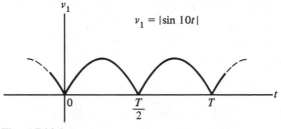

Figure P14-4

14-5. The Fourier series of the square wave shown in (a) of the figure is

$$f(t) = \frac{4A}{\pi} \cos \omega_0 t - \frac{4A}{3\pi} \cos 3\omega_0 t + \frac{4A}{5\pi} \cos 5\omega_0 t - \ldots$$

where $\omega_0 = 2\pi/T$. The Fourier series for the sawtooth function $g(t)$ is

$$g(t) = \frac{2B}{\pi} \sin \omega_0 t - \frac{2B}{2\pi} \sin 2\omega_0 t + \frac{2B}{3\pi} \sin 3\omega_0 t - \ldots$$

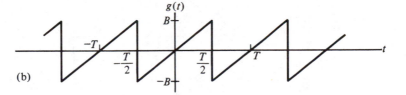

(a)

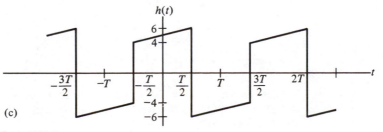

(b)

(c)

Figure P14-5

as shown in (b) of the figure. Find the magnitude and phase angles of the first two harmonics of the function $h(t)$ sketched in (c) of the figure.

14-6. The Fourier series for a square wave is given in Prob. 14-5. Make use of that information to obtain an expression for the Fourier series for the waveform shown in the figure.

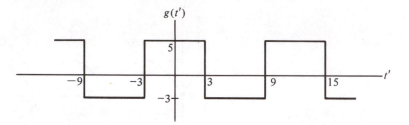

Figure P14-6

14-7. Given that the Fourier series for the function shown in (a) of the figure is

$$f(x) = 0.5 + 0.637 \sin x + 0.212 \sin 3x + 0.127 \sin 5x + \ldots$$

write the Fourier expansion for the function given in (b) of the figure, including the same harmonics as given for the original $f(x)$.

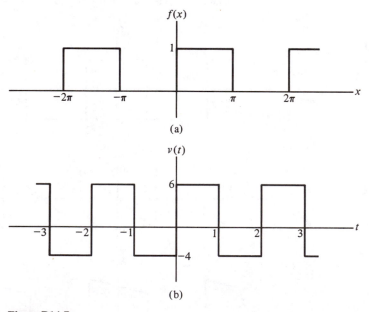

Figure P14-7

14-8. The waveform shown in the figure consists of segments of sine waves of alternating amplitude. Determine the Fourier series for the waveform up to the fourth harmonic.

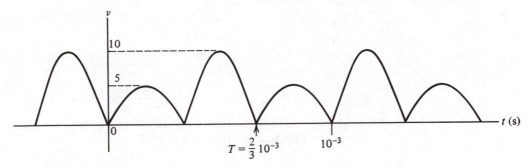

Figure P14-8

14-9. The following functions specify the envelope of the line spectrum of a periodic function. Find the critical points and the bandwidth for each case.

(a) $10e^{-f^2/500}$

(b) $\dfrac{1}{1 + 0.02(|f| - 100)^2}$

(c) $\dfrac{5}{|1 + 0.003f|}$

14-10. Repeat Prob. 14-9 for the following functions:

(a) $\dfrac{1}{1 + af^2}$

(b) $\dfrac{1}{3 + 0.2(|f| - 25)^2}$

(c) $7e - \dfrac{(f - 150)^2}{256}$

14-11. The Fourier series for the periodic function $f(x)$ is given by

$$f(x) = \frac{2}{\pi} \sum_{k=1}^{\infty} \frac{(-1)^{k+1}}{k} \sin kx$$

Using this relationship, draw (with the scales of the plot clearly labeled) the function whose Fourier series is given by

$$g(t) = a_0 + \sum_{k=1}^{\infty} \left(a_k \sin \frac{k\pi t}{6} + b_k \cos \frac{k\pi t}{6} \right)$$

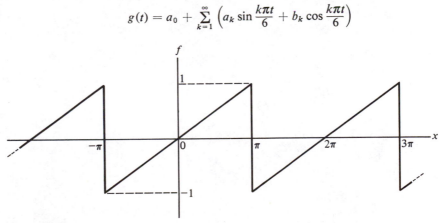

Figure P14-11

where

$$a_0 = 10$$

$$a_k = \frac{(-1)^{k+1} \times 10}{k\pi} \cos \frac{\pi k}{2}$$

$$b^k = \frac{(-1)^{k+1} \times 10}{k\pi} \sin \frac{\pi k}{2}$$

14-12. Find the Fourier coefficients of the waveform shown in the figure for which $T = 10$ and the function is value 5, 2, or 0.

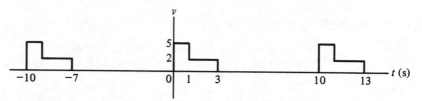

Figure P14-12

15

CAPACITOR–OP AMP CIRCUITS

15-1 FREQUENCY RESPONSE

Frequency is a word we commonly use in describing our hearing experiences. We know that a tuba or a bassoon produces low-frequency tones, that a flute or a piccolo produces high-frequency tones. The term "frequency response" is also commonplace: high-fidelity (hi-fi) amplifiers are described as having good low-frequency response, sometimes good high-frequency response. We know that a chirping or a buzzing are high-frequency sounds and that they can be suppressed by using a filter to reject the unwanted frequencies. Similarly, if a particular frequency is objectionable, a 60-Hz hum for example, it can be suppressed using a notch filter. In tuning a radio, we are selecting one band of frequencies, that occupied by the radio station we wish to receive, and rejecting broadcasters on all other frequencies. We thus understand that a radio tuner is a band-pass filter. These concepts that are familiar at the audio-frequency range are common in all frequency ranges. In optics, for example, we know that filters can reject all colors but red, or that Polaroid lenses reject objectionable light, making seeing easier.

In this chapter we begin to formalize the terms we have just used. When we use "frequency response" we will ordinarily mean a function or a plot that describes the magnitude and phase as a function of the frequency of a sine wave, ω or f. If the transfer function or gain function is $T(s)$, we let $s = j\omega$, and plot $|T(j\omega)|$ and $\theta(j\omega)$, the phase of $T(j\omega)$. These plots are useful in engineering practice because these magnitude and phase functions may be routinely

measured using standard instrumentation: a sine-wave generator and an oscil-
loscope, for example. Because the range of values of ω is often large, it is
common to use a logarithmic frequency scale. Gain also sometimes has a
large range of values, leading to a logarithmic gain scale. The logarithmic
and semilogarithmic coordinate systems we will use are shown in Fig. 15-1.
The coordinate system of Fig. 15-1(b) is the one we use later in the chapter
for Bode plots.

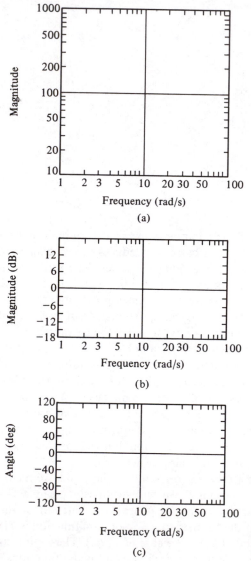

Figure 15-1. Coordinate systems to be used: (a) log-log coordinates, (b) semi-log coordinates for magnitude, and (c) semilog coordinates for phase.

The RC circuit shown in Fig. 15-2 is simply a voltage divider. The voltage-divider equation in terms of the impedances of the circuit is

$$\frac{V_2}{V_1} = T(s) = \frac{Z_2}{Z_1 + Z_2} \tag{15-1}$$

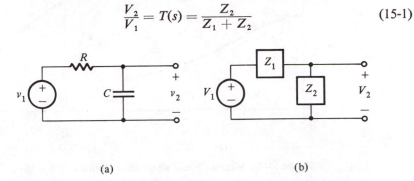

(a)   (b)

Figure 15-2. (a) Low-pass filter with (b) elements characterized by their impedances.

where, as shown previously, $Z_1 = R$ and $Z_2 = 1/Cs$. Then

$$T(s) = \frac{1/Cs}{R + 1/Cs} = \frac{1/RC}{s + 1/RC} \tag{15-2}$$

To obtain the magnitude and phase functions, we let $s = j\omega$ and write

$$T(j\omega) = \frac{1}{RC}\frac{1}{j\omega + 1/RC} \tag{15-3}$$

From this equation, we obtain the magnitude and the phase,

$$|T(j\omega)| = \frac{1/RC}{[\omega^2 + (1/RC)^2]^{1/2}} \tag{15-4}$$

and

$$\theta = -\tan^{-1}\omega RC \tag{15-5}$$

Determining a few values of these functions will suffice to show the general characteristics of the magnitude and the phase.

ω	$|T(j\omega)|$	$\theta(\omega)$
0	1	0
$1/RC$	$1/(2)^{1/2}$	$-45°$
$5/RC$	$\sim 1/5$	$-78.7°$
∞	0	$-90°$

As a frequency response, this is a low-pass filter, since high frequencies are rejected. From the plot shown in Fig. 15-3, we see that the magnitude decreases slowly with increasing ω.

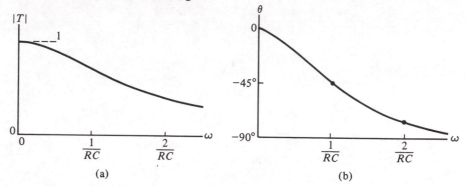

Figure 15-3.   Magnitude and phase characteristics of Eq. 15-3.

If we need a more rapid decrease with ω, we might be tempted to connect two of the circuits of Fig. 15-2 in cascade, as shown in Fig. 15-4. This is a ladder circuit which may be routinely analyzed, beginning at the output and working to the input. The result of such an analysis is the following:

$$\frac{V_2}{V_1} = \frac{(1/RC)^2}{s^2 + 3(s/RC) + (1/RC)^2} \tag{15-6}$$

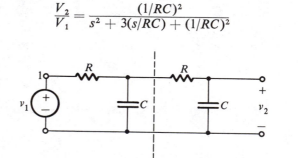

Figure 15-4.   Two circuits as in Fig. 15-2(a) connected in cascade without isolation.

We cite this result to make it clear that it is not equivalent to the simple squaring of Eq. 15-2, which would be

$$\frac{V_2}{V_1} = \frac{(1/RC)^2}{(s + 1/RC)(s + 1/RC)} = \frac{(1/RC)^2}{s^2 + 2s/RC + (1/RC)^2} \tag{15-7}$$

which might have been anticipated from the multiplication of transfer functions as $T = T_1 T_2$.

Now Eq. 15-7 is trivial to obtain, whereas Eq. 15-6 requires a tedious ladder analysis. True, both are low-pass responses and both do have magnitude responses which decrease more rapidly than that of Eq. 15-4.

How can we make Eq. 15-7 valid? The answer is to make use of the *voltage follower* shown in Fig. 15-5. This circuit is a special case of the non-inverting op-amp circuit considered earlier. Since at the input of the op amp, $V_x = 0$ and by the way in which it is connected $V_1' = V_+$ and $V_2' = V_-$, then $V_2' = V_1'$ or $V_2'/V_1' = 1$. We see that the voltage follower isolates the two RC circuits, so that current in one circuit marked 2 does not affect the current in the circuit marked 1. Hence, for this cascade connection, $T = T_1 T_2$ is valid and Eq. 15-7 is the proper transfer function.

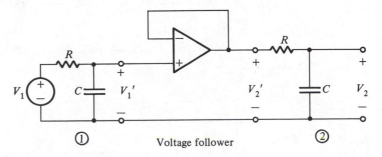

Figure 15-5. Same circuits as in Fig. 15-4 but with the isolation of a voltage follower.

Looking back now to the circuit of Fig. 15-4 and the response of Eq. 15-6, we see that the currents in the ladder are interacting, whereas the currents in Fig. 15-5 are isolated. Both circuits may be used as filters, but the approximation of Eq. 15-7 is valid only with isolation. The voltage follower is used extensively in circuit design to provide this isolation.

The response of Eq. 15-7 may be found from the values previously found for Eqs. 15-4 and 15-5 by squaring those magnitudes and doubling the angles. The result is shown in Fig. 15-6.

The circuit of Fig. 15-5 is a combination of passive circuits and op-amp circuits connected in cascade or tandem. The same result may be obtained

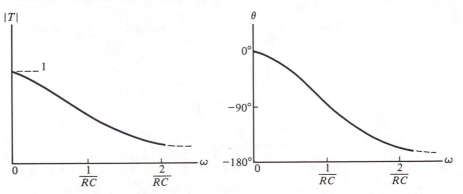

Figure 15-6. Magnitude and phase response of the circuit of Fig. 15-5.

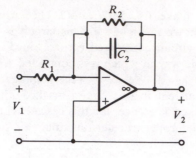

Figure 15-7. Lossy integrator circuit.

by a single module, that shown in Fig. 15-7. For the inverting amplifier, we found that $T = -R_2/R_1$, which we now generalize by replacing the R's by Z's to give

$$T = -\frac{Z_2}{Z_1} \tag{15-8}$$

For the circuit given, we see that $Z_1 = R_1$ and

$$Z_2 = \frac{1}{(1/R_2) + C_2 s} \tag{15-9}$$

so that

$$T = -\frac{1/R_1 C_2}{s + 1/R_2 C_2} \tag{15-10}$$

Except for the negative sign, this equation is equivalent to Eq. 15-2 and so has the magnitude and phase characteristic already given in Fig. 15-3. Note that by cascading two of these modules of Fig. 15-7, we may obtain the response of Eq. 15-7, which is shown in Fig. 15-6.

Two variations of the op-amp circuit of Fig. 15-7 are important. If $R_2 = \infty$, meaning that R_2 is removed, we have the circuit given in Fig. 15-8. First observe that

$$i_{C_2} = C_2 \frac{dv_C}{dt} = C_2 \frac{dv_2}{dt} \tag{15-11}$$

and that $i_{R_1} = v_1/R_1$. Since KCL gives us $i_{C_2} = -i_{R_1}$, we obtain

$$\frac{v_1}{R_1} = -C_2 \frac{dv_2}{dt} \tag{15-12}$$

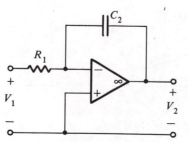

Figure 15-8. Integrator circuit.

or integrating to obtain v_2, assuming C_2 is uncharged initially,

$$v_2 = \frac{-1}{R_1 C_2} \int_0^t v_1 \, dt \tag{15-13}$$

which means that the output is proportional to the integral of the input. Hence, this circuit is known as an op-amp *integrator*. We will use it often.

Now the transfer function for the integrator is given by Eq. 15-10 with $R_2 = \infty$, and so is

$$T = \frac{-1}{R_1 C_2 s} \tag{15-14}$$

and we obtain the magnitude and phase functions by letting $s = j\omega$. Hence,

$$|T(j\omega)| = \frac{1}{R_1 C_2 \omega} \tag{15-15}$$

and

$$\theta = 90° \tag{15-16}$$

Thus the frequency response for the integrator is that shown in Fig. 15-9.

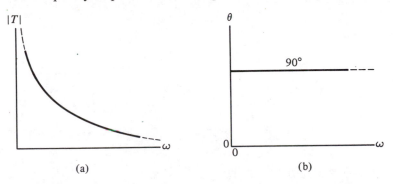

(a)   (b)

Figure 15-9.   Magnitude and phase response corresponding to Eq. 15-14.

The other circuit of interest which is related to that given in Fig. 15-7 is shown in Fig. 15-10. For this circuit $i_R = v_2/R_2$ and $i_C = C_1 \, dv_1/dt$, and equating i_C and $-i_R$ gives

$$v_2 = -R_2 C_1 \frac{dv_1}{dt} \tag{15-17}$$

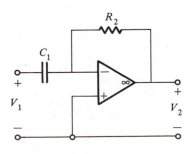

Figure 15-10.   Differentiating circuit.

This circuit is then a *differentiator*, and the output is the derivative of the input. This circuit is suitable for use only with inputs that do not change abruptly with time, for a large value of the derivative would give a large output, v_2, which would damage the op amp.

The frequency response for the differentiator is found directly from Eq. 15-9, and since $Z_2 = R_2$, and $Z_1 = 1/C_1 s$, we have

$$T = -R_2 C_1 s \qquad (15\text{-}18)$$

If we let $s = j\omega$ once more, then

$$|T(j\omega)| = R_2 C_1 \omega \qquad (15\text{-}19)$$

and

$$\theta = -90° \qquad (15\text{-}20)$$

These magnitude and phase response functions are shown in Fig. 15-11.

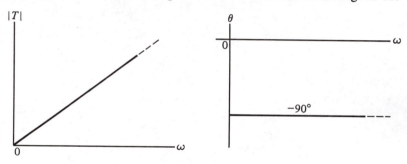

Figure 15-11.   Magnitude and phase responses for the circuit of Fig. 15-10.

15-3 POLES AND ZEROS OF $T(s)$

Analysis of the circuits in the last section gave a transfer function, $T(s)$, in the following forms:

$$T_1 = K_1 s \qquad (15\text{-}21)$$

$$T_2 = \frac{K_2}{s} \qquad (15\text{-}22)$$

$$T_3 = \frac{K_3}{s + s_1} \qquad (15\text{-}23)$$

$$T_4 = \frac{K_4}{(s + s_2)^2} \qquad (15\text{-}24)$$

and

$$T_5 = \frac{K_5}{(s + s_3)(s + s_4)} \qquad (15\text{-}25)$$

We seek a general form that will cover all these equations and others yet to come. That form is

$$T(s) = K\frac{(s + z_1)(s + z_2)\ldots(s + z_n)}{(s + p_1)(s + p_2)\ldots(s + p_m)} = K\frac{p(s)}{q(s)} \qquad (15\text{-}26)$$

where K is a real constant sometimes called the *scale factor*, and

$$-z_1, -z_2, -z_3, \ldots, -z_n \qquad (15\text{-}27)$$

are known as the *zeros* of $T(s)$ or the roots of the equation, $p(s) = 0$, and

$$-p_1, -p_2, -p_3, \ldots, -p_m \qquad (15\text{-}28)$$

are known as the *poles* of $T(s)$ or the roots of the equation, $q(s) = 0$. The factors of the form $(s + z_j)$ are called zero factors, while those of the form $(s + p_j)$ are known as pole factors. Poles and zeros of the form shown by Eq. 15-26 are called simple or distinct. If a pole or zero is repeated r times, it is said to be of multiplicity r. Often, we say simply a double pole, a double zero, and so on. Poles and zeros are plotted in the complex plane. using a 0 for each zero, and $\times$ for each pole. All of the $T(s)$ we have considered in this chapter have zeros and poles located on the negative real axis.

EXAMPLE 15-1

Describe the following $T(s)$ in terms of its poles and zeros:

$$T(s) = \frac{s(s + 2)}{(s + 1)(s + 3)} \qquad (15\text{-}29)$$

Solution

$T(s)$ has poles at -1 and -3, and zeros at the origin and at -2. Then all are on the negative real axis of the s-plane (see Fig. 15-12).

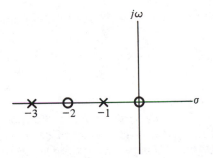

Figure 15-12. Poles and zeros for $T(s)$ in Example 15-1.

Poles and zeros for the forms of T given in Eqs. 15-21 through 15-25 are summarized in Table 15-1.

Table 15-1

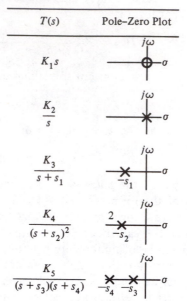

$T(s)$	Pole–Zero Plot
$K_1 s$	
$\dfrac{K_2}{s}$	
$\dfrac{K_3}{s+s_1}$	
$\dfrac{K_4}{(s+s_2)^2}$	
$\dfrac{K_5}{(s+s_3)(s+s_4)}$	

15-4 MAGNITUDE AND PHASE FROM POLE–ZERO PLOTS

In this section we show that the magnitude and phase responses may be visualized or computed in terms of the pole and zero locations for $T(s)$. This will be found to offer conceptual advantages, especially in the design process, where specifications frequently relate to pole and zero locations.

We begin with the $T(s)$ of Eq. 15-18, which is the transfer function of the differentiator,

$$T = \frac{V_2}{V_1} = -R_2 C_1 s \tag{15-30}$$

First, let us simplify by considering a normalized transfer function

$$T = s \tag{15-31}$$

Letting $s = j\omega$ yields

$$T(j\omega) = j\omega \tag{15-32}$$

This phasor is shown *superimposed* on the s-plane in Fig. 15-13. This is a convenience because we wish to determine the magnitude and phase of $T(j\omega)$ as ω increases—meaning for larger values of ω as we move up the imaginary axis. Figure 15-13 shows this movement for successively larger values of ω. Clearly, the magnitude of this phasor is the value of ω at which it is drawn, and the phase of the phasor is always 90° for all ω. Thus, we obtain the magnitude and phase variation with frequency that is depicted in Fig. 15-13(b) and (c).

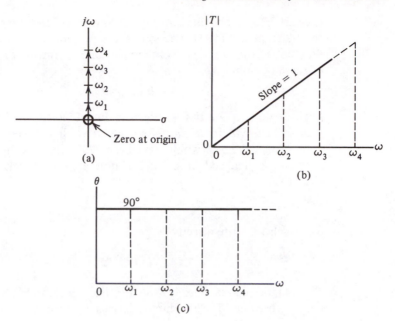

Figure 15-13. Graphical method for constructing the magnitude and phase plots.

Now if we return to Eq. 15-30, we see that the magnitude of the transfer function when $s = j\omega$ is that given in Fig. 15-13(b) multiplied by R_2C_1 as shown in Fig. 15-14(a). At a specific frequency, ω_2, the magnitude is $R_2C_1\omega_2$, and the slope of the straight line

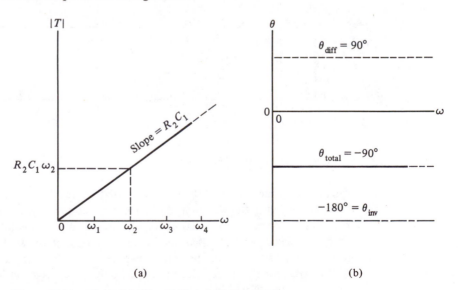

Figure 15-14. Plots of (a) Eq. 15-33 and (b) Eq. 15-34.

is clearly R_2C_1. The phase of Eq. 15-30 is modified by the -1 multiplier, which results from the inverter action of the op-amp circuit. Call the phase of the differentiator θ_{diff} and that of the invertor θ_{inv}; the net or total phase is the sum of these two phases,

$$\theta_{\text{total}} = \theta_{\text{diff}} + \theta_{\text{inv}} \tag{15-34}$$

Each of these quantities is shown in Fig. 15-14(b), and the resultant phase is seen to be $-90°$ for all ω. Note that if we had taken θ_{inv} to be $180°$ rather than $-180°$, the sum would have been $270°$, which is the same as $-90°$.

The normalized transfer function that will relate to the integrator circuit is

$$T = \frac{1}{s} \tag{15-35}$$

for which we write directly,

$$|T(j\omega)| = \frac{1}{\omega}, \qquad \theta = -90° \tag{15-36}$$

This magnitude response is shown in Fig. 15-15 and is recognized as a rectangular hyperbola. The magnitude is seen to approach an infinite value as fre-

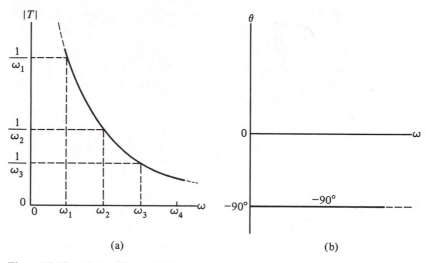

(a)   (b)

Figure 15-15.   Plots of Eqs. 15-36.

quency approaches zero. This is visualized in terms of the s-plane location of the pole with the aid of Fig. 15-16. As frequency increases, we move up the imaginary axis. The actual magnitude is the reciprocal of the distance from the origin, and the phase is the negative of that shown by each successive phasor. We will eventually consider pole–zero plots containing both poles and zeros, and this difference in role of poles and zeros with respect to magnitude and phase will become more evident.

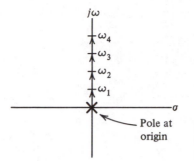

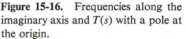

Figure 15-16. Frequencies along the imaginary axis and $T(s)$ with a pole at the origin.

For the actual integrator transfer function given by Eq. 15-14, we must divide the magnitude function of Eq. 15-36 by R_1C_2. Taking into account the -1 sign that results from inverting action of the op-amp circuit, the phase of the integrator is seen to be

$$\theta_{\text{total}} = \theta_{\text{int}} + \theta_{\text{inv}} = -90° - 180° = +90° \qquad (15\text{-}37)$$

for all $\omega > 0$.

So much for poles and zeros at the origin. Let that zero move into the left-half of the s-plane along the negative real axis to a position $-\sigma_1$ shown in Fig. 15-17. Thus zero corresponds to the zero factor $(s + \sigma_1)$, so the transfer function is

$$T(s) = s + \sigma_1 \qquad (15\text{-}38)$$

With $s = j\omega$,

$$T(j\omega) = j\omega + \sigma_1 \qquad (15\text{-}39)$$

We can determine the magnitude and phase functions employing our knowledge of complex numbers from Chapter 9; hence, the magnitude is the sum of the squares of the components, from which we extract the square root,

$$|T(j\omega)| = \sqrt{\omega^2 + \sigma_1^2} \qquad (15\text{-}40)$$

and the phase is the inverse tangent of the imaginary part divided by the real part, or

$$\theta = \tan^{-1} \frac{\omega}{\sigma_1} \qquad (15\text{-}41)$$

However, we wish to find a geometrical interpretation that is easy to visualize. In Fig. 15-17, the real and imaginary components of $T(j\omega)$ of Eq. 15-39 are shown in the s-plane. The real part is shown as a phasor directed from O to D, and the imaginary part as a phasor from O to B. The addition of these two phasors gives the resultant phasor directed from O to C. This resultant phasor has the magnitude and phase of $T(j\omega)$. From the figure we can see that this magnitude and phase do indeed have the values given in Eqs. 15-40 and 15-41, since the magnitude is the square root of the sum of the squares of the right triangle $ODCO$, and the tangent of the angle of line OC is the ratio of DC and OD.

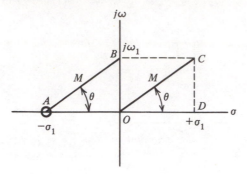

Figure 15-17. Geometry for the case of a zero at $s = -\sigma_1$.

Also shown on the phasor diagram of Fig. 15-17 is the phasor directed from A to B, from the position of the zero at $-\sigma_1$ to a point on the imaginary axis, $j\omega_1$. It is important that we recognize that phasors AB and OC have the same magnitude and angle, shown as M and θ. If they are identical, we can use either. But phasor AB has more meaning to us because it originates at the zero position, as was previously the case when the zero was located at the origin, and it will be used for both zeros and poles.

Next, we let frequency have successively larger values, ω_1 ω_2, ω_3, in Fig. 15-18(a). At each frequency, the phasor drawn from the zero has the

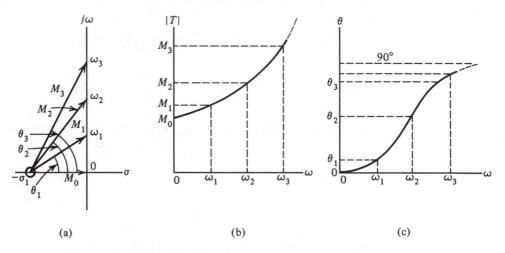

(a)   (b)   (c)

Figure 15-18. Magnitude and phase functions for Eqs. 15-40 and 15-41.

magnitude of $T(j\omega)$, and the phasor forms the corresponding angle with respect to the real axis. In (b) of the figure, we see how this magnitude increases with increasing frequency, and in (c) how the angle increases from 0 to 90°. With experience, you will visualize the curves of (b) and (c) in Fig. 15-18 in your mind while looking only at (a).

Suppose that there is a pole at $-\sigma_1$ of the s-plane, in place of the zero of Figs. 15-17 and 15-18, so that the transfer function is

$$T(s) = \frac{1}{s + \sigma_1} \tag{15-42}$$

The same construction shown in Fig. 15-17 applies to the denominator, so is the same variation in magnitude and phase with increasing frequency that is shown in Fig. 15-19(a). However, the magnitude $|T(j\omega)|$ is now the reciprocal

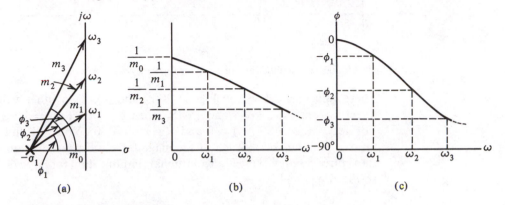

(a)   (b)   (c)

Figure 15-19. Plots corresponding to those of Fig. 15-18 when a pole replaces a zero at $s = -\sigma_1$.

of this phasor length, and the angle of $T(j\omega)$ is the negative of the angle drawn with respect to the real axis. Thus,

$$T(j\omega_k) = \frac{1}{m_k} \underline{/-\phi_k} \tag{15-43}$$

in polar form. The variation of the magnitude and phase of this $T(j\omega)$ is that shown in (b) and (c) of Fig. 15-19, showing that the magnitude decreases with increasing frequency, and the phase angle has a characteristic inverse tangent shape with increasing frequency, going from $0°$ to $-90°$.

Let us reexamine the plots of Eq. 15-43 in greater detail with the objective of making sketching of the magnitude and phase functions easier to accomplish. The magnitude of Eq. 15-42 with $s = j\omega$ is the reciprocal of Eq. 15-40, or

$$|T(j\omega)| = \frac{1}{\sqrt{\omega^2 + \sigma_1^2}} \tag{15-44}$$

and the phase is the negative of Eq. 15-41,

$$\theta = -\phi = -\tan^{-1}\frac{\omega}{\sigma_1} \tag{15-45}$$

Let ω_1 be the frequency at which $\omega_1 = \sigma_1$, as shown in Fig. 15-20(a). Under this condition, Eqs. 15-44 and 15-45 become

$$|T(j\omega_1)| = \frac{1}{\sqrt{2}\,\sigma_1} \tag{15-46}$$

and

$$\theta(j\omega_1) = -45° \tag{15-47}$$

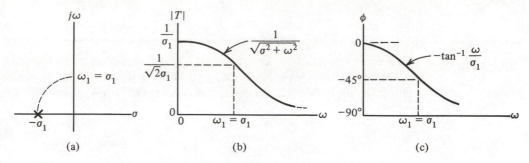

Figure 15-20.   Values of magnitude and phase at the frequency $\omega_0 = \sigma_1$.

These values are shown in Fig. 15-20(b) and (c). Two other values are seen directly from Eqs. 15-44 and 15-45: when $\omega = 0$, $T(j0) = 1/\sigma_1$ and $\phi = 0°$. For very large ω, $T(j\infty) = 0$ and $\phi(j\infty) = -90°$. You will find that you can make a fair start of a sketch of the magnitude and phase functions using only these three points! There is an additional simplification for large ω. Let $\omega = k\sigma_1$ so that Eq. 15-44 is

$$|T(j\omega)| = \frac{1}{\sigma_1\sqrt{k^2 + 1}} \approx \frac{1}{k\sigma_1} \qquad (15\text{-}48)$$

The error in the approximation is 5% for $k = 3$ and less than 2% for $k = 5$.

We turn next to transfer functions with more than one pole. An example of such a function is

$$T(s) = \frac{5}{(s + 1)(s + 5)} \qquad (15\text{-}49)$$

We may think of this function as the product of two other functions,

$$T_1(s) = \frac{1}{s + 1} \qquad \text{and} \qquad T_2 = \frac{5}{s + 5} \qquad (15\text{-}50)$$

such that

$$T(j\omega) = T_1(j\omega)T_2(j\omega) = |T_1||T_2|\underline{/-\phi_1 - \phi_2} \qquad (15\text{-}51)$$

The poles are displayed in the complex plane in Fig. 15-21(a), and the phasor diagram in terms of which we may visualize the magnitude and phase variation

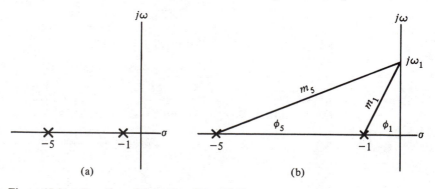

Figure 15-21.   Location of the poles of Eq. 15-49.

is shown in (b) of the figure. As the point marked $j\omega_1$ moves up the imaginary axis, we see that $1/m_1$ and $1/m_5$ grow small decreasing from their initial values of 1 and $\frac{1}{5}$. The phase starts at $\omega = 0$ at $0°$ for each phasor, and ϕ_1 and ϕ_5 continue to increase until each is at $90°$. The complete plots of the magnitude and phase of $T(j\omega)$ are shown in Fig. 15-22. Note there the usefulness of the

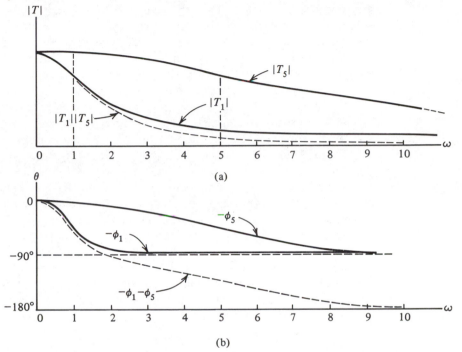

Figure 15-22. Construction of the magnitude and phase of Eq. 15-49 by multiplication of magnitudes and summation of phase.

rule given in Fig. 15-20 at the frequencies $\omega = 1$ and 5, where each magnitude has reduced to $1/\sqrt{2}$ of its initial value, and the phase is $-45°$ for each.

Figure 15-22 also suggests the ease with which phases are added in comparison with multiplication required in finding the total magnitude. We can replace multiplication by addition if we deal with logarithmic quantities, since

$$\log|T_1||T_2| = \log|T_1| + \log|T_2| \tag{15-52}$$

This observation serves to introduce the next topic, Bode plots.

15-5 BODE PLOTS

The *logarithmic gain or transfer function magnitude* is defined in terms of a new unit called the *decibel* (dB), as follows:

$$M = 20 \log_{10}|T(j\omega)| \quad \text{dB} \tag{15-53}$$

Frequently, M is referred to simply as *gain in dB*. It should be observed that the gain (magnitude) function

$$|T| = \left|\frac{V_2}{V_1}\right|$$

may take on values, as a function of $j\omega$, which may be greater than 1, equal to 1, or less than 1. Then the gain in dB is (by Eq. 15-53) positive, zero, or negative, respectively. Further, a gain of zero represents a dB gain of $-\infty$ and a gain of infinity represents a dB gain of $+\infty$. Other important values are 20 dB corresponding to a gain of 10, and positive or negative multiples of 20 dB corresponding to positive or negative powers of 10. Finally, a gain of 6 dB corresponds approximately to a gain of 2.

Plots of the magnitude gain function M in dB and the phase function $\theta(j\omega)$, usually in degrees, versus ω are typically required for a wide range of ω. For these reasons the plots are usually made using a logarithmic scale for ω. Thus, the number of units of the abscissa, for a given ω, are equal to $\log_{10}\omega$; for example, $\omega = 10$ corresponds to $\log_{10} 10 = 1$ unit, $\omega = 10^{\pm k}$ corresponds to $\pm k$ units, with $\omega = 1$ nominally corresponding to 0 units. Coordinates for plotting gain M in dB and θ in degrees versus ω (on a $\log_{10}\omega$ scale) are shown in Fig. 15-23. These are semilogarithmic coordinates and special (semilog) graph papers are available for making such plots.

Plots of M and θ versus ω (on a $\log_{10}\omega$ scale) are known as Bode plots after Hendrik Bode, formerly of the Bell Laboratories. The usefulness of this particular choice will become apparent as we deal with design. We begin by considering transfer functions with a zero and then a pole at the origin, as was done in Section 15-4. Let

$$T(s) = s = \sigma_0\left(\frac{s}{\sigma_0}\right) \tag{15-54}$$

where σ_0 is a normalizing factor. From Eq. 15-53,

$$M = 20\log\left|\sigma_0\left(\frac{j\omega}{\sigma_0}\right)\right| = 20\log\sigma_0 + 20\log\frac{\omega}{\sigma_0} \tag{15-55}$$

Now $20\log\sigma_0$ dB is a constant which we will presently store and then add it to M in completing the Bode plot. Let the remaining term be M_1, which does vary with frequency, in fact with the logarithm of frequency,

$$M_1 = 20\log\frac{\omega}{\sigma_0} \qquad \text{dB} \tag{15-56}$$

Clearly, a plot of M_1 as a function of $\log \omega/\sigma_0$ will be a straight line with a positive slope. What will be the value of that slope? And in what units?

Figure 15-23 shows the frequency coordinate of the Bode plot, in which ω is shown on a logarithmic scale and u on a linear scale. Two points are shown as u_1 and u_2, for which

$$u_1 = \log\omega_1 \qquad \text{and} \qquad u_2 = \log\omega_2 \tag{15-57}$$

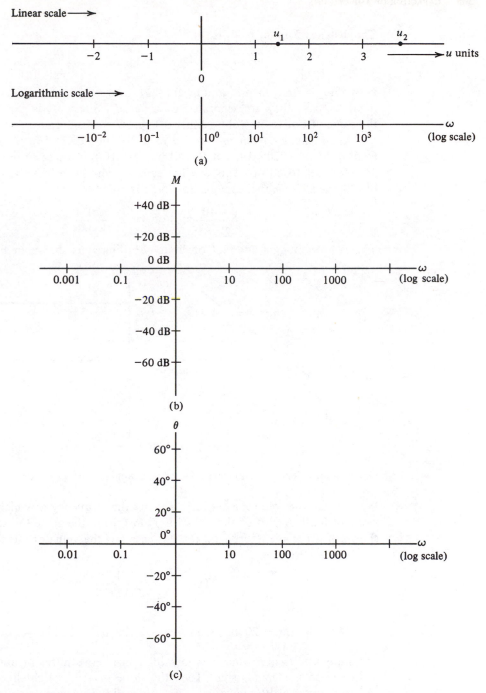

Figure 15-23. (a) Comparison of linear and logarithmic scales, (b) semilog coordinates for magnitude, and (c) semilog coordinates for phase.

The distance from u_1 to u_2 is

$$u_2 - u_1 = \log \omega_2 - \log \omega_1 = \log \frac{\omega_2}{\omega_1} \tag{15-58}$$

Now if $\omega_2 = 2\omega_1$, then u_1 and u_2 are said to be separated by an *octave*, whereas if $\omega_2 = 10\omega_1$, they are separated by a *decade*. Since M_1 is expressed in dB, the slope of Eq. 15-56 will be either dB/octave or dB/decade.

As ω changes from $\omega = \sigma_0$ to $\omega = 2\sigma_0$ in Eq. 15-56, M_1 will increase from 0 dB to 6.0206 dB. Let us call it 6 dB. If ω changes to $\omega = 10\sigma_0$, then M_1 increases to 20 dB. This will be repeated with each increase of an octave or a decade. Thus, the slope of Eq. 15-56 is either

$$6 \frac{dB}{octave} \quad \text{or} \quad 20 \frac{dB}{decade} \tag{15-59}$$

The complete magnitude plot on Bode coordinates is shown in Fig. 15-24(a).

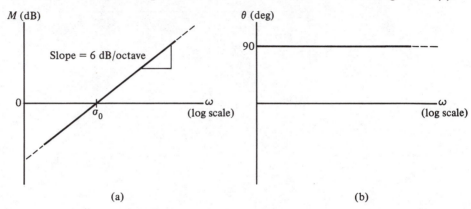

(a)   (b)

Figure 15-24.   Slope of 6 dB/octave on semilog coordinates.

We now see the role for σ_0, which is to set the frequency at which M_1 has the value of 0 dB. A plot of the phase of $T(j\omega)$ using the same log scale for ω is shown in (b) of the figure, and for this case has the value of 90° for all ω.

We next consider

$$T(s) = \frac{1}{s} = \frac{1}{\sigma_0(s/\sigma_0)} \tag{15-60}$$

for which

$$M = 20 \log |T(j\omega)| = -20 \log \sigma_0 - 20 \log \frac{\omega}{\sigma_0} \tag{15-61}$$

Our study can be short since the only difference between this M and that found in Eq. 15-55 is a minus sign! Then the slope of the straight line is negative and is either -6 dB/octave or -20 dB/decade. The complete Bode plot for Eq. 15-61 is shown in Fig. 15-25.

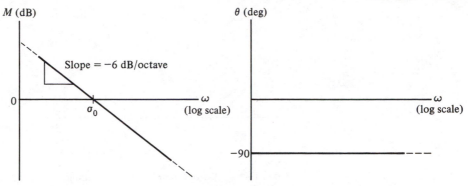

Figure 15-25. Slope of -6 dB/octave on semilog coordinates.

Now we move on to a zero at $-\sigma_0$ in the s-plane. The normalized form for the transfer function is

$$T(j\omega) = \sigma_0 \left(1 + j\frac{\omega}{\sigma_0}\right) \tag{15-62}$$

For this function, M is

$$M = 20 \log \sigma_0 + 20 \log \left|1 + j\frac{\omega}{\sigma_0}\right| \tag{15-63}$$

and, as before, we store $20 \log \sigma_0$ for the time being, and consider the remainder, which is

$$M_1 = 20 \log \left|1 + j\frac{\omega}{\sigma_0}\right| \quad \text{dB} \tag{15-64}$$

We first determine the low- and high-frequency asymptotes for M_1. For low frequencies,

$$M_1 \approx 20 \log 1 = 0 \text{ dB}, \qquad \omega \ll \sigma_0 \tag{15-65}$$

while for high frequencies,

$$M_1 \approx 20 \log \frac{\omega}{\sigma_0} \text{ dB}, \qquad \omega \gg \sigma_0 \tag{15-66}$$

The high-frequency asymptote is familiar as Eq. 15-56, which is the equation of a straight line having a slope of 6 dB/octave, and having the value $M_1 = 0$ dB at the frequency $\omega = \sigma_0$. The two asymptotic lines meet $\omega = \sigma_0$ as shown in Fig. 15-26. This asymptotic cradle holds the actual response, which we obtain from Eq. 15-64 by determining the magnitude of the complex number,

$$M_1 = 20 \log \sqrt{1 + \left(\frac{\omega}{\sigma_0}\right)^2} = 10 \log \left[1 + \left(\frac{\omega}{\sigma_0}\right)^2\right] \tag{15-67}$$

We could simply plot M_1 by determining enough points for an accurate plot. Instead, we give a simple rule for the construction of this M_1. Solving Eq.

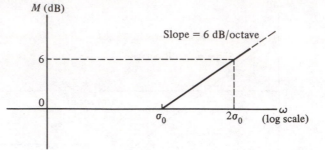

Figure 15-26. Asymptotic plots corresponding to Eqs. 15-65 and 15-66.

15-67 for ω at which the magnitude has the value M_1 dB, we obtain

$$\omega = \sigma_0 \sqrt{10^{M_1/10} - 1} \tag{15-68}$$

for which values can be routinely found using a calculator. You may wish to verify the values tabulated in Table 15-2. The values found are shown in Fig. 15-27. These five points may be routinely located using the asymptotic lines as references, and a smooth curve drawn through these points.

Table 15-2

ω	True Magnitude (dB)	Asymptotic Magnitude (dB)	Difference (dB)
$\sigma_0/2$	1	0	1
$0.7648\sigma_0$ $= \sigma_0/1.308$	2	0	2
σ_0	3	0	3
$1.308\sigma_0$	4.3	2.3	2
$2\sigma_0$	7	6	1

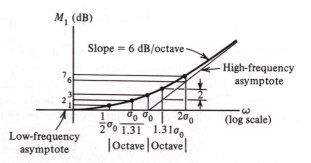

Figure 15-27. Specific frequencies for which the actual response can be easily calculated from the asympototic response.

A step-by-step procedure for accomplishing the things we have just described is as follows:

1. Given $T(j\omega)$ as in Eq. 15-62, determine σ_0. Let this frequency be called the *break frequency*.
2. From σ_0 on the frequency scale, draw the two asymptotic straight lines—one of zero slope extending to lower values of frequency, one of 6 dB/octave slope extending to higher frequencies.
3. At the break frequency, the actual response is displaced 3 dB from the intersection of asymptotes.
4. At frequencies one octave below and one octave above the break frequency, the difference in the actual and asymptotic curve is 1 dB.
5. If greater accuracy is required, locate the frequencies at which the difference given in Table 15-2 is 2 dB.
6. Join the points you have just found as a smooth curve, and the plot is complete.

There is no similar simple procedure for plotting the phase function on Bode coordinates. This phase function is found from Eq. 15-62 and is

$$\theta = \tan^{-1}\frac{\omega}{\sigma_0} \qquad \text{rad} \qquad\qquad (15\text{-}69)$$

Since we usually plot phase in degrees rather than radians, we scale this equation by multiplying by $360/2\pi = 57.3$,

$$\theta = 57.3 \tan^{-1}\frac{\omega}{\sigma_0} \qquad \text{degrees} \qquad\qquad (15\text{-}70)$$

However, most calculators perform this multiplication automatically, so it is an issue of little concern. One approximation is depicted in Fig. 15-28 as the dashed line, which has the value of 45° at $\omega = \sigma_0$ and has a slope of 45° per decade.

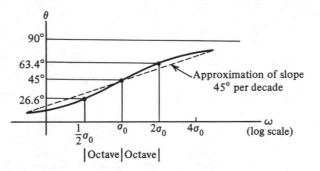

Figure 15-28. Frequencies at which the actual phase response is easily found from a linear approximation to phase.

For a pole of $T(s)$ at $s = -\sigma_0$, the transfer function is

$$T(s) = \frac{1}{s + \sigma_0} \qquad (15\text{-}71)$$

Letting $s = j\omega$, we find M to be

$$M = -20 \log \sigma_0 - 20 \log \left| 1 + j\frac{\omega}{\sigma_0} \right| \qquad (15\text{-}72)$$

which is the negative of the M for the corresponding zero factor, Eq. 15-63. Hence, we reach the same conclusion we did in comparing a zero and pole at the origin: The Bode plot for the magnitude of $T(s)$ given by Eq. 15-71 is the negative for the corresponding $T(s)$ with a zero at $s = -\sigma_0$, and the desired result obtained when M_1 of Fig. 15-27 is tipped over. And the same conclusion applies to the phase.

We can now make Bode plots for a pole or a zero at any location on the negative real axis of the s-plane. We will show next that the results we have will suffice to make a Bode plot for any number of poles and zeros along the negative real axis. We will do so by an example, and then generalize.

Figure 15-29 shows a cascade connection of two RC–op amp circuits.

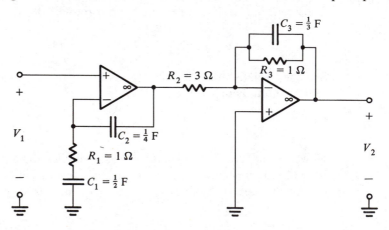

Figure 15-29.   Cascade connection of two op-amp circuits.

Since the two circuits are isolated, the two sections may be analyzed separately and the results combined. The objective in studying this circuit is to prepare a Bode plot for the magnitude of V_2/V_1. For the first module,*

$$T_1 = 1 + \frac{Z_2}{Z_1} = 1 + \frac{1/C_2 s}{R_1 + 1/C_1 s} = \frac{s + 6}{s + 2} \qquad (15\text{-}73)$$

---

*These equations were derived in Chapter 5 using R_1 and R_2. We have let R_1 become Z_1 and R_2 become Z_2, as discussed in Chapter 9.

and

$$T_2 = -\frac{Z_2}{Z_1} = \frac{-1}{R_1}\left(\frac{R_3/C_3 s}{R_3 + 1/C_3 s}\right) = \frac{-1}{s+3} \qquad (15\text{-}74)$$

so that

$$T = T_1 T_2 = \frac{-(s+6)}{(s+2)(s+3)} \qquad (15\text{-}75)$$

We first prepare this equation to be in standard form,

$$T(s) = \frac{-(1+s/6)}{(1+s/2)(1+s/3)} \qquad (15\text{-}76)$$

On the Bode coordinates of Fig. 15-30, the break frequencies of 2, 3, and 6 are shown. For the first two, the plots break downward, while for that breaking

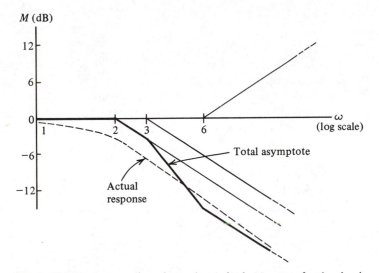

Figure 15-30. Asymptotic and actual magnitude responses for the circuit of Fig. 15-29.

at $\omega = 6$, the curve is breaking upward. The figure shows the total asymptotic curve, obtained by adding the other three. This is especially easy, since all of these are straight lines. The actual response is also shown on the figure. This is the manner in which the circuit of Fig. 15-29 behaves as frequency increases. The characteristic may be described as a low-pass filter.

To generalize, if $T(s)$ has poles and zeros on the negative real axis, then

$$T(s) = K\frac{(s+\sigma_1)(s+\sigma_2)\dots(s+\sigma_n)}{(s+\sigma_a)(s+\sigma_b)\dots(s+\sigma_m)} \qquad (15\text{-}77)$$

We divide each pole and zero factor by the appropriate σ_j, and then take the logarithm of the individual magnitudes,

$$M = 20 \log |T(j\omega)| = 20 \log K\frac{\sigma_1\sigma_2\cdots\sigma_n}{\sigma_a\sigma_b\cdots\sigma_m} + 20 \log \left|1 + j\frac{\omega}{\sigma_1}\right|$$

$$+ 20 \log \left|1 + j\frac{\omega}{\sigma_2}\right| + \ldots - 20 \log \left|1 + j\frac{\omega}{\sigma_a}\right| \qquad (15\text{-}78)$$

$$- 20 \log \left|1 + j\frac{\omega}{\sigma_b}\right| - \ldots$$

The only difference in numerator and denominator factors is a sign. Because of the normalization used, each of the factors is drawn from the break frequency on the $M = 0$-dB line. These are then added and the total asymptotic response thus determined. If greater accuracy is required, the actual responses for each of the factors in Eq. 15-78 is determined and these are added.

There remains discussion of the first factor in Eq. 15-78, call it M_0,

$$M_0 = 20 \log K\frac{\sigma_1\sigma_2\cdots\sigma_n}{\sigma_a\sigma_b\cdots\sigma_m} \qquad \text{dB} \qquad (15\text{-}79)$$

This is a constant, of course, but it may be included in the Bode plot in two different ways: (1) It can shift the remaining plot either upward or downward, the shape of the response remaining the same. (2) The M coordinates can be changed, changing the value previously marked 0 dB to a new value of M_0 dB. The second of these is preferred, of couse, since it is so easy to accomplish!

One last point about Bode plots: we may start from an experimentally determined response, cut and try until suitable break frequencies are found, and then write an expression for $T(s)$ in the form of Eq. 15-77. In Fig. 15-31, the dashed curve represents an experimentally determined response. Through several trials, the asymptotic response also shown in the figure as the solid lines is found. With this asymptotic response, the break frequencies can be read off, and $T(s)$ becomes

$$T(s) = K\frac{(s + 1)(s + 3)}{(s + \frac{1}{2})(s + 6)} \qquad (15\text{-}80)$$

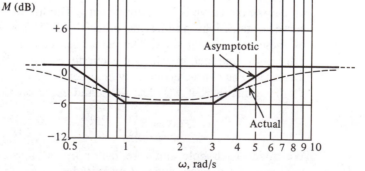

Figure 15-31.   Bode plot for Eq. 15-80.

The value of K can be determined from the value of the asymptotic curve at some value of frequency. Thus, when $s = 0$, $T(0) = 1$ (corresponding to 0 dB), so for this example $K = 1$.

PROBLEMS

15-1. Given the asymptotic Bode plot shown in the figure. Find the transfer function, $T(s)$, and determine all constants.

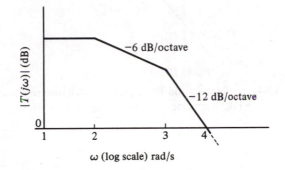

Figure P15-1

15-2. Repeat Prob. 15-1 for the asymptotic Bode plot of the figure. You are to determine all constants in $T(s)$.

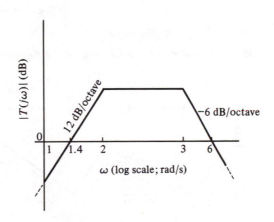

Figure P15-2

15-3. Given the asymptotic Bode plot shown in the figure. You are to write the corresponding transfer function, $T(s)$, and determine all constants.

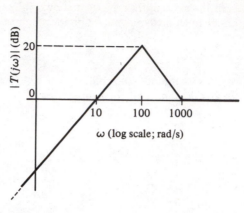

Figure P15-3

15-4. The asymptotic Bode plot of the figure represents a bandpass filter. From the plot, determine $T(s)$, including all constants.

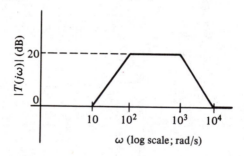

Figure P15-4

15-5. The asymptotic Bode plot of the figure represents a band-elimination filter.

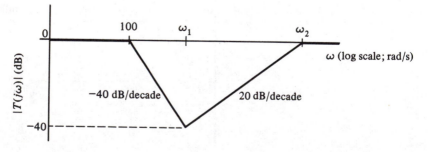

Figure P15-5

First, determine the numerical value for ω_1 and ω_2, and then write the equation for $T(s)$ with all numerical values determined.

15-6. Given the asymptotic Bode plot shown in the figure. Find the corresponding transfer function $T(s)$, and determine all constants.

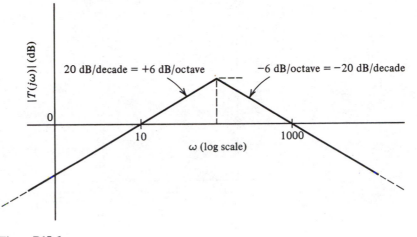

Figure P15-6

15-7. The kind of plot of $|T(j\omega)|$ shown is sometimes called a notch filter. For this figure, determine the corresponding $T(s)$, including the determination of all constants.

Figure P15-7

15-8. The asymptotic Bode plot of the figure represents a bandpass amplifier. For this response, determine the transfer function, $T(s)$, including the evaluation of all constants.

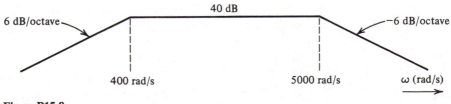

Figure P15-8

15-9. Given the transfer function asymptotic Bode plot shown in the figure, determine $T(s)$, including all constants.

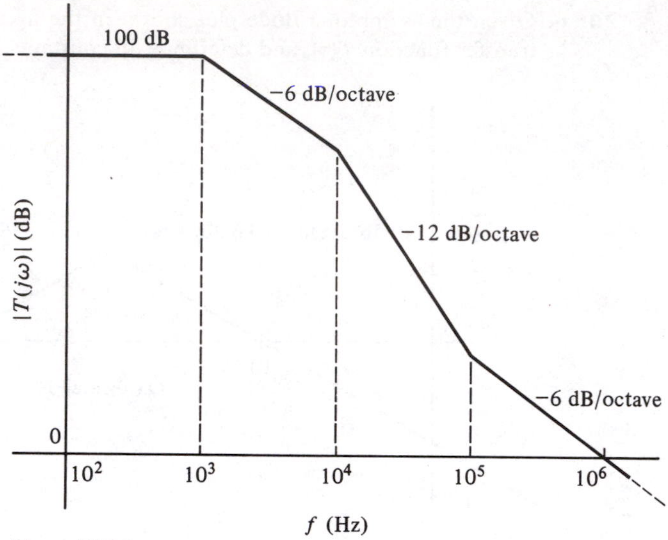

Figure P15-9

15-10. For the following $T(s)$, plot the straight-line asymptotes for the Bode plot and plot the phase as a function of frequency using logarithmic coordinates for frequency.

$$T(s) = 50\left[\frac{1 + 0.025s}{s(1 + 0.05s)}\right]$$

15-11. Repeat Prob. 15-10 for the following transfer function:

$$T(s) = \frac{1000s}{(1 + 0.01s)(1 + 0.0025s)}$$

15-12. For the following $T(s)$, plot the straight-line asymptotes for the Bode plot and also plot the phase as a function of frequency using logarithmic coordinates for frequency.

$$T(s) = 180\left[\frac{s(1 + 0.01s)}{(1 + 0.05s)(1 + 0.0001s)}\right]$$

15-13. Repeat Prob. 15-12 for the following transfer function:

$$T(s) = \frac{1,000,000s(s + 1000)}{(s + 3,000,000)(s + 20,000)(s + 10)}$$

16

CIRCUIT DESIGN

The process of circuit design involves many different trade-offs. Cost, accuracy, reliability, size, power dissipation, fabrication ease, and so on, are some of the factors involved. Several successive designs are generally necessary to arrive at an acceptable circuit. Our interests here are to arrive at a good circuit that meets design requirements but not to be concerned with the other items, such as cost, size, and so on. Even so, we must be careful to use reasonable and practical elements and values. For example, a 1-F capacitor is not an acceptable value because capacitors do not come in these sizes; nor is 1 Ω an acceptable value. Resistors should be at least 100 Ω or greater and capacitors at most 0.001 F or less to be practical.

We also concentrate on inductorless circuits in this chapter. Inductors and transformers are important in many power circuits, but they are not essential to signal-processing circuits discussed here. Inductors are also not very reliable, nor are they suited to modern electronic technology. Fortunately, op amps, resistors, and capacitors are entirely sufficient to provide any desired circuit behavior. For these reasons we concentrate on op-amp circuits in this chapter.

Op-amp circuits also provide a major design convenience, the ability to cascade units in tandem to obtain the overall transfer function. Each unit or module can have a simple transfer function that is easy to design. The overall function then is the product of the module transfer functions and the design is complete. The isolation between units is usually quite good. In some cases it may be desirable to choose the component values so as to make the input impedance high enough, and in other cases it may be preferable to insert an isolator between the modules. We assume in our discussion that the isolation between modules is perfect, with the understanding that either isolating modules are inserted or the input impedance is high enough.

Our design process will involve very simple op-amp modules whose transfer functions will generally have a single zero and a single pole. This means that the overall transfer function for the cascade of such modules will have poles and zeros on the negative real axis of the complex plane since each RC–op amp module will have that constraint. We will be concerned only with designs of the magnitude of transfer functions, such as the low pass, the band pass, and other filters. The process is similar but more complex for the phase functions. It then follows that the asymptotes in the magnitude Bode plots as discussed in Chapter 15 will have slopes of 6 db/octave (or the equivalent of 20 db/decade) on the $\log_{10} \omega$ scale. Multiples of 6-db/octave slope may be obtained by multiple poles or zeros in the overall transfer function.

The design process thus constitutes approximating the ideal or desired magnitude transfer function by an asymptotic Bode plot with slopes of 6 db/octave or, at times, multiples of 6 db/octave. From this approximate plot of the magnitude transfer function, we write down the corresponding transfer function in terms of its poles and zeros. We then split this function as a rule into terms with a single pole and perhaps a single zero. Each of the terms is realized by means of an op-amp module. The module may be either a noninverting or an inverting one, since we are interested only in the magnitude function. These modules are then cascaded together to obtain the desired design.

16-2 OP AMP MODULES

In our design process, one of the steps is the realization of the single module from the single pole/zero transfer-function term. We will do this by a simple, process of compiling the different possible circuits and their transfer functions. All we need to do then is to look up the table for the desired modular transfer function and find the corresponding op-amp circuit. We will compile some inverting and some noninverting versions for our modules. The noninverting op amp usually has a higher input impedance and may be preferred in general. We will, however, not show any preference in the choice of circuits in our discussion.

The circuit in Fig. 16-1 is the general inverting op amp with impedances Z_1 and Z_2 to be selected to achieve various transfer functions for the modules. We will list the impedances for the different cases and the corresponding pole–zero locations. The transfer function for the circuit in term of the impedances is given by

$$T(s) = \frac{V_2}{V_1} = -\frac{Z_2}{Z_1} \tag{16-1}$$

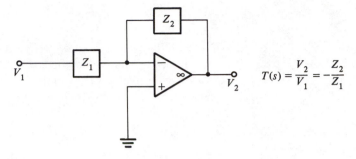

$$T(s) = \frac{V_2}{V_1} = -\frac{Z_2}{Z_1}$$

Figure 16-1. General form of the inverting op-amp circuit.

1. $$Z_2 = \frac{1}{Cs}; \qquad Z_1 = R; \qquad T = -\frac{1/RC}{s} \qquad (16\text{-}2)$$

$$\text{pole at } s = 0 \qquad (16\text{-}3)$$

2. $$Z_2 = R; \qquad Z_1 = \frac{1}{Cs}; \qquad T = -RCs \qquad (16\text{-}4)$$

$$\text{zero at } s = 0 \qquad (16\text{-}5)$$

Let Z_A and Z_B be as defined in Fig. 16-2. We designate the R's and C's involved with the subscript 2 or 1, depending on whether

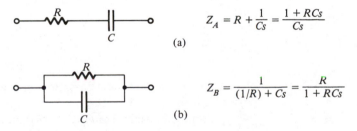

$$Z_A = R + \frac{1}{Cs} = \frac{1 + RCs}{Cs}$$

(a)

$$Z_B = \frac{1}{(1/R) + Cs} = \frac{R}{1 + RCs}$$

(b)

Figure 16-2. Two passive circuits that will be used in the circuit of Fig. 16-1 to substitute for Z_1 and Z_2.

they appear in the feedback path or not. For example, if $Z_2 = Z_A$, the resistance and the capacitance in Z_A will be designated as R_2 and C_2.

3. $$Z_2 = Z_B; \qquad Z_1 = R_1; \qquad T = -\frac{R_2}{R_1(1 + R_2 C_2 s)} \qquad (16\text{-}6)$$

$$\text{pole at } s = \frac{-1}{R_2 C_2} \qquad (16\text{-}7)$$

4. $$Z_2 = R_2; \qquad Z_1 = Z_A; \qquad T = -\frac{R_2 C_1 s}{1 + R_1 C_1 s} \qquad (16\text{-}8)$$

$$\text{zero at } s = 0, \qquad (16\text{-}9)$$

$$\text{pole at } s = -\frac{1}{R_1 C_1} \qquad (16\text{-}10)$$

5. $Z_2 = Z_B$: $Z_1 = Z_B$; $T = -\dfrac{R_2(1 + R_1C_1s)}{R_1(1 + R_2C_2s)}$ (16-11)

$$\text{zero at } s = -\frac{1}{R_1C_1}\qquad(16\text{-}12)$$

$$\text{pole at } s = -\frac{1}{R_2C_2}\qquad(16\text{-}13)$$

6. $Z_2 = Z_A$: $Z_1 = Z_A$; $T = -\dfrac{C_1(1 + R_2C_2s)}{C_2(1 + R_1C_1s)}$ (16-14)

$$\text{zero at } s = -\frac{1}{R_2C_2}\qquad(16\text{-}15)$$

$$\text{pole at } s = -\frac{1}{R_1C_1}\qquad(16\text{-}16)$$

7. One useful exception to the single pole–zero module occurs in this case. We include it here because it is no more complicated than the above and yet in a single module takes the place of two modules.

6. $Z_2 = Z_B$: $Z_1 = Z_A$; $T = -\dfrac{R_2C_1s}{(1 + R_2C_2s)(1 + R_1C_1s)}$

$$\qquad(16\text{-}17)$$

$$\text{zero at } s = 0$$

$$\text{poles at } s = -\frac{1}{R_2C_2}\qquad(16\text{-}18)$$

$$\text{and at } s = -\frac{1}{R_1C_1}\qquad(16\text{-}19)$$

The circuits for these cases together with their pole–zero locations are shown in Fig. 16-3.

We now consider some cases of noninverting op-amp circuits. The general circuit is shown in Fig. 16-4 and the transfer function is

$$T(s) = \frac{V_2}{V_1} = 1 + \frac{Z_2}{Z_1}\qquad(16\text{-}20)$$

8. $Z_2 = \dfrac{1}{C_2s}$: $Z_1 = Z_A$; $T = \left[1 + \dfrac{C_1/C_2}{1 + R_1C_1s}\right]$ (16-21)

$$\text{zero at } s = -\left(\frac{1}{R_1C_1} + \frac{1}{R_1C_2}\right)\qquad(16\text{-}22)$$

$$\text{pole at } s = -\frac{1}{R_1C_1}\qquad(16\text{-}23)$$

9. $Z_2 = R_2$: $Z_1 = Z_A$; $T = \left[1 + \dfrac{R_2C_1s}{1 + R_1C_1s}\right]$ (16-24)

$$\text{zero at } s = -\frac{1}{R_1C_1 + R_2C_1}\qquad(16\text{-}25)$$

$$\text{pole at } s = -\frac{1}{R_1C_1}\qquad(16\text{-}26)$$

Circuit	Pole–Zero Location	$-T(s)$

1.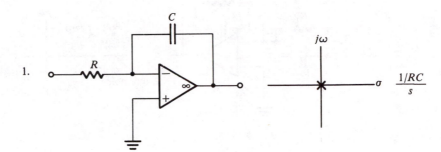

$$\dfrac{1/RC}{s}$$

2.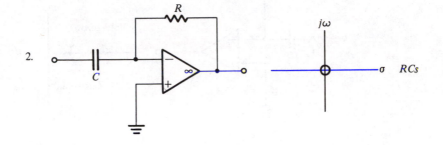

RCs

3.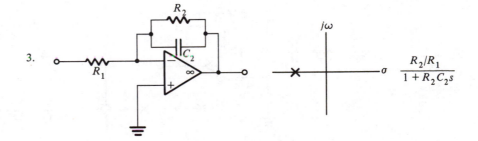

$$\dfrac{R_2/R_1}{1 + R_2 C_2 s}$$

4.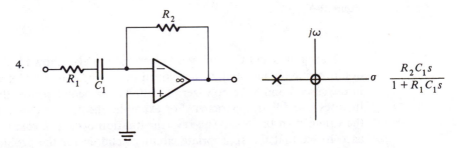

$$\dfrac{R_2 C_1 s}{1 + R_1 C_1 s}$$

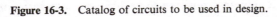

Figure 16-3. Catalog of circuits to be used in design.

5.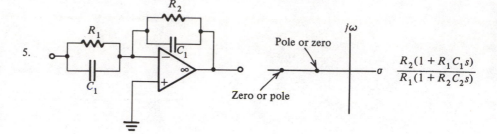

$$\frac{R_2(1 + R_1 C_1 s)}{R_1(1 + R_2 C_2 s)}$$

6.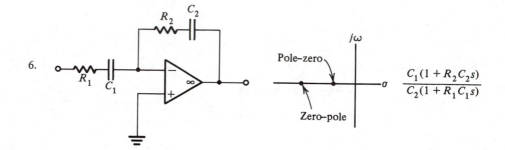

$$\frac{C_1(1 + R_2 C_2 s)}{C_2(1 + R_1 C_1 s)}$$

7.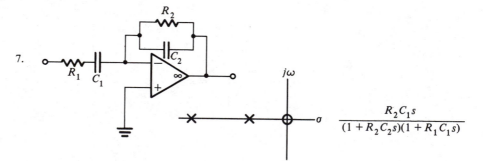

$$\frac{R_2 C_1 s}{(1 + R_2 C_2 s)(1 + R_1 C_1 s)}$$

Figure 16-3. (*Cont.*)

The circuits and the pole–zero locations are also shown in Fig. 16-3. Other noninverting circuits may be developed in similar manner. Notice that whereas in the cases 5 and 6 the pole–zero locations can be independently chosen, those in cases 8 and 9 are restricted. For example, the zero in case 8 is further from the origin than the pole. The reverse situation occurs in case 9. Care must be used to see that the appropriate circuit is chosen for the problem at hand. In any case, we are now in a position to realize any pole–zero combination.

8.

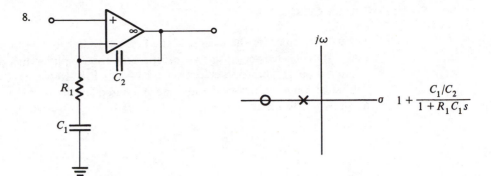

$$1 + \frac{C_1/C_2}{1 + R_1 C_1 s}$$

9.

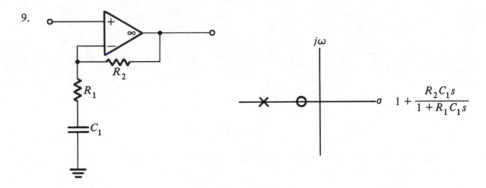

$$1 + \frac{R_2 C_1 s}{1 + R_1 C_1 s}$$

Figure 16-3. (*Cont.*)

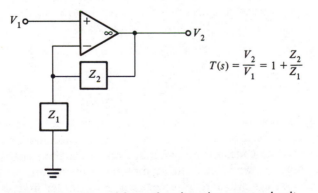

$$T(s) = \frac{V_2}{V_1} = 1 + \frac{Z_2}{Z_1}$$

Figure 16-4. General form of noninverting op amp circuit.

Let us now consider the design of the low-pass filter or amplifier. There is no real difference between the two except that the amplifier has passband gain greater than 1. The characteristic is shown in Fig. 16-5 with the dotted line. The ideal magnitude plot is then approximated by the 6-db/octave asymptote as shown. The break frequency is at ω_1. It is easy to write down the transfer

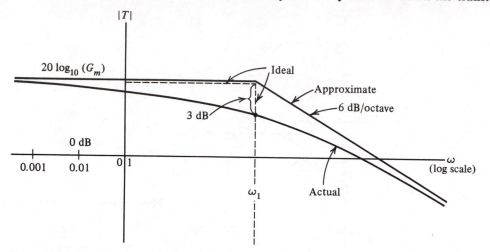

Figure 16-5. Actual and asymptotic form of the Bode plots developed in Chapter 15.

function that will have the desired Bode characteristic. It should have a single pole at $s = -\omega_1$. The gain G_m, which is the gain in the passband as represented by the passband asymptote, is obtained by the proper choice of the elements.

The desired transfer function is then written down as follows:

$$T(s) = \pm \frac{K_1}{s + \omega_1} \tag{16-27}$$

or

$$= \pm \frac{K_2}{1 + s/\omega_1} \tag{16-28}$$

where Eq. 16-28 is the same as Eq. 16-27 and is obtained simply by dividing the numerator and the denominator by ω_1. The transfer function has a pole at $s = -\omega_1$ and the constant $K_2 = K_1/\omega_1$ determines the passband gain. Recall that in Chapter 15 we plotted the asymptotes by considering $|T(j\omega)|$ and considering the values when $\omega \gg \omega_1$ or $\omega \ll \omega_1$. In the first case the magnitude of Eq. 16-27 reduces to that obtained by reglecting ω_1 and in the second case to that obtained by neglecting s. Similarly, for Eq. 16-28 we either neglect 1 or s/ω_1, respectively.

The low-frequency gain ($\omega \ll \omega_1$) is then K_1/ω_1 or, equivalently, K_2. This is required to be G_m. The transfer function may now be written as

$$T(s) = \pm \frac{G_m \omega_1}{s + \omega_1} \tag{16-29}$$

$$= \pm \frac{G_m}{1 + s/\omega_1} \tag{16-30}$$

The $\pm$ sign in the above equations may be chosen arbitrarily since our interest is only in the magnitude.

We now choose the module from our compilation in Fig. 16-3 to match the desired transfer function. The circuit that will have the desired characteristic is found in Fig. 16-3 case 3. The transfer function is

$$T(s) = - \frac{R_2/R_1}{1 + R_2 C_2 s} \tag{16-31}$$

Comparing Eqs. 16-31 and 16-30, we obtain

$$R_2 C_2 = \frac{1}{\omega_1} \tag{16-32}$$

and

$$\frac{R_2}{R_1} = G_m \tag{16-33}$$

Equations 16-32 and 16-33 provide us with a complete design of the desired low-pass filter with a specified break point ω_1 and the low-frequency gain G_m. Recall from Chapter 15 that the actual gain at the break frequency ω_1 is 3 dB below the corner value since the breakpoint asymptote has a negative slope. The actual curve for the magnitude of T is shown in the figure.

Alternatively, we could have selected the circuit with a single pole and no finite zero as required by the asymptotic Bode plot and written down Eq. 16-31 directly. Then, since the breakpoint is at ω_1 (requiring a pole at $s = -\omega_1$), make $1/R_2 C_2 = \omega_1$. The passband gain $R_2/R_1 = G_m$ is obtained by letting $|s = j\omega|$ be very small. Of course, this approach is no different from the previous one. We merely skip some steps but remember all the implications. With experience, such a direct design approach comes naturally.

EXAMPLE 16-1

Design a low-pass filter with a passband of $1/2\pi$ kHz and a low-frequency gain of 40 dB.

Solution

$$20 \log_{10} G_m = 40 \text{ dB}; \qquad G_m = 100$$

and

$$\omega_1 = 2\pi \cdot \frac{1}{2\pi} \times 10^3 \text{ rad/s}$$

$$= 10^3$$

By Eqs. 16-32 and 16-33,

$$R_2C_2 = \frac{1}{\omega_1} = 10^{-3}$$

and

$$G_m = \frac{R_2}{R_1} = 100$$

Let us choose

$$R_1 = 1 \text{ k}\Omega$$

Then

$$R_2 = 100 \text{ k}\Omega$$

and

$$C_2 = \frac{10^{-3}}{10^5} = 0.01 \ \mu\text{F}$$

The design is shown in Fig. 16-6 and the gain at $\omega_1 = 10^3$ rad/s is 37 dB.

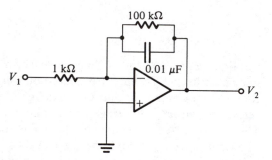

100 kΩ

1 kΩ

0.01 μF

V_1

V_2

Figure 16-6. Low-pass filter designed in Example 16-1.

16-4 THE BAND-PASS CIRCUIT

The ideal band-pass filter characteristic is shown by the dashed curve in Fig. 16-7. The critical frequencies are ω_1 and ω_2 with the passband between them and the stop band outside. The midband gain is designated G_m. The Bode asymptotic plot to approximate the ideal is also shown. Figure 16-8 shows the different asymptotes needed to construct the asymptotic Bode plot of Fig. 16-7. A zero at very low value of ω is needed to provide the first positive slope up to the cutoff frequency ω_1. A pole at $s = -\omega_1$ will provide a negative slope asymptote starting at ω_1, and the sum of the positive and negative slopes will yield the desired constant asymptote in the midband. A second pole at $s = -\omega_2$ will provide the negative slope asymptote starting at ω_2. The only question is where the zero should be located at the low end. A good choice is to place the

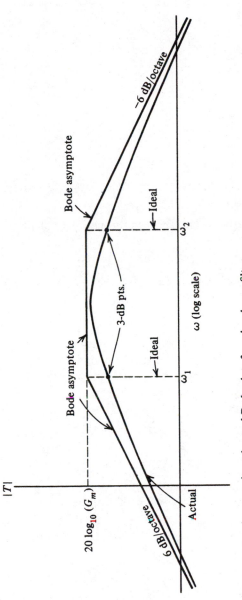

Figure 16-7. Asymptotic and actual Bode plots for a band-pass filter.

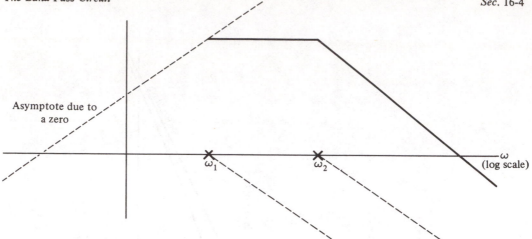

Figure 16-8. Breaking the asymptotic Bode plot into component parts from which $T(s)$ can be written.

zero at $s = 0$, since that will allow the attenuation of the entire low-frequency end. The transfer function of the circuit should be

$$T(s) = \pm \frac{Ks}{(s + \omega_1)(s + \omega_2)} \qquad (16\text{-}34)$$

where the constant K must be such that the midband value is obtained. The midband value of the magnitude of T, $|T(j\omega)|$, is obtained by considering the asymptotes in the midband region. In the midband $|s = j\omega|$ is greater then ω_1 and we may safely neglect ω_1 in the term $(s + \omega_1)$. The second term in the denominator $(s + \omega_2)$ becomes just ω_2 since $|s|$ is smaller than ω_2 in the midband region. Thus, in the midband

$$|T| = \frac{K}{\omega_2} \qquad (16\text{-}35)$$

$$= G_m \qquad (16\text{-}36)$$

giving us the value of K to be

$$K = G_m \omega_2 \qquad (16\text{-}37)$$

and the transfer function is

$$T(s) = \pm \frac{G_m \omega_2 s}{(s + \omega_1)(s + \omega_2)} \qquad (16\text{-}38)$$

$$= \pm \frac{(G_m/\omega_1)s}{(1 + s/\omega_1)(1 + s/\omega_2)} \qquad (16\text{-}39)$$

Equations 16-38 or 16-39 represent the transfer function to be realized. We could split the transfer function above into single pole–zero terms and realize these terms by modules in Fig. 16-3. These modules are then cascaded to obtain the overall transfer function design. However, case 7 provides us with a single module with a zero at $s = 0$ and two poles on the negative real axis, as required by Eq. 16-39. This module then provides us with an efficient design. The transfer function of the case 7 module is

$$T(s) = -\frac{R_2 C_1 s}{(1 + R_2 C_2 s)(1 + R_1 C_1 s)} \tag{16-40}$$

Comparing Eqs. 16-40 and 16-39, we see we have a choice in equating either $R_2 C_2$ or $R_1 C_1$ to say, ω_1. The rest will follow. Let us choose

$$R_1 C_1 = \frac{1}{\omega_1} \tag{16-41}$$

Then

$$R_2 C_2 = \frac{1}{\omega_2} \tag{16-42}$$

and

$$R_2 C_1 = \frac{G_m}{\omega_1} \tag{16-43}$$

so that

$$R_2 C_1 \omega_1 = G_m \tag{16-44}$$

or by substituting Eq. 16-41 into Eq. 16-44, we obtain

$$\frac{R_2}{R_1} = G_m \tag{16-45}$$

The design equations are thus Eqs. 16-41, 16-42, and 16-45.

EXAMPLE 16-2

Design a band-pass filter with passband between 1 and 100 kHz. The midband gain is required to be 40 dB.

Solution

$$\omega_1 = 2\pi \times 10^3 \text{ rad/s}$$

$$\omega_2 = 2\pi \times 10^5 \text{ rad/s}$$

$$G_m = 100 \text{ since } 20 \log_{10} G_m = 40 \text{ dB}$$

From Eqs. 10-41, 10-42, and 10-45,

$$R_1 C_1 = \frac{1}{2\pi \times 10^3}$$

$$R_2 C_2 = \frac{1}{2\pi \times 10^5}$$

$$\frac{R_2}{R_1} = 100$$

Choose

$$R_1 = 100 \ \Omega$$

Then

$$R_2 = 10 \text{ k}\Omega$$

$$C_1 = \frac{1}{2\pi} \times 10^{-5} = 1.59 \ \mu\text{F}$$

$$C_2 = \frac{1}{2\pi} \times 10^{-9} = 1.59 \times 10^{-10} = 159 \text{ pF}$$

The complete design is shown in Fig. 16-9. The gain at 1 kHz and at 100 kHz will be 37 dB, which is 3 dB below the midband gain of 40 dB.

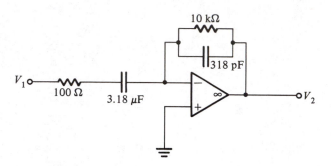

Figure 16-9.   Band-pass filter designed in Example 16-2.

16-5 THE BAND-ELIMINATION CIRCUIT

The ideal characteristic of a band-elimination filter is shown in Fig. 16-10. The asymptotes approximating the ideal and the shape of the actual transfer function are also shown. $20 \log_{10} (G_m)$ is the midband gain in dB and $20 \log_{10} G_0$ is the passband gain. The difference between $20 \log_{10} G_m$ and $20 \log_{10} G_0$ is the midband rejection of the filter. If $G_0 = 1$, the rejection is just $20 \log_{10} G_m$.

The transfer function for the circuit is seen to require a pole at some $s = -\omega_0$ with $\omega_0 < \omega_1$ to provide the first $- 6$-dB/octave asymptote. A zero at $s = -\omega_1$ then results in the flat midband asymptote. A second zero at $s = -\omega_2$ provides the $+ 6$-dB/octave asymptote, and finally the pole at $s = -\omega_\infty$ results in a flat asymptote for the high-frequency end. These asymptotes are shown in Fig. 16-11. The transfer function is

$$T(s) = \pm \frac{K(s + \omega_1)(s + \omega_2)}{(s + \omega_0)(s + \omega_\infty)} \tag{16-46}$$

The midband gain is when $\omega_1 < \omega < \omega_2$. The magnitude of $s = j\omega$ in the midband is thus greater than ω_1 and ω_0 but less than ω_2 and ω_∞. The transfer function magnitude in the midband is

$$K\frac{\omega_2}{\omega_\infty} = G_m \tag{16-47}$$

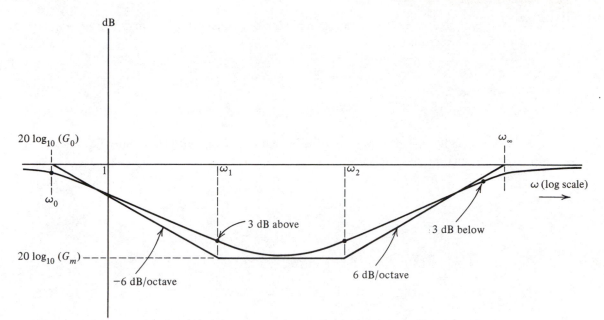

Figure 16-10. Asymptotic and actual Bode plot for the band-elimination filter.

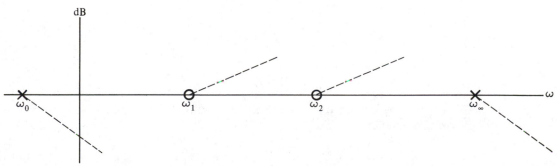

Figure 16-11. Breaking the asymptotic Bode plot for the band-elimination filter into component parts from which $T(s)$ can be written.

so that

$$K = G_m \frac{\omega_\infty}{\omega_2} \tag{16-48}$$

The low-frequency magnitude G_0 is obtained when $|s|$ is smaller than ω_0 or since the high-frequency magnitude is the same, it is obtained when $|s|$ is greater than ω_∞. Thus.

$$K \frac{\omega_1 \omega_2}{\omega_0 \omega_\infty} = K = G_0 \tag{16-49}$$

The first term is obtained from Eq. 16-46 when $|s|$ is smaller than ω_0 (and also

$\omega_1, \omega_2, \omega_\infty$). The second term is obtained when $|s| > \omega_\infty$ (and also $\omega_0, \omega_1, \omega_2$). Then from Eq. 16-49,

$$\frac{\omega_1\omega_2}{\omega_0\omega_\infty} = 1 \qquad (16\text{-}50)$$

or

$$\frac{\omega_2}{\omega_\infty} = \frac{\omega_0}{\omega_1} \qquad (16\text{-}51)$$

By substituting the value of $K = G_0$ from Eq. 16-49 into Eq. 16-48, we obtain

$$G_0 = G_m\frac{\omega_\infty}{\omega_2} \qquad (16\text{-}52)$$

or

$$\frac{\omega_2}{\omega_\infty} = \frac{G_m}{G_0} \qquad (16\text{-}53)$$

Finally, from Eqs. 16-51 and 16-53, we have

$$\frac{\omega_0}{\omega_1} = \frac{\omega_2}{\omega_\infty} = \frac{G_m}{G_0} \qquad (16\text{-}54)$$

Subject to the constraint of Eq. 16-54, the transfer function is from Eqs. 16-49 and 16-46,

$$T(s) = \pm\frac{G_0(s + \omega_1)(s + \omega_2)}{(s + \omega_0)(s + \omega_\infty)} \qquad (16\text{-}55)$$

$$= \pm G_0\frac{\omega_1\omega_2}{\omega_0\omega_\infty}\frac{(1 + s/\omega_1)(1 + s/\omega_2)}{(1 + s/\omega_0)(1 + s/\omega_\infty)} \qquad (16\text{-}56)$$

$$= \pm G_0\frac{(1 + s/\omega_1)(1 + s/\omega_2)}{(1 + s/\omega_0)(1 + s/\omega_\infty)} \qquad (16\text{-}57)$$

To realize this transfer function, we must first express it as a product of single pole–zero terms. Thus,

$$T(s) = \left[\pm G_0\frac{(1 + s/\omega_1)}{(1 + s/\omega_0)}\right] \times \left[\pm\frac{(1 + s/\omega_2)}{(1 + s/\omega_\infty)}\right] \qquad (16\text{-}58)$$

Each term in the brackets may be realized by one of the modules in Fig. 16-3. The candidates are cases 5 and 6 generally and 8 and 9 with a constraint on the relative pole–zero locations.

We will choose the module of case 5 for our design discussion, but the others can serve just as well. The transfer functions we wish to realize are the terms in Eq. 16-58. The terms realized as modules should have the transfer functions given by

$$T_1(s) = \pm G_0\frac{1 + s/\omega_1}{1 + s/\omega_0} \qquad (16\text{-}59)$$

and

$$T_2(s) = \pm\frac{1 + s/\omega_2}{1 + s/\omega_\infty} \qquad (16\text{-}60)$$

The transfer function of the circuit in case 5, Fig. 16-3, is

$$T_m(s) = -\frac{R_2(1 + R_1 C_1 s)}{R_1(1 + R_2 C_2 s)} \qquad (16\text{-}61)$$

Comparing Eqs. 16-61 and 16-59, we have for the first module

$$\frac{R_2^{\textcircled{1}}}{R_1^{\textcircled{1}}} = G_0 \qquad (16\text{-}62)$$

$$R_1^{\textcircled{1}} C_1^{\textcircled{1}} = \frac{1}{\omega_1} \qquad (16\text{-}63)$$

$$R_2^{\textcircled{1}} C_2^{\textcircled{1}} = \frac{1}{\omega_0} \qquad (16\text{-}64)$$

Similarly, by comparing Eq. 16-61 with Eq. 16-60, we obtain the design equations for the second module:

$$\frac{R_2^{\textcircled{2}}}{R_1^{\textcircled{2}}} = 1 \qquad (16\text{-}65)$$

$$R_1^{\textcircled{2}} C_1^{\textcircled{2}} = \frac{1}{\omega_2} \qquad (16\text{-}66)$$

$$R_2^{\textcircled{2}} C_2^{\textcircled{2}} = \frac{1}{\omega_\infty} \qquad (16\text{-}67)$$

We must, of course, remember that the design specifications of ω_1, ω_2, G_m, and G_0 and the constraints of Eq. 16-54 give us the values of ω_0 and ω_∞. The two modules realized above are then cascaded to obtain the overall transfer function, as shown in Fig. 16-12.

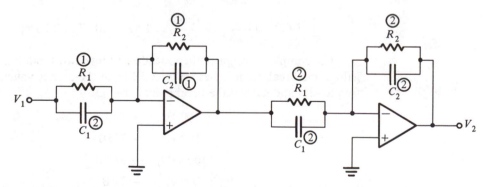

Figure 16-12.   General form of the two-module circuit having a design based on Eq. 16-61.

EXAMPLE 16-3

Design a band-elimination filter with a midband rejection of -40 dB and the stop band between 10 and 100 kHz.

Solution

$$\omega_1 = 2\pi \times 10^4$$

$$\omega_2 = 2\pi \times 10^5$$

$$\frac{G_m}{G_0} = \frac{1}{100}; \quad \text{assume } G_0 = 1, \quad \text{then } G_m = \frac{1}{100}$$

By Eq. 16-54,

$$\omega_0 = \omega_1 G_m = 2\pi \times 10^2$$

$$\omega_\infty = \frac{\omega_2}{G_m} = 2\pi \times 10^7$$

Then by Eqs. 16-62 to 16-64, the first module parameters are

$$\frac{R_2^{①}}{R_1^{①}} = 1; \quad \text{let } R_2^{①} = R_1^{①} = 1 \text{ k}\Omega$$

$$R_1^{①} C_1^{①} = \frac{1}{\omega_1} = \frac{10^{-4}}{2\pi}; \quad C_1^{①} = \frac{10^{-4}}{2\pi \times 10^3} = \frac{10}{2\pi} \times 10^{-8} = 1.59 \times 10^{-8} \text{ F}$$

$$= 0.016 \ \mu\text{F}$$

$$C_2^{①} = \frac{1}{\omega_0 R_1^{①}} = \frac{10^{-5}}{2\pi} = 1.59 \ \mu\text{F}$$

The second module parameters are given by Eqs. 16-65 to 16-67,

$$\frac{R_2^{②}}{R_1^{②}} = 1; \quad R_2^{②} = R_1^{②} = 1 \text{ k}\Omega$$

$$C_1^{②} = \frac{1}{\omega_2 R_1^{②}} = \frac{10^{-8}}{2\pi} = 0.0016 \ \mu\text{F}$$

$$C_2^{②} = \frac{1}{\omega_\infty R_2^{②}} = \frac{10^{-10}}{2\pi} = 1.59 \times 10^{-11} = 15.90 \text{ pF}$$

The complete design is the cascade of these two modules. The actual gain at the break frequencies are ± 3 dB from the corner values. Thus, the frequencies and gain at the breakpoints are

$$100 \text{ Hz}, \quad -3\text{dB}$$

$$10 \text{ kHz}, \quad -37\text{dB}$$

$$100 \text{ kHz}, \quad -37 \text{ dB}$$

$$10 \text{ MHz}, \quad -3 \text{ dB}$$

16-6 THE NOTCH FILTER

The band-elimination circuit of Section 16-5 may be specialized in a couple of different ways. The first one is the notch filter, which attempts to eliminate a very narrow band of frequencies around some specific frequency, say ω_1. For

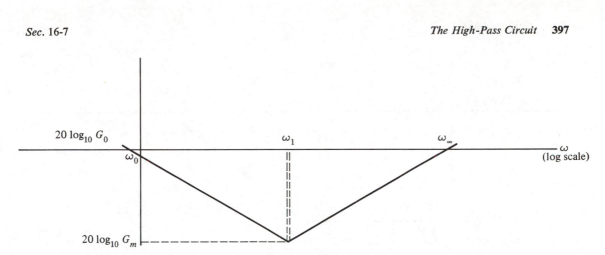

Figure 16-13. Asymptotic Bode plot for a special case of the band-elimination filter known as a notch filter.

example, the 60-Hz hum in many circuits may be eliminated by a notch filter with the notch centered at 60 Hz. The notch characteristic is shown in Fig. 16-13. We have assumed 6-dB/octave slopes for the asymptotes. This would result in a rather shallow notch. A sharper notch may be obtained by assuming slopes of multiples of 6 dB/octave with the corresponding multiple poles and zeros.

The design of a notch is no different from the band-elimination filter. Here we have $\omega_1 = \omega_2$. If we assume $G_0 = 1$, then by Eq. 16-54,

$$\frac{\omega_0}{\omega_1} = \frac{\omega_1}{\omega_\infty} = G_m \qquad (16\text{-}68)$$

The design specifications are usually given in terms of ω_1 and G_m (or $20 \log_{10} G_m$ dB). The other parameters of the transfer function are then

$$\omega_0 = G_m \omega_1 \qquad (16\text{-}69)$$

and

$$\omega_\infty = \frac{\omega_1}{G_m} \qquad (16\text{-}70)$$

The rest of the design is routine. There is a double zero $s = -\omega_1$ since now $\omega_2 = \omega_1$. The rest follows as before.

16-7 THE HIGH-PASS CIRCUIT

The high-pass filter eliminates the low frequencies from ω_1 to ω_2. The specialization here is that $\omega_1 = 0$. Therefore, ω_0 is also zero. The desired characteristic shown in Fig. 16-14 requires a transfer function

$$T(s) = \pm G_0 \frac{s + \omega_2}{s + \omega_\infty} \qquad (16\text{-}71)$$

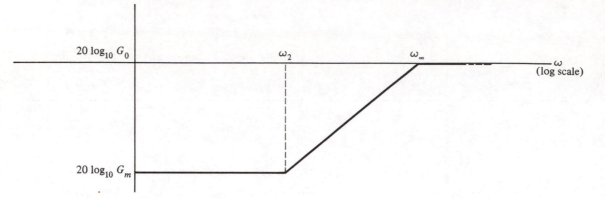

Figure 16-14. Asymptotic Bode plot for a high-pass filter.

Here G_0 is the magnitude at very high values of $|s|$. Then the value of G_m is the magnitude at very low values of $|s|$, or

$$\frac{\omega_2}{\omega_\infty} = \frac{G_m}{G_0} \tag{16-72}$$

The rest of the design is straightforward as before and only one module is necessary in this case.

EXAMPLE 16-4

Design a filter to eliminate all frequencies up to $1/2\pi \times 10^3$ Hz with a rejection of -40 dB.

Solution

$$\omega_2 = 10^3$$
$$G_m = \tfrac{1}{100}; \qquad G_0 = 1$$
$$\omega_\infty = 10^5$$

The design parameters for the circuit of Fig. 16-15 are given by Eqs. 16-65 to 16-67.

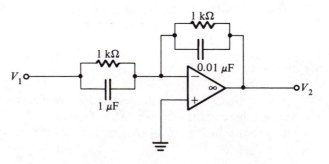

Figure 16-15. Circuit designed in Example 16-4.

$$\frac{R_2}{R_1} = 1; \qquad R_2 = R_1 = 1 \text{ k}\Omega$$

$$R_1 C_1 = \frac{1}{\omega_2} = 10^{-3}; \qquad C_1 = 1 \ \mu\text{F}$$

$$C_2 = \frac{1}{\omega_\infty R_2} = 10^{-8} = 0.01 \ \mu\text{F}$$

PROBLEMS

16-1. The circuit shown in the figure has the following element values: $R_2 = 10 \text{ k}\Omega$, $R_1 = 1 \text{ k}\Omega$, $C_2 = 1 \ \mu\text{F}$, and $C_1 = 10 \ \mu\text{F}$. For these values, determine the voltage-ratio transfer function, V_2/V_1, and plot the asymptotic Bode magnitude response.

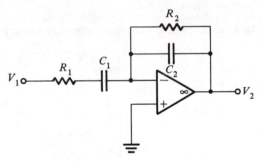

Figure P16-1

16-2. For the two circuits shown in the figure, determine the voltage-ratio transfer functions

$$T(s) = \frac{V_2(s)}{V_1(s)}$$

as a quotient of polynomials, with each polynomial normalized so that the coefficient of the highest-order term is unity. Sketch the pole and zero locations for $T(s)$ and relate these locations to the values of the elements in the circuit.

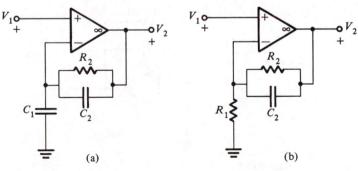

Figure P16-2

16-3. For the cascade-connected circuit shown in the figure, complete an asymptotic Bode plot for the magnitude function $|V_2/V_1|$. Carefully identify the low- and high-frequency asymptotes and designate the values of important slopes.

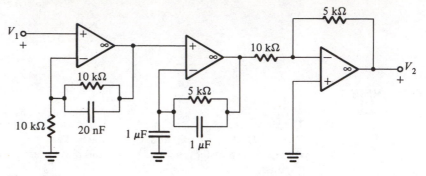

Figure P16-3

16-4. Repeat Prob. 16-4 for the cascade-connected circuit shown in the figure.

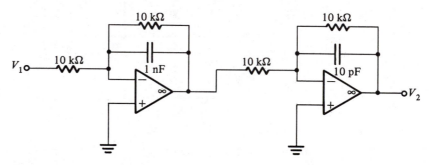

Figure P16-4

16-5. The accompanying figure shows two different circuits connected in cascade. For this circuit, complete an asymptotic Bode plot for the magnitude of the voltage-ratio transfer function. Carefully identify the low- and high-frequency asymptotes and important slopes.

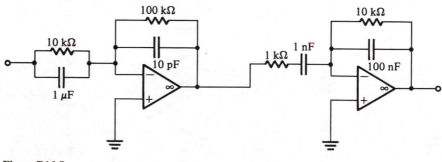

Figure P16-5

cascade circuit shown in the figure, the elements have the following
$R_1 = 1\ M\Omega$, $R_2 = 100\ k\Omega$, $R_3 = 100\ k\Omega$, $R_4 = 10\ M\Omega$, $C_2 = 10\ nF$,
10 μF. For these values, determine the voltage-ratio transfer function
e asymptotic Bode magnitude function, carefully identifying low-
quency asymptotes and important slopes.

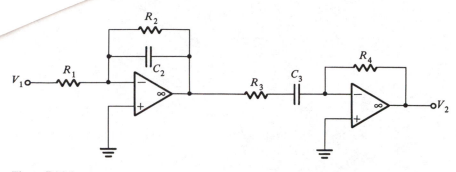

Figure P16-6

16-7. The circuit shown in the figure consists of a cascade connection of several simple
section. The first stage is an inverting stage in which the input accomplishes the
addition of V_0 and V_1, the second stage is a passive RC circuit, and the third
stage is a noninverting circuit. In the entire circuit, all resistors and capacitors
have the same values. (a) Let $V_0 = 0$ and find the transfer function $T = V_2/V_1$.
(b) Let $V_0 = V_2$, and show that the system then oscillates as discussed in
Chapter 8. Determine the frequency of oscillation.

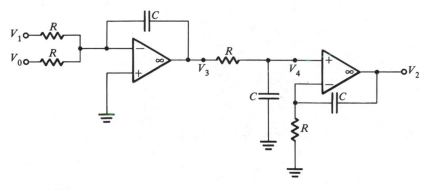

Figure P16-7

16-8. For the op-amp circuit shown in the figure, (a) show that

$$\frac{V_2}{V_1} = \frac{s - 1/RC}{s + 1/RC}$$

and (b) that the phase relating V_2 to V_1 is

$$\theta = -2 \tan^{-1} \omega RC$$

(*Hint:*   Make use of superposition.)

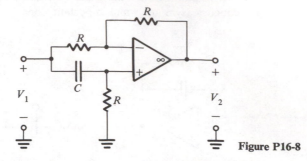

Figure P16-8

16-9. For this problem, you are to design an op-amp circuit that exactly satisfies the specification function

$$T(s) = \frac{(1 + 0.01s)(1 + 0.0001s)}{(1 + 0.0002s)(1 + 0.005s)}$$

16-10. Design an op-amp circuit using a single op amp, two resistors, the smaller of these having a value of 2 kΩ, satisfying the Bode plot specifications shown in the accompanying figure.

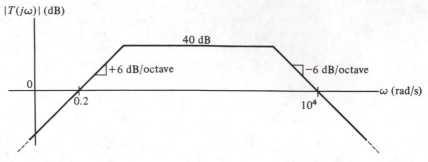

Figure P16-10

16-11. Design an op-amp circuit with the transfer function given in the figure. You are permitted to use two op amps, but all resistors except one must have the resistance value of 10 kΩ.

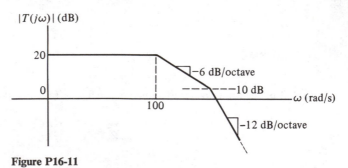

Figure P16-11

16-12. Find the component values for a noninverting *RC*–op amp circuit with only one op amp having the Bode plot magnitude characteristics shown in the accompanying figure.

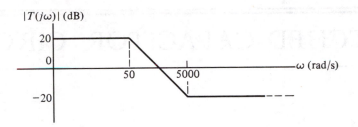

Figure P16-12

I7

SWITCHED-CAPACITOR CIRCUITS

In this chapter we consider circuits that are made by substituting an arrangement of switches and capacitors for resistors. Such circuits find application when the realization is in integrated form.

17-1 THE MOS SWITCH

A special class of integrated circuits are made using MOS (metal–oxide–semi-conductor) technology. An important element in circuit design is the *transistor*, which is represented in Fig. 17-1(a) using MOS technology. The transistor is a three-terminal device having terminals which are identified as source, gate, and drain. The circuit symbol is shown in part (b) of the figure. In one specialized use of this transistor, the voltage between the gate and the source, v_{GS}, is either zero or a value larger than a threshold value, v_{cr}, typically 1 or 2 V. In this mode of operation, for reasons discussed below, the device is known as an MOS switch.

The voltage that controls the switching action is shown on the figure as v_{GS}, and the path of interest is that between S and D, with equivalent resistance R_{eq}. When the voltage v_{GS} is less than v_{cr}, then R_{eq} is quite large, perhaps 100 to 1000 MΩ. When v_{GS} is greater than v_{cr}, then R_{eq} is relatively small, of the order of 10 kΩ.

Thus, the ratio of the two resistance values is quite large, of the order of 100,000. By switching between the two values of v_{GS}, we are able to switch between equivalent resistances of a very large relative ratio. In most practical circuit applications, a ratio of resistance values switched is much more important

404

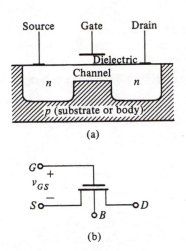

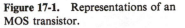

Figure 17-1. Representations of an MOS transistor.

in a switch than the absolute values involved. For these reasons, the MOS transistor used in this mode is called the MOS switch. The switch is in the on state when $v_{GS} > v_{cr}$, and it is in the off state when $v_{GS} < v_{cr}$. We will speak of (relatively) low resistance in the on state and high resistance in the off state.

A representation of these facts is shown in Fig. 17-2 together with an idealized model for the MOS switch: either a short circuit (low resistance) in the on state or an open circuit (high resistance) in the off state. This is shown in Fig. 17-3 with the switch open or closed depending on the value of v_{GS}. Such a switch is also known as a *single-pole single-throw* (SPST) switch.

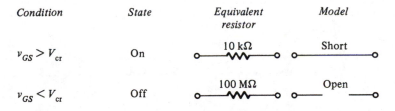

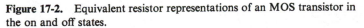

Figure 17-2. Equivalent resistor representations of an MOS transistor in the on and off states.

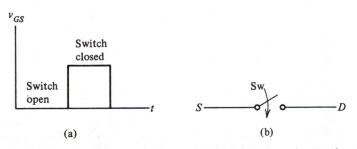

Figure 17-3. Voltage from grid to source as it relates to the transistor behaving as a switch.

Although the actual MOS switch has substantial resistance even in the on state, we will use the ideal model to simplify our discussion considerably.

The voltage waveform that is used to activate the MOS switch is shown in Fig. 17-4. Here we have changed variables by letting v_{GS} be ϕ to conform to

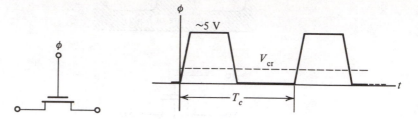

Figure 17-4.   Voltage waveform generated by a clock that is applied to the transistor.

common usage. This waveform is generated by a clock, which is an important part of digital systems. It is a pulse train which is periodic and of period T_c, as shown in the figure. The quantity $f_c = 1/T_c$ is known as the *clock frequency* of the pulse train. In digital systems the clock provides a timing standard to start or stop operations. Here we use it to turn on and then off the MOS switch in a repetitive manner.

Figure 17-5 shown the waveforms of a two-phase clock of the type we will need in our discussion. It is required that ϕ_1 and ϕ_2 have the same frequency and also that they be nonoverlapping, so that when ϕ_1 is on, ϕ_2 will be off,

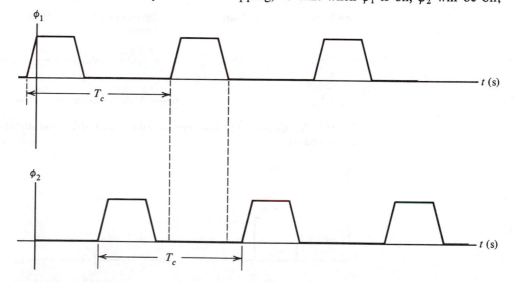

Figure 17-5.   Voltages of a two-phase clock generator. Observe that when ϕ_1 is on, ϕ_2 is off, and vice versa.

and vice versa. Here we use the term "on" to mean relative high voltage, and the term "off" to mean low voltage.

If two MOS switches are connected in series as shown in Fig. 17-6 and then driven by the two waveforms of Fig. 17-5, there will never be a direct connection from 1 to 2, since one of the two switches will always be open. This is shown in Fig. 17-7(a) in terms of the models using SPST switches. The switch Sw_1 is driven by the clock ϕ_1 and Sw_2 by ϕ_2. Thus when Sw_1 is open, Sw_2 will be closed, and when Sw_1 is closed, Sw_2 will be open.

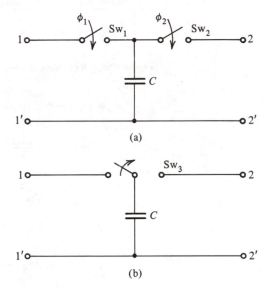

(a)

(b)

Figure 17-7. The two switches operating in a SPST mode as in (a) are equivalent to one switch operating the SPDT mode, as shown in (b).

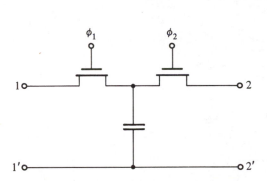

Figure 17-6. Two transistor switches are connected to a capacitor as shown to form the basic switched-capacitor circuit.

The net effect of the two switches operated by the two-phase clock is to alternately switch the capacitor connection between terminals 1 and 2. The term "switched capacitor" is applied to this type of operation.

The switching action described above is equivalent to that shown in part (b) of the figure. The switch Sw_3 is switched at the clock frequency, allowing, as before, a connection by the capacitor C either to 1 or to 2. Such a switch is called a *single-pole double-throw* (SPDT) switch. This kind of switch is thus implemented by means of two SPST switches operated by a two-phase clock.

Such a scheme for switching is easily extended to four MOS switches as shown in Fig. 17-8. With the two clock voltages ϕ_1 and ϕ_2 connected as shown, we now have a *double-pole double-throw* (DPDT) switch. Such a switch is useful; two applications are illustrated in Fig. 17-9. In (a) of that figure, the switches are closed in positions a, and the capacitor charges to the value $v_C = v_{1a}$

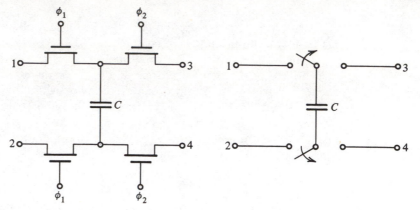

Figure 17-8.   Switching scheme involving four MOS transistors.

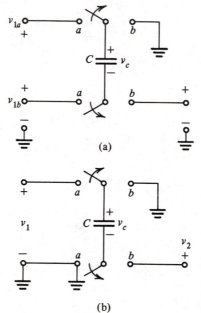

Figure 17-9.   Switching scheme producing a voltage difference as in (a) and an inverted (multiplied by -1) voltage in (b).

$- v_{1b}$. When the switches are moved to positions b, then

$$v_2 = v_{1b} - v_{1a} \tag{17-1}$$

or the difference of two voltages is formed. If $v_{1b} = 0$ and $v_{1a} = v_1$ as in Fig. 17-9(b), then

$$v_2 = -v_1 \tag{17-2}$$

or an inverter has been formed such that the output is the negative of the input.

This completes a catalog of MOS switching configurations. We next turn to the use of such switches and switched capacitor circuits.

Figure 17-7(b) is repeated as Fig. 17-10 and shows a capacitor C_R with a periodic SPDT switch. Analysis of such a circuit is possible by examining its behavior when the switch is in one position and then in the other position.

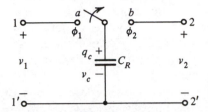

Figure 17-10. Basic switched-capacitor circuit.

We will do so with a view to determining a simple approximation to the circuit behavior rather than an exact behavior.

Let us first assume that the input voltage v_1 is constant and the switch is initially in the position a. Then the circuit reduces to that of Fig. 17-11(a), where we have shown a switch resistance R_1 for the on state. Ideally, this resistance is close to zero, but a realistic value in our case is 10 kΩ. We will examine the behavior with the realistic value first.

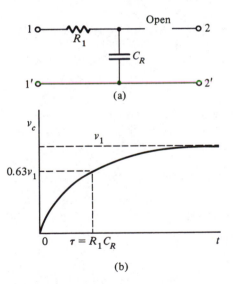

Figure 17-11. Voltage of capacitor C_R as a function of time. The time constant is very short but finite.

If v_1 is constant, the voltage of the capacitor will increase as shown in Fig. 17-11(b) with the time constant $\tau = R_1 C_R$. We have stated that R_1 is about 10^4, and a typical value for C_R is 1 pF, then v_C will reach 63% of its final value in time equal to one time constant,

$$\tau = R_1 C_R = 10^4 \times 10^{-12} = 10 \text{ ns} \qquad (17\text{-}3)$$

409

The capacitor will thus be charged to the voltage v_1 in about four time constants or 40 ns. This is extremely fast relative to the type of signals we usually deal with. For most practical purposes, then, we can say the capacitor is charged instantly to the input voltage v_1. This is the case for an ideal switch.

The ideal switch assumption is thus quite reasonable if all our signals are slowly varying with time relative to the time constant of 10 ns. We will assume in all that follows that this assumption is valid at all times. We also assume that the clock period is small enough so that the signals do not change during one period of the clock. Thus, even if the input voltage v_1 is a function of time, the capacitor will be assumed charged to v_1 instantly as if we had an ideal switch connection.

If the switch is now changed to position b in Fig. 17-10 the capacitor will discharge instantly (or, more realistically, with the same time constant) until it reaches the output voltage v_2, which we also assume is either a constant or a slowly varying signal, as discussed above.

The capacitor is, thus, first charged to v_1 by the input signal, and then discharged to v_2 at the output end, in one period of the two-phase clock. Moreover, this process is repeated in each period of the clock. There is thus a net charge transferred to the output side. The charge transferred by the capacitor in one period, to the terminal marked 2 will be

$$q = C_R(v_1 - v_2) \qquad (17\text{-}4)$$

and this will be accomplished in time T_c, the period of the clock.

Under our assumptions, the signals v_1 and v_2 hardly change during this one period of the clock. So during this time interval, the current is simply

$$i(t) = \frac{\Delta_q}{\Delta_t} \simeq \frac{C_R(v_1 - v_2)}{T_c} \qquad (17\text{-}5)$$

Alternatively, we could obtain the same current at terminal 2, during the same time interval, if we place an appropriate resistor between terminals 1 and 2 as in Fig. 17-12. Then

$$i(t) = \frac{1}{R_c}(v_1 - v_2) \qquad (17\text{-}6)$$

By equating the Eqs. 17-5 and 17-6, we see that the size of such an equivalent resistor to give the same value of current, during this same time interval, is

$$R_C = \frac{v_1 - v_2}{i} = \frac{T_c}{C_R} = \frac{1}{f_c C_R} \qquad (17\text{-}7)$$

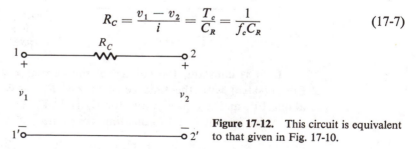

Figure 17-12.   This circuit is equivalent to that given in Fig. 17-10.

We see that the equivalent resistor value is independent of the signal values involved during the period of interest. It is, therefore, possible to replace the switched-capacitor circuit during each period by the same equivalent resistor value given in Eq. 17-7. In other words, a switched capacitor as shown in Fig. 17-10 is (approximately) equivalent to a series resistor, as shown in Fig. 17-12.

It is this equivalence between the switched capacitor and the resistor that gives us a simple way to analyze and design switched-capacitor circuits. In design, we first design the circuits using resistors, capacitors, and op amps. We then replace the resistors by equivalent switched capacitors. In analysis of switched-capacitor circuits, we merely replace the switched capacitors by their equivalent resistors.

Care must be exercised during design to make sure that the range of resistor values in the preliminary design will yield practical values of capacitors in the equivalent switched-capacitor version. Let us use the typical values given previously to determine the range of values of R_c that might be realized. Using $C_R = 1$ pF, and a typical value of $f_c = 100$ KHz, we see, by Eq. 17-7, that R_C is 10 MΩ. This value is in a useful range. Other values may be realized by adjusting capacitor values or the clock frequency.

For the approximation of Eq. 17-5 to be valid, it is necessary that the clock period be small enough so that the signals v_1 and v_2 are effectively constant during this interval. This implies that the signals are varying very slowly in time with respect to the clock. Such a restriction can be translated to the frequency domain if we recall that the Fourier spectrum of slowly varying signals is concentrated primarily at the lower-frequency levels.

Thus, we must require that the switching frequency f_c be much larger than the significant spectrum frequencies of $v_1(t)$ and $v_2(t)$. This turns out to be the case for voice-frequency filters and the switched capacitor may be regarded as a direct replacement for the resistor. If this is not the case, other more complex methods must be used to analyze the behavior of switched-capacitor circuits.

In the treatment of this chapter, we assume that Eq. 17-5 is valid. Our objective is to present the basic concepts on which switched-capacitor filters are designed.

17-3 ANALOG OPERATIONS

In this section we are concerned with the four analog operations on voltages: addition, subtraction, multiplication, and integration. We are concerned with accomplishing these operations with switched-capacitor circuits, but in each case we will show the corresponding analog circuits using resistors, capacitors, and op amps as derived in earlier chapters.

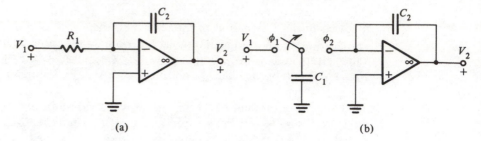

(a)   (b)

Figure 17-13. The circuit of (b) is the switched-capacitor equivalent of that shown in (a), which was studied in Chapters 5 and 15.

We begin with the integrator circuit of Fig. 17-13, for which the transfer function is

$$\frac{V_2}{V_1} = \frac{-1}{R_1 C_2 s} \tag{17-8}$$

Here v_1 and v_2 are the transforms (or phasors) and s is the complex frequency variable.

The corresponding switched-capacitor circuit has a transfer function which is obtained by substituting R_1 from Eq. 17-7 such that

$$\frac{V_2}{V_1} = -f_c \left(\frac{C_1}{C_2}\right) \frac{1}{s} \tag{17-9}$$

Both circuits are inverting, and both represent integrators.

It is significant that Eq. 17-9 involves the ratio C_1/C_2. Capacitors can be formed in MOS technology with accuracy, so the ratio can be realized with the same accuracy.

If addition of two voltages is required in addition to integration, then the circuit of Fig. 17-14 may be used, paying special attention to maintain the same switching sequence for the two inputs. For this circuit

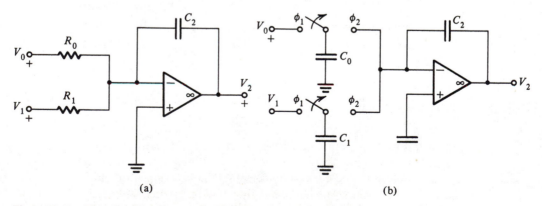

(a)   (b)

Figure 17-14. The circuit of (b) is the switched-capacitor equivalent of that shown in (a) used for summing and integrating.

$$V_2 = \frac{-1}{R_0 C_2 s} V_0 + \frac{-1}{R_1 C_2 s} V_1 \tag{17-10}$$

To find the corresponding equation for the switched-capacitor circuit of Fig. 17-14(b), we substitute the equivalent values of the switched capacitors, from Eq. 17-7, for the resistors:

$$V_2 = -f_c \left(\frac{C_0}{C_2}\right)\frac{1}{s} V_0 - f_c \left(\frac{C_1}{C_2}\right)\frac{1}{s} V_1 \tag{17-11}$$

For the special case $C_0 = C_1 = C$,

$$V_2 = -f_c \left(\frac{C}{C_2}\right)\frac{1}{s}(V_0 + V_1) \tag{17-12}$$

so that the circuit integrates the sum of two voltages and multiplies by a constant. It is interesting to observe that for $f_c = 100 \text{ KHz}$ and $C = 10^{-12}$, the multiplier magnitude is 10^6, providing a pretty high gain!

The circuit shown in Fig. 17-15(a) is the familiar single-pole circuit, also known as a lossy integrator, for which

$$\frac{V_2}{V_1} = \frac{-1/R_1 C_2}{s + 1/R_3 C_2} \tag{17-13}$$

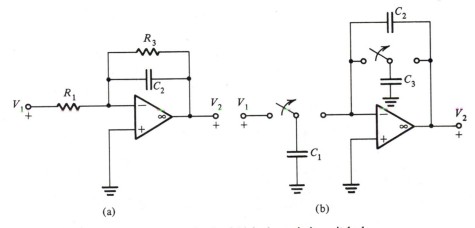

(a)   (b)

Figure 17-15. The lossy integrator circuit of (a) is shown in its switched-capacitor form in (b).

The corresponding switched-capacitor circuit is shown in (b) and the gain function is

$$\frac{V_2}{V_1} = \frac{-f_c(C_1/C_2)}{s + f_c(C_3/C_2)} \tag{17-14}$$

for which the dc gain is C_1/C_3.

Frequently, we come across circuits involving an input from some other part of the circuit. The analysis in such a case is straight forward using the superposition principle. Suppose that v_0 is the voltage being fed back from some other part in the circuit, as shown in Fig. 17-16. Then we may use the

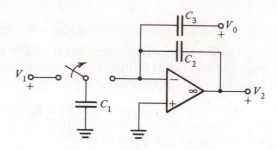

Figure 17-16. One voltage is multiplied and added to the integral of the other.

concept of superposition to write

$$V_2 = -f_c \left(\frac{C_1}{C_2} \right) \frac{1}{s} V_1 - \frac{C_3}{C_2} V_0 \qquad (17\text{-}15)$$

We have considered thus far a variety of inverting circuits. We next consider some noninverting circuits. The first of these is the noninverting integrator circuit of Fig. 17-17(a), for which

$$\frac{V_2}{V_1} = \frac{1}{R_1 C_2} \frac{2}{s} \qquad (17\text{-}16)$$

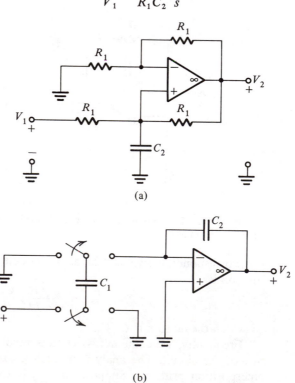

(a)

(b)

Figure 17-17. Two forms of noninverting integrator circuit.

The corresponding switched-capacitor circuit is shown in Fig. 17-17(b). It is seen to be a combination of the inverting integrator of Fig. 17-13(b) and the inverting switching arrangement of Fig. 17-9 described by Eq. 17-2. Combining Eq. 17-2 with Eq. 17-9, we have

$$\frac{V_2}{V_1} = f_c \left(\frac{C_1}{C_2}\right) \frac{1}{s} \tag{17-17}$$

If we use the other switching arrangement shown in Fig. 17-9 in combination with the standard integrator circuit of Fig. 17-13(b) as shown in Fig. 17-18(b), then the transfer function becomes

$$V_2 = f_c \left(\frac{C_1}{C_2}\right) \frac{1}{s} (V_1 - V_0) \tag{17-18}$$

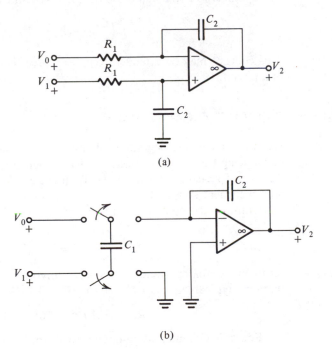

(a)

(b)

Figure 17-18. Two forms of a circuit that integrates the difference of two voltages.

The circuit forms a difference of voltages, multiplies by $f_c C_1 / C_2$, and integrates. This is a useful circuit with many applications. The corresponding *RC*–op amp circuit is shown in (a) of Fig. 17-18, and for this circuit

$$V_2 = \frac{1}{R_1 C_2} \frac{1}{s} (V_1 - V_0) \tag{17-19}$$

As in Fig. 17-16, a circuit with several inputs, one of which is fed back from another part of the circuit, as shown in Fig. 17-19, may be analyzed

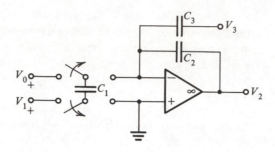

Figure 17-19. Switched-capacitor circuit described by Eq. 17-20.

by superposition. We make use of Eq. 17-18 and write

$$V_2 = f_c\left(\frac{C_1}{C_2}\right)\frac{1}{s}(V_1 - V_0) - \frac{C_3}{C_2}V_3 \tag{17-20}$$

This concept may be extended to cases where there are more than three inputs to a circuit.

17-4 RANGE OF CIRCUIT ELEMENT SIZES

We observed in the beginning of this chapter that the original motivation in introducing the switched capacitor was to achieve MOS integrated-circuit realizations of resistors at voice-band frequencies. These factors place constraints on the element sizes to be used, as we shall see.

Consider the op-amp circuit shown in Fig. 17-13, for which

$$\frac{V_2}{V_1} = \frac{-1}{R_1C_2}\frac{1}{s} \tag{17-21}$$

As in many analog circuits, this result involves the product R_1C_2. In earlier chapters, typical values of $C = 0.01$ μF and $R = 10$ kΩ. For these values,

$$R_1C_2 = 10^{-8} \times 10^4 = 10^{-4} \text{ s} \tag{17-22}$$

For a MOS integrating-circuit realization, suppose that a typical value of capacitor available is 10 pF. Then to give the same RC product, the value of R must be

$$R_1 = \frac{10^{-4}}{C_2} = \frac{10^{-4}}{10^{-11}} = 10 \text{ M}\Omega \tag{17-23}$$

We must realize this resistance using a switched capacitor using the equivalence of Eq. 17-7. Then to achieve this value of resistance using again a capacitor of 10 pF, the frequency of the clock switch must be

$$f_c = \frac{1}{R_1C_R} = \frac{1}{10^7 \times 10^{-11}} = 10 \text{ kHz} \tag{17-24}$$

So in MOS circuits, we deal with a different range of element values: small capacitors, large resistors, and a modest frequency range of switching.

We will not pursue here the generation of the two-phase signals shown in Fig. 17-5, which operate at the clock frequency. We will merely state that the clock frequency is generally tunable (i.e., it is adjustable over a range). The range is quite wide and it is reasonable to assume a range up to 2 MHz. The phases ϕ_1 and ϕ_2 can be thus generated with the f_c range up to 2 MHz.

We mentioned earlier that the clock frequency must be much higher than the highest significant frequency components in the signal spectrum. In filters with a cut off at some frequency, say f_0, it is safe to have the clock frequency at least 10 times the cutoff. That is,

$$f_c > 10 f_0 \qquad (17\text{-}25)$$

A typical value for f_c may well be around 100 kHz.

Let us now turn to the various values of equivalent resistors that may be realized by typical values of switched capacitors. Let us assume a typical range of the capacitor values to be

$$0.1 \text{ pF} < C < 100 \text{ pF} \qquad (17\text{-}26)$$

The limits above for realizable capacitors give the range of equivalent resistors that can be simulated with the given clock frequencies. For example:

$$f_c = 100 \text{ kHz} \qquad 0.1 \text{ M}\Omega < R < 100 \text{ M}\Omega \qquad (17\text{-}27)$$

$$f_c = 1 \text{ kHz} \qquad 10 \text{ M}\Omega < R < 10{,}000 \text{ M}\Omega \qquad (17\text{-}28)$$

These numbers indicate the wide range of values of equivalent resistance available to the designer and the fact that such values may be achieved by either adjusting the capacitor or tuning the clock frequency.

17-5 FIRST-ORDER FILTERS

There are several general approaches for the design of switched-capacitor filters. Conceptually, the simplest approach is the one introduced, in an earlier chapter, for the design of the op-amp filters. It consists of cascading a number of first- and/or second-order sections to achieve some specified overall response. We have already seen how to design such op-amp circuits and filters. Here we show some examples of design where we first obtain the op-amp circuits and then replace the resistors by their equivalent switched capacitors.

The simplest section used in the cascade-connection method of design is the first-order low-pass filter shown in Fig. 17-20(a) for which the transfer function is

$$T(s) = \frac{-1/R_1 C_2}{s + 1/R_1 C_2} \qquad (17\text{-}29)$$

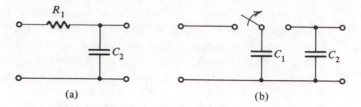

(a)   (b)

Figure 17-20.   Simple low-pass filter shown in two realizations.

The switched capacitor equivalent is shown in Fig. 17-20(b). If we substitute $R_1 = 1/f_c C_1$, then Eq. 17-29 becomes

$$T(s) = \frac{-f_c(C_1/C_2)}{s + f_c(C_1/C_2)} \qquad (17\text{-}30)$$

From this equation, we see that the bandwidth, given by the half-power frequency of the response, of this circuit is

$$\omega_{\text{hp}} = f_c \frac{C_1}{C_2} \qquad (17\text{-}31)$$

Thus, we see that the filter can be tuned (i.e., its bandwidth adjusted) by either the clock frequency or the ratio of capacitors.

Another first-order filter section is that shown in Fig. 17-21(a) with the

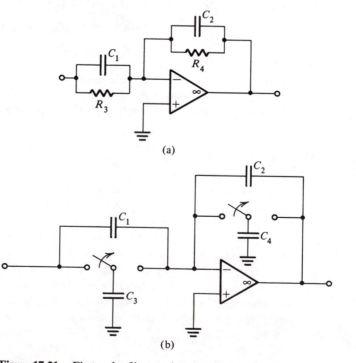

(a)

(b)

Figure 17-21.   First-order filter section described by Eq. 17-33.

switched-capacitor equivalent shown in (b) of the figure. The transfer function is

$$T(s) = -\frac{C_1 s + 1/R_3 C_1}{C_2 s + 1/R_4 C_2} \qquad (17\text{-}32)$$

Then the corresponding equation for the switched-capacitor case is

$$T(s) = -\frac{C_1 s + f_c(C_3/C_1)}{C_2 s + f_c(C_4/C_2)} \qquad (17\text{-}33)$$

Suppose that the filter specifications we wish to realize are indicated by the asymptotic Bode plot of Fig. 17-22, which is seen to be a band-elimination filter. The transfer function is

$$T(s) = \frac{(s + 10^3)(s + 10^4)}{(s + 10^2)(s + 10^5)} \qquad (17\text{-}34)$$

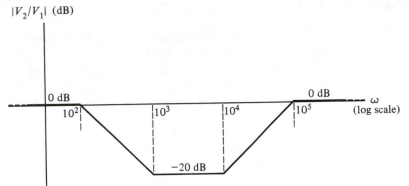

Figure 17-22. Asymptotic Bode plot for Eq. 17-34.

We first break $T(s)$ into the product of factors having the appearance of Eq. 17-33.

$$T(s) = \frac{s + 10^3}{s + 10^2} \frac{s + 10^4}{s + 10^5} \qquad (17\text{-}35)$$

We will equate coefficients of these factors with those of Eq. 17-33. Let us select as our clock frequency $f_c = 10{,}000$ Hz. For convenience we select $C_1 = C_2$ so that $C_1/C_2 = 1$. If we let $C_1 = C_2 = 10$ pF for both sections, then for the first section

$$C_3 = 1 \text{ pF} \qquad \text{and} \qquad C_4 = 0.1 \text{ pF} \qquad (17\text{-}36)$$

while for the second section

$$C_3 = 10 \text{ pF} \qquad \text{and} \qquad C_4 = 100 \text{ pF} \qquad (17\text{-}37)$$

We observe that the range of capacitor values satisfies the limits imposed by Eq. 17-26. The final filter design is shown in Fig. 17-23.

The approach is straightforward and can be followed for all types of filter design. We will not discuss each type of filter design separately but leave it to the reader. The problems cover a wide range of possibilities.

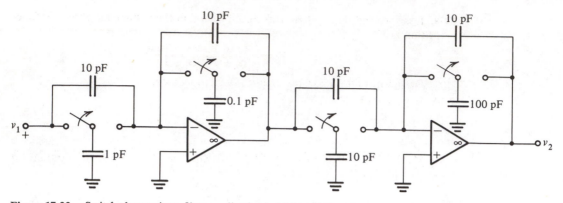

Figure 17-23. Switched-capacitor filter realization which satisfies the specifications of Eq. 17-34.

We conclude this chapter with a few observations. The switched capacitor has become an important circuit element in the integrated-circuit technology. The switched capacitor–resistor equivalence is approximate and valid only for the signal assumptions made; that is, the signals are slowly varying or, equivalently, their spectrum contains primarily low-frequency components. Of course, all circuit models are valid only for the assumptions made, as we have seen in the case of other elements, such as op amps, linear resistors, and so on. With these assumptions in mind and realizing the ubiquitous nature of integrated circuits, the switched capacitor must be included as a major circuit element in electronic circuits.

PROBLEMS

The following problems are design problems in the sense that there is no unique answer. For convenience, it is suggested that a clock frequency of 10 kHz be used and that capacitor values be restricted to the range 0.1 to 10 pF. These are guidelines, and the designer may well have other objectives in mind, resulting in a different design.

17-1. Redesign the circuit given in Fig. 16-6 using switched-capacitor methods.

17-2. Redesign the circuit of Fig. 16-9 to make use of switched-capacitor techniques.

17-3. Redesign the circuit given in Fig. 16-15 using switched-capacitor methods.

17-4. Reconsider the problem given as Prob. 16-4. Design a solution to the specifications given.

17-5. Consider the specifications originally given in Prob. 16-5. Complete a design using switched-capacitor methods, using the guidelines given at the top of this page.

17-6. Repeat Prob. 17-5 for the specifications of Prob. 16-6.

17-7. Repeat Prob. 17-5 for the specifications of Prob. 16-10.

17-8. Repeat Prob. 17-5 for the specifications of Prob. 16-11.

17-9. It is specified that a filter be designed such that the voltage-ratio transfer function is

$$T(s) = \frac{(s + 10^4)(s + 10^5)}{(s + 10^3)(s + 10^6)}$$

Design this filter using switched-capacitor methods.

17-10. Repeat Prob. 17-9 for the specifications

$$T(s) = \frac{(s + 10^2)(s + 10^5)}{(s + 10^3)(s + 10^4)}$$

Design the filter using switched-capacitor methods and the range of values specified at the beginning of this set of problems.

INDEX

A

Active elements, 48, 52, 84
Admittance, 242
Algebraic sum, 39
Ammeter, 10
Ampere, 2
Analog addition, 133
Analysis, 1
  ac, 220
Apparent power, 296
Available power, 115
Average power, 291

B

Band-elimination circuits, 314, 392
Bandpass circuits, 388
Bandwidth, 311
Battery, 8
Bode plots, 365
Branch, 52
  constraints, 146
  currents, 52

Branch (*Contd.*)
  voltages, 52
Break frequency, 371
Bridged-*T*, 109

C

Capacitor, 147
  discharge, 163
  op amp circuits, 349
  quiescent state, 163
  series and parallel, 155
Charge, 1
Circuit, 1
Circuit functions, 265
Circuit model, 17
Clock frequency, 406
Closed loop, 1, 26
Coefficient of coupling, 189
Complex exponentials, 230, 257
Complex frequency, 260
Complex numbers, 224
  conjugate, 298
Conceptual scheme, 1

Condenser, 147
Conductance, 14, 107, 244
Conjugate, 227, 298
Conjugate match, 298
Conservation laws, 4
    charge, 4
    energy, 48
Continued fraction, 106
Continuous spectrum, 309, 339
Controlled sources, 49, 65
Coulomb, 2
Coupled coils, 189
Current, 1
    conventional, 3
    electron, 3
Current divider circuit, 42, 273
Cycles per second, 215

D

Damped oscillations, 203
Damped sinusoid, 258
Datum, 9
Dc analysis, 220
Decade, 368
Dependent source, 49
Differential amplifier, 135
Differential equations, 158
    homogeneous, 158
    nonhomogeneous, 165
Differentiator circuit, 356
Discrete spectrum, 307
Distortion, 314
Dot convention, 191
Driving-point functions, 125

E

Effective value, 295
Energy, 6, 45
    electric, 153
    magnetic, 179
Envelope, 309
Equivalent circuit, 58, 73

Equivalent Circuit (*Contd.*)
    elements, 58
    impedance, 271
    resistance, 84
Equivalent impedance, 271
Euler's identity, 226
Even symmetry, 329
Excitation, 113, 163
Exponential representation, 226

F

Faraday's law, 179
Feedback, 116
Filters, 313
Floating elements, 9
Flux linkages, 178
Forced response, 163
Fourier integral, 339
Fourier series, 305
    analysis, 321
    coefficients, 326
    truncated, 325
Fourier transform, 340
Frequency, 216
    break, 371
Frequency response, 349
Frequency spectrum, 302, 321

G

Gain, 36, 125, 266
Gauss elimination, 97
Generalized exponential, 256
Generalized impedance, 260
Ground, 9

H

Half-wave symmetry, 329
Harmonics, 302, 307
Henry, 179
Hertz (Hz), 215
Highpass circuits, 397

Homogeneous differential equation, 158

I

Ideal transformer, 192
Imaginary part, 224
Impedance, 235, 242
Impedance function, 265
Impulse, 152
Independent sources, 52, 75, 111
Inductors, 178
   series and parallel, 180
Input impedance, 243
Input port, 124
   resistance, 76, 85, 125
Input resistance, 84
Integrator circuit, 355, 412
Internal node, 59
Inverse transform, 340
Inverting circuits, 129, 412

J

Joule (J), 45
Jump discontinuity, 331

K

Kirchhoff laws for phasors, 234
Kirchhoff's current law, 39
Kirchhoff's voltage law, 26

L

Lag, 216
Ladder circuits, 106
Lead, 216
Linear elements, 154
Linearity, 154
Linear resistor, 14
Line spectrum, 307
Load, 73
Loop analysis, 98

Loop current, 29, 99
Lossy integrator, 354
Lowpass filter, 351, 386

M

Magnetic coupling, 189
Magnetic flux, 178
Magnitude and phase, 214, 265, 358
Maximum power transfer, 297
Metal-oxide-semiconductor (MOS), 404
Mixed sources circuits, 103
Model, 14
Modules for design, 380
MOS switch, 404
Mutual inductance, 190, 241

N

Node, 26, 39, 68
Node analysis, 92
Node-to-datum voltage, 34, 92
Node voltage, 93
Nonhomogeneous differential
   equations, 165
Noninverting op amp circuit, 385
Nonlinear resistor, 14
Norton equivalent circuit, 75, 278
Notch filter, 322, 396

O

Octave, 368
Odd symmetry, 329
Ohm's law, 12
Op amp circuits, 125
   differentiator, 171
   gain, 126
   integrator, 170
   inverting, 128
   noninverting, 129
Open circuit, 18, 73, 148, 165
Operational amplifier (op amp), 124

Oscillations, 209
Output resistance, 125
Overshoot, 208

P

Parallel connection, 20
　current sources, 60
　resistors, 58
Passive elements, 48
Period, 217
Periodic functions, 302
Phase angle, 216, 265
Phase distortion, 314
Phase equalization, 314
Phasor, 222
　addition, 223
　currents, 234
　diagram, 238
　voltage, 234
Polar form, 224
Polarity, 7
Poles and zeros, 356
Power, 292
　apparent, 296
　available, 115
　average, 292
　factor, 297
　sign, 46
　transfer, 113, 297
Pulse train, 310
　spectrum, 335

Q

Quiescent state, 159

R

RC circuits, 157
Reactance, 244
　elements, 146
　function, 266
Real part, 224

Rectangular form, 224
Reference direction, 2, 9
　current, 10
　voltage, 15
Resistance, 12, 244
　input, 84
Resistors, 12
　parallel, 23, 59
　series, 22, 58
Ringing, 203
Rise time, 202
Rms value, 295
Root-mean-square, 296
Rotating phasor, 257
Rotational symmetry, 229

S

Self inductance, 190
Series connection, 20
　impedance, 271
　resistors, 58
　voltage sources, 60
Settling time, 208
Short circuit, 18, 74, 179
Signal processing, 214, 312, 321
(sin x) /x, 337
Sinusoid, 214
　addition, 218
　steady state, 220
SI units, 4
Source shifting, 68
Source transformation, 63, 277
Spectrum, 307
　analyzer, 327
　continuous, 309
　line, 307
　shaping, 333
s plane 260
Steady state, 163, 220
　analysis, 211
　response, 163
Steinmetz's analog, 222
Step response, 201

Sum of phasors, 231
Superposition, 110, 154, 284
Suppressed nodes, 59
Susceptance, 244
    functions, 266
Switched-capacitor filters, 404
    first-order, 417
Symmetry of signals, 329
System function, 265

T

Taps, voltage, 38
Terminal pair, 124
Thévenin equivalent circuit, 73, 278
Time constant, 159, 185
Time delay, 202
Transfer function, 36, 125, 266
Transform, 339
Transformer, 187, 192
Transient response, 163
Truncated Fourier series, 325
Turns ratio, 193
Two-port circuits, 124

U

Unilateral, 126
Unit output methods, 108

V

Voltage, 6, 101
Voltage divider, 36, 273
    taps, 38
Voltage follower, 131, 353
Voltage source, 49, 68
Voltmeter, 8

W

Watt-hour, 45
Waveform, 214
    shaping, 314
Window panes, 100

Z

Zero resistance, 19
Zeros, 256